Ecofictions, Ecorealities, and Slow Violence in Latin America and the Latinx World

Ecofictions, Ecorealities, and Slow Violence in Latin America and the Latinx World brings together critical studies of Latin American and Latinx writing, film, visual, and performing arts to offer new perspectives on ecological violence. Building on Rob Nixon's concept of "slow violence," the contributions to the volume explore the processes of environmental destruction that are not immediately visible yet expand in time and space and transcend the limits of our experience. Authors consider these forms of destruction in relation to new material contexts of artistic creation, practices of activism, and cultural production in Latin American and Latinx worlds. Their critical contributions investigate how writers, cultural activists, filmmakers, and visual and performance artists across the region conceptualize, visualize, and document this invisible but far-reaching realm of violence that so tenaciously resists representation.

The volume highlights the dense web of material relations in which all is enmeshed and calls attention to a notion of agency that transcends the anthropocentric, engaging a cognition envisioned as embodied, collective, and relational. *Ecofictions, Ecorealities, and Slow Violence* measures the breadth of creative imaginings and critical strategies from Latin America and Latinx contexts to enrich contemporary ecocritical studies in an era of heightened environmental vulnerability.

Ilka Kressner received her PhD in Spanish from the University of Virginia. She is currently Associate Professor of Hispanic Studies at the University at Albany, State University of New York.

Ana María Mutis received her PhD in Spanish at the University of Virginia. She is currently Assistant Professor in the Department of Modern Languages and Literatures at Trinity University.

Elizabeth M. Pettinaroli received her PhD in Spanish Literature at the University of Virginia. She is currently Associate Professor of Spanish Literature and Chair of Latin American and Latinx Studies at Rhodes College.

Routledge Studies in World Literatures and the Environment

Human Minds and Animal Stories
How Narratives Make Us Care About Other Species
Wojciech Małecki, Piotr Sorokowski, Bogusław Pawłowski, and Marcin Cieński

Climate and Crises
Magical Realism as Environmental Discourse
Ben Holgate

Ecocriticism and the Semiosis of Poetry
Holding on to Proteus
Aaron Moe

Christina Rossetti's Environmental Consciousness
Todd O. Williams

Ecoprecarity
Vulnerable Lives in Literature and Culture
Pramod K. Nayar

The Environment on Stage
Scenery or Shapeshifter?
Julie Hudson

Ecofictions, Ecorealities, and Slow Violence in Latin America and the Latinx World
Edited by Ilka Kressner, Ana María Mutis, and Elizabeth M. Pettinaroli

For more information about this series, please visit: www.routledge.com/Routledge-Studies-in-World-Literatures-and-the-Environment/book-series/ASHER4038

Ecofictions, Ecorealities, and Slow Violence in Latin America and the Latinx World

Edited by Ilka Kressner, Ana María Mutis and Elizabeth M. Pettinaroli

Routledge
Taylor & Francis Group
NEW YORK AND LONDON

First published 2020
by Routledge
52 Vanderbilt Avenue, New York, NY 10017

and by Routledge
2 Park Square, Milton Park, Abingdon, Oxon, OX14 4RN

Routledge is an imprint of the Taylor & Francis Group, an informa business

First issued in paperback 2021

Library of Congress Cataloging-in-Publication Data
A catalog record for this book has been requested

ISBN: 978-0-367-42671-2 (hbk)
ISBN: 978-1-03-208482-4 (pbk)
ISBN: 978-1-003-00177-5 (ebk)

Typeset in Sabon
by Apex CoVantage, LLC

Contents

List of Illustrations viii
Acknowledgments ix

Introduction 2
ILKA KRESSNER, ANA MARÍA MUTIS, AND ELIZABETH M. PETTINAROLI

PART I
Bad Living: Mutations, Monsters and Phantoms 37

1 Monsters and Agritoxins: The Environmental Gothic in Samanta Schweblin's *Distancia de rescate* 39
ANA MARÍA MUTIS

2 Toxic Nature in Contemporary Argentine Narratives: Contaminated Bodies and Ecomutations 55
GISELA HEFFES

3 The Ruins of Modernity: Synecdoche of Neoliberal Mexico in Roberto Bolaño's *2666* 74
DIANA ALDRETE

PART II
Econarratives and Ecopoetics of Slow Violence 93

4 The Representation of Slow Violence and the Spatiality of Injustice in *Y tu mamá también* and *Temporada de patos* 95
LAURA BARBAS-RHODEN

5 The Voice of Water: Spiritual Ecology, Memory, and Violence in *Daughter of the Lake* and *The Pearl Button* 114
IDA DAY

6 From Polluted Swan Song to Happy Armadillos: The Cold War's Slow Violence in Nicaragua 128
JACOB G. PRICE

PART III
Protracted Degradation and the Slow Violence of Toxicity 145

7 Collateral Damage: Nature and the Accumulation of Capital in Héctor Aguilar Camín's *El resplandor de la madera* and Jennifer Clement's *Prayers for the Stolen* 147
ADRIAN TAYLOR KANE

8 Violence, Slow and Explosive: Spectrality, Landscape, and Trauma in Evelio Rosero's *Los ejércitos* 162
CARLOS GARDEAZÁBAL BRAVO

9 The Environmentalism of Poor Women of Color in Mayra Santos-Febres's *Nuestra Señora de la Noche* 180
CHARLOTTE ROGERS

PART IV
Materialities, Performances, and Ecologies of Praxis 197

10 Slow Violence in a Digital World: Tarahumara Apocalypse and Endogenous Meaning in *Mulaka* 199
LAUREN WOOLBRIGHT

11 Slow Violence in the Scientific Ecosystem: Decolonial Ecocriticism on Science in the Global South 218
THAIANE OLIVEIRA

12 Bodies, Transparent Matter, and Immateriality: Compagnie Käfig's Ecodance Performances 236
ILKA KRESSNER

13 *Llubia Negra*: Fetishism of Form, Temporalities of Waste, and Slow Violence in Cartonera Publishing of the Triple Frontier (Paraguay, Brazil, Argentina) 257
ELIZABETH M. PETTINAROLI

Contributors 277
Index 281

Illustrations

0.1	Pedro Ruiz: *Total Eclipse of the Heart* (2006)	1
5.1	Nélida Ayay in *Daughter of the Lake* (2015)	118
10.1	Terégori Esqueltico, the Lord of Death, as he appears in a cutscene (in-game cinematic passage during which the player does not have control of the player-character)	210
10.2	Wa'ruara Gu'wi, the Seeló boss in the town of Paquime	213
11.1	The entirety of the Brazilian scientific output on climate change	228
11.2	Network of coauthorship of Brazilian and international research on climate change	229
12.1	*Agwa* (2014)	242
12.2	*Agwa*, final moments	243
12.3	*Pixel* (2015)	250
13.1	*Llubia Negra: 11 narradores paraguayos y non-paraguayos* (2009)	259
13.2	Various publications by Yiyi Yambo *cartonera* publishing house, using local textiles and materials such as *ñandutí*—a traditional Paraguayan embroidered lace knitted by local women	260
13.3	Photo collage by Douglas Diegues, *Llubia Negra: 11 narradores paraguayos y non-paraguayos* (2009)	270
13.4	Yiyi Yambo collective	272

Acknowledgments

Novel and creative responses to ecological slow violence on the part of activists, writers, and theorists in the Latin American and Latinx worlds inspired this collaborative project, and the interest and enthusiasm of many individuals helped our ideas develop into what became this volume. For their generous contributions of chapters and their conscientious work toward the volume's completion, we would like to thank the authors: Diana Aldrete, Laura Barbas-Rhoden, Ida Day, Carlos Gardeazábal Bravo, Gisela Heffes, Adrian Taylor Kane, Thaiane Oliveira, Jacob G. Price, Charlotte Rogers, and Lauren Woolbright.

Our project's first collective open forum took place in the context of the panel "The Dimensions of Disaster: Scale, Circulation, and the Specular Economy in Latin American Disaster Writing," held at the International Congress of the Latin American Studies Association (LASA), in Barcelona, Spain, in May 2018. Our thanks go to Mark Anderson for spearheading this effort and for facilitating discussions on the topics, works, and methodologies. The VII Annual SARAS Conference (South American Institute for Resilience and Sustainability Studies) in December of 2017 in Maldonado, Uruguay, provided several of the authors in this collaboration with a place to bring ecocritical perspectives into dialogue with the environmental humanities in the region and to rethink our contributions from truly trans- and interdisciplinary perspectives. We are grateful to Nestor Mazzeo, Jorge Marcone, Eduardo Gudynas, George Handley, Víctor Vich, Rachel Price, Patrícia Vieira, Zelia M. Bora, Andrea Casals, and Jesse Lee Kercheval, among others, for enriching dialogues. Douglas Diegues and Fernando Villaraga opened a window onto *cartonera* publishing practices and welcomed us to share in activists' and writers' innovative responses to slow violence.

Colleagues have been critical, inspiring, and passionate interlocutors about ecocriticism in Latin American and Latinx worlds. At UAlbany and Trinity University, thanks go to Alejandra Aguilar Dornelles, Jesús Alonso-Regalado, Alejandra Bronfman, Selma Cohen, Luis Cuesta, Timothy Sergay, Carmen Serrano, and Heather Sullivan. Pablo J. Davis shared his expertise in and love for Latin America and always found time for stimulating transdisciplinary conversations. For their generous, valuable, and timely assistance with bibliography, logistics, materials, and contacts, we are in debt to Darlene Brooks, Kenan Padgett, Charlie Kenny, and Larry Ahokas at Rhodes College and Jason Hardin at Trinity University. Clavelina and Marian Rodríguez Monin were an indispensable guide to research along the Triple Frontier and are the best ambassadors the region could have.

Our institutions, The University at Albany, State University of New York; Rhodes College; and Trinity University supported this project, seeing its mission as mirroring their own commitments to engaged inquiry into humankind's ecological future and a just society. At Routledge, our thanks go to Michelle Salyga and Bryony Reece for their immediate interest in and steadfast support of our project from day one. Our experience could only be described as a smooth editorial ride. We are grateful to Scott Garner for his keen editorial eye and to Robin Bissett and Diona Espinosa for their skillful help with the formatting and proofreading of some of the chapters of this volume.

We wish to express our appreciation to the artists, practitioners, and publishers who generously gave permission to include reprints of their works here: to Pedro Ruiz for allowing us to include a reprint of his stimulating painting *Total Eclipse of the Heart* from the collection *Glifosatos 2* (2006) at the beginning of our introduction; to director Ernesto Cabellos for use of an image from his film *Daughter of the Lake* (2015), included in Ida Days' chapter; to Adolfo Aguirre and Lienzo Games for their permission to include two images of the video game *Mulaka* (2018), discussed by Lauren Woolbright in her chapter here; and to photographer Benoîte Fanton for granting permission to reprint three of her photographs of Compagnie Käfig's performances of *Agwa* (2008; images taken in 2014) and of *Pixel* (2014; image taken in 2015) that are included in Ilka Kressner's chapter. While *cartonera* publishing embraces the notion of "copyleft" and permission is thereby not necessary, we acknowledge Yiyi Yambo's generously providing us with materials and stories.

Lastly but in so many ways most importantly, we happily recognize our debt of gratitude to our families, including our partners Pablo J. Davis, Julio Estevez-Bretón, and Sebastián Serrano for their generous support, for sharing in our intellectual excitement, and for the crucial everyday help that made it possible for us to find the time and energy to make *Ecofictions, Ecorealities, and Slow Violence in Latin America and the Latinx World* happen. They are our inspiration and what makes this work worthwhile.

Figure 0.1 Pedro Ruiz: *Total Eclipse of the Heart* (2006).

Source: Reprinted with permission by artist.

Introduction

Ilka Kressner, Ana María Mutis, and Elizabeth M. Pettinaroli

In Pedro Ruiz's painting *Total Eclipse of the Heart* (2006), an expansive black-and-white horizontal landscape is cut by a long white trail left by a plane. From the title of the series, *Glifosatos 2*, we learn that the white line approaching the indigenous village is not the harmless vapor exhaust from the aircraft but a stream of glyphosate, the herbicide used by the Colombian government in its aerial drug crop eradication program. This program, conducted under U.S. oversight, was intended to control illicit drugs at the source by spraying glyphosate-based herbicides over opium poppy and coca plants. In 2015, the Ministry of Health banned the aerial spraying of glyphosate over illegal crops after the International Agency for Research on Cancer (IARC), an agency of the World Health Organization, declared that this pesticide might be carcinogenic. Aside from killing food crops, contaminating water supplies, and threatening the rainforests and wildlife, glyphosate has been reported to produce adverse effects on humans that include non-Hodgkins lymphoma, skin problems, and an increased risk of miscarriages and premature births (Massey 281; Erler 84). However, since the 2016 peace agreement with the FARC has led to an increase in the number of illicit crops, the Colombian government is now considering a return to the aerial spraying of glyphosate.

We have chosen Ruiz's painting to open this volume because it quite clearly showcases Rob Nixon's notion of slow violence. Nixon's conceptualization, developed in *Slow Violence and the Environmentalism of the Poor*, refers to a specific form of environmental destruction that unfolds gradually over long periods of time and is dispersed beyond national boundaries, ecosystems, and elements. Unlike catastrophic ecological violence, which mediatically rivets public attention, this slow violence is characterized by a pernicious invisibility; it "remain[s] outside our flickering attention span—and outside the purview of a spectacle-driven corporate media" (6). While Nixon examines works by authors and environmental activists in an anglophone, postcolonial context, the essays of the present collection critically rethink Nixon's arguments in the geographical and material contexts of the Latin American and Latinx worlds and in relation to traditional and experimental areas of cultural production.

They explore how writers, cultural activists, filmmakers, video gamers, performance artists, and scientists from the region document, conceptualize, and visualize this specific form of violence that so often resists representation.

At first sight, Ruiz's painting does not strike us as a work about violence, much less ecological violence. Still, for someone observing the painting more carefully, there is something quite unsettling about the growing dominance of the white line from the left to the right side of the landscape frame. The line of colorless poison that cuts the scene and approaches the village becomes utterly threatening when we read the title of the series and are referred to glyphosate's harmful effects. In the words of writer William Ospina ("Pedro Ruiz y la caricia del veneno"), the beautiful landscapes painted by Ruiz "no están hechos para que nuestra buena conciencia se regodee en el deleite del mundo natural" ("are not made for our good conscience to revel in delight of the natural world"); rather, the menacing presence of that white brushstroke symbolizes the toxic arrow that quietly wounds people and nature:

> Es el hilo de muerte que deja a su paso el artefacto blanco, el pájaro mecánico, el juguete del viento. Y basta ese objeto minúsculo en el aire, y basta ese trazo blanco en el cielo, para que los paisajes pierdan su inocencia, y nuestra morada en la tierra se llene de un horror sutil y de una pavorosa advertencia. Esos livianos artefactos del viento, esas avionetas tejidas por la destreza humana, pasan sembrando muerte indiscriminada sobre todas las cosas. Es la muerte lineal, la muerte sutil, la muerte etérea, la muerte impalpable.
>
> ("Pedro Ruiz y la caricia del veneno")
>
> It is the thread of death that the white artifact leaves behind, the mechanical bird, the toy of the wind. And that tiny object in the air, and that white stroke in the sky, are enough for the landscapes to lose their innocence, and to fill our abode on earth with a subtle horror and a frightening warning. Those light artifacts of the wind, those planes woven by human dexterity, sow indiscriminate death on all things. It is a linear death, a subtle death, an ethereal, impalpable death.

This disquieting effect is exacerbated only by the paratext, the title of the painting. By using a pop song from the eighties, the artist appeals to our levity, perhaps even frivolity. Ospina suggests that the titles of Ruiz's collection refer to the love songs heard on the radio by the pilots of these aircrafts as they sprayed their poison, in which case the titles are meant to highlight their indifference and callousness ("Pedro Ruiz y la caricia del veneno"). But the title itself—*Total Eclipse of the Heart*—also speaks to the blindness of their, and our, hearts. The phrase "eclipse of the heart" invites meditation on the connection of invisibility to emotional indifference, highlighting both the hidden nature of the violence depicted in the painting and the grave consequences of our inability to see it. In addition, more indirectly, the title might also even refer to the force of an atmosphere to act on living bodies and to cast a shadow over the beat of life.

Nixon's Concept of Slow Violence

Nixon shows us that we need to broaden the notion of violence from its instantaneous and immediately visible guises to incremental ones that expand in time and space, transcend our humanly imaginable time frames, and broaden clearly demarcated geographical areas into regions that are much less clearly delineated. What is at stake is to understand this specific form of attritional violence and look beyond the apparent unspectacularity of its ecocidal unfolding to detect the causes that have been and are currently jeopardizing the livelihoods, cultures, and futures of millions of people and ecozones in the Latin American and Latinx Global South. Here, one of the major challenges, as Nixon reminds us, is representational (3). How can we convey this form of violence beyond a spectacular metonymical headline image? Several essays in our volume significantly discuss forms of agrochemical violence in areas of heavy agrarian activity where the spraying of herbicides and insecticides led to long-term contamination of drinking water (including eutrophication and other forms of toxic contamination) and resulted in the loss of biodiversity, cellular mutation, and ultimately aquatic and rural dead zones. In order to examine and communicate slow violence, "apprehension is a critical word . . ., a crossover term that draws together the domains of perception, emotion, and action" (Nixon 14). All essays assembled here thus touch on strategies to apprehend forms of "long emergencies" (3)—intellectually and with the help of our senses and emotional connections, as well as with a general perceptual openness that necessarily transcends disciplinary forms of expertise. By selecting certain spatiotemporal contexts and using specific images, symbols, and narrative sequences, they engage in new forms of witnessing of this violence in connection to questions of spatial justice through art and aim at engaging directly in action that resists amnesia related to ecological violence.

For Nixon, "the most conceptually ambitious and influential figures within the ecocritical turn have been [Lawrence] Buell and [Ursula] Heise, who deserve special credit for the reach and rigor of their innovative work, which has powerfully reshaped the priorities of literary studies and the environmental humanities more broadly" (33). Both these thinkers are Americanists, while Nixon's "training is in postcolonial studies, and as such, the 'elsewheres' that fringe their work constitutes [his] intellectual foreground" (33) in an anglophone context. While Buell's and Heise's research remains foundational for us as well, our scholarly background resonates with Nixon's insofar as we all work in Latin American and Latinx cultural history and artistic expression. The deep impact of the Spanish and Portuguese colonial legacy and its contemporary guises are a conceptual cornerstone for our scholarly approaches. For more than 500 years, Latin America has been providing soil, natural resources, and cheap labor to the Global North. But in addition, as Walter Mignolo reminds us, it has also developed an impressive range of counter discourses, practices of dissent, and social movements that may serve as inspiration to understand and learn from the region (Mignolo, *La idea* 74). We therefore study elaborations of space and place in Latin American and Latinx expression, representations of violence (ecocidal and others) in film and literature, and conceptions of alternative spaces in literary, visual, and performing arts, often from a comparative perspective. We conceive the "elsewheres" mentioned by Nixon as transnational zones of relation that go beyond the division of Spanish-speaking Ibero-America and Lusophone Brazil and that include artistic responses in some of the many indigenous languages as well. For instance, this volume discusses the epistemologies of water by indigenous populations in Peru, Bolivia, and Chile and videogames on and created by Tarahumara indigenous people living in Chihuahua, Mexico.

Among the key topics discussed by Nixon that strongly resonate with the explorations in the chapters of this book is the potential of imagination to make the "unapparent appear" (15). While he refers to the power of "imaginative writing" (15), our authors explore a variety of media that include novel writing (both narrative and lyrical), filmmaking, dancing, and video gaming, as well as imaginative uses of materialities such as cardboard or plastic cups. We hope to show the wealth of Latin American and Latinx experimental art making as forms of creative activism that challenge the notion of environmental degradation as inevitable and unveil neocolonial and neoliberal tactics that prescribe passivity and inaction as the only possible attitude.

Slow Violence's Impact

Since its publication in 2011, Rob Nixon's *Slow Violence* has inspired innovative explorations and perspectives by scholars in an astonishing range of contexts, including literary and artistic production, regulatory bodies of knowledge in international law and ethics, and spheres of cultural practice and technological control. Furthermore, scholars have embraced Nixon's concept as a way to put their own methodological approaches to the test within their fields of study. Netflix's highly popular television series *Cromo* (Argentina) has received scholarly attention to address the effects of slow violence on disenfranchised citizens living under the organizing logic of a politics of death that affects the individual, nation, and environment (Caña Jiménez). The topics of migration in films and of bodies of water in ecopoetry from Latin America have been given a bird's-eye treatment from an ecocritical perspective that addresses slow violence (Pérez-Melgosa; López). Chinese science fiction narratives dedicated to such topics as climate change, terraforming, and environment degradation engage the notion of critically examining the effects of the electronics recycling industry and its less visible environmental and occupational impacts on nature and humans. Authors have explained and explored the pernicious way in which economics, technological developments, and government policies intersect to transform the ecology, environment, and climate (Liu and Li). The effects of protracted aggression and life under siege in the Gaza Strip have also been examined through this concept as a way of rendering visible the obliteration of "normalcy" by means of dromocolonization (the colonization of humanity through technoscientific acceleration) in the self-governing Palestinian territory. By bringing these forms of violence to light through poetic (rather than visceral) realism, writer-activists have also worked to reclaim agency by resisting the reductive logics and subjectification of human rights discourses and reductive arguments of the anglophone media (Hesse). Scholars in film studies have looked into the representation by filmmakers of innocence and absence in childhood in order to illuminate the scope of slow violence and the ways it complicates or obstructs the psychic apprehension of environmental harm (Cecire). Slow violence has also been deployed to shed light on cinematic figurations that are avisual or affective (rather than strictly representational) in films on narcotrafficking (Llamas-Rodriguez).

Additionally, the covert dynamics of slow violence at play in the regulation and implementation of international law and ethics has found plenty of examples in the industry of mining.[1] Case studies of mountain communities in Peru reveal that theorizations of corporate exploitation legitimize capitalism's violence and, in particular, the slow violence that degrades local environments (Gamu and Dauvergne). Nixon's broader redefinition of violence to account for nonphysical impacts occurring over broad temporal scales has also made it possible to go beyond the injuries that extraction has wrought on the environment and to recognize, for example, the impact of slow violence on human rights defenders in Tanzania (Holterman).

Relatedly, Nixon's discussion of strategies of occlusion has provided a framework for broaching situations of insecurity, precarity, and disorder too gradual to achieve recognition as crises. For example, cultural anthropologists have looked into industrial practices and risk perceptibility in the context of a steady stream of impositions plaguing locals in Chile (Ureta) and in the instance of Baltimore, Maryland, with its placement of dumps and the building of trash incinerators (Ahmann). Analysis of slowing, speeding up, reordering, and strategically punctuating time by parties in conflict unveils creative forms of temporal orchestration and moral manipulation. Selective mobilization of the understanding of climate change has also been identified as a strategy aimed at obscuring the causalities of slow violence—a case of how the politically motivated management of information can hinder equity and transparency (O'Lear). And geographers have looked into the rise of cashless technologies as a form of slow violence in which financial tactics are used to undermine the provision of care for asylum seekers in the UK and the retention of welfare benefits for Aboriginal citizens in Australia's Northern Territory; such practices enact new forms of border securitization that slowly but permanently block political participation by those with precarious claims to citizenship (Coddington). In environmental media studies, slow violence has even led critics to reconsider theories of technology by interrogating such phenomena as the temporalities of the global e-waste recycling trade and the speed, acceleration, and simultaneity brought by information and communications technologies (ICTs). In these contexts, planned obsolescence functions as a type of slow violence that structures the environmental politics of the information age (LeBel).

Nixon's project has also inspired multiple disciplines to reappraise their methodologies, themselves considered as epistemological tools that may effect slow violence. From the perspective of the social sciences, researchers have examined the severing of social problems from their frameworks in relation to social movements and attempted to situate the temporal, spatial, and experiential aspects of suffering within a comparative framework that considers the phenomenological structure of social problems (Skotnicki). This vein of self-critique has also reached the field of archeology through a consideration of its place within capitalism; in this light, archeological practice emerges as a slow exercise of extraction (Hutchings and La Salle). Recent studies have brought new perspectives to the analysis of gender policies (Pérez) and to the exploration of slow violence in neoliberal attacks against public education (Giroux and Proasi). The gesture of self-examination echoes Anne McClintock's condemnation of formalism through "fetishism of the form," as well as Nixon's embrace of environmental humanities perspectives that join history and the literary (quoted in Nixon 31). In a similar vein, the concept of slow violence has helped address the temporal dimensions of problems such as social stigma (Barnwell) and systems of organized exploitation of children ranging from informal economy to outright criminal enterprise (Iom). Finally, scholars inspired by Nixon's forensic examination of the long-term effects of environmental damage recently came together at a symposium at the University of Hertfordshire, Hatfield, to explore the visualization of slow violence, as well as its far-reaching consequences in areas such as health, economics, education, and migration.

Slow Violence in Latin America

In the early sixteenth century, the first implementation of Western colonialism on a global scale significantly marked the history of what we know today as Latin America, and is therefore of particular import for understanding the contexts that facilitated the emergence of contemporary slow violence and the logics that continue to sustain it. The conquest and colonization of the New World took place via the framework of an ontological shift within the European worldview that transformed the relationship between man and the environment, and that fostered new modes of intervention in the world.[2] The spoils unfolding from conquest were part of man's mandate to understand the world's complexity and its mechanisms, an animist-organic perspective in which the working of nature came to be a deliberate activity perceptible to humans (Ingegno 244–5).

In the iconographic and conceptual distancing of "human" from "earth," Tim Ingold identifies a deep cognitive schism between a global ontology of detachment and the local ontology of engagement over which it prevails. World and nature lie beneath an observer who now envisions them as a globe and surface awaiting human intervention. This novel ontological approach, a radical departure from those of the past, assumes the human refashioning of nature-as-object as responding to properties immanent in nature itself. Ingold perceives, in this teleological paradox, links between that evolving image of the world as a globe and the modern conception of environment as an object of management rather than a dwelling, and as "an object of appropriation for collective humanity" (214.) Despite their totalizing efforts, these global perspectives remain illusory and incomplete, open to the possibility of critique of man's role on earth. Discussion on nature also served, in all its contingency, as a form and a forum for the debate of differing views of dominion and humanity.

It is precisely in this context where Iberian projects of colonization anchored the legitimization of Western logics of extraction wielded by new technologies. In *The Matter of Empire*, Orlando Bentancor reminds us of the metaphysical foundations of imperial instrumental reason and its role in providing a framework for writings on natural law that were so central to Spain's justification of its empire. Western metaphysical instrumentalism served as the basis for regarding nature as open to technical manipulation and is the foundation for the contemporary positing of nature as a repository of inert raw materials to be manipulated through the complexities of active technological intervention (8–15). This notion of "nature" as form and matter served to justify the extractive ambitions of the Iberian Monarchy from the island of Hispaniola to the indomitable Andes and from the territorial conquests in the Pacific to a dream of subjugating China. Francisco de Vitoria's lectures, venturing into debates about sovereignty over indigenous peoples by a particular employment of the principle of the law of nations (*ius gentium*) (Bentancor 52–3), laid the legal groundwork for the Monarchy's assertion of a worldwide right of extraction and trade of metals. And José de Acosta, in his *De procuranda indorum* (1588), put forward a project to rationalize the control of nature and the practice of mining through the supremacy of technology. In this compendium, Acosta presented a program that articulated mechanisms and technologies for labor systems, as well as methods for the administration and disciplining of peoples into the ostensible condition of civilization according to the presence of civility and "policy."

The condition of "policy" was hotly debated, as evidenced by the polemic on the nature of the Amerindian, most famously the philosophical and legal debates at Valladolid and the robust body of literature that preceded and followed it.[3] One side of the debate continued to justify imperial rule over the New World through appeal to arguments of anthropologic asymmetry between civilized and barbarous peoples as put forward by Juan Ginés de Sepúlveda. Others, such as the cosmographer and royal chronicler Alonso de Santa Cruz, challenged empire by invoking the explanatory power of Christian universalism upheld by José de Acosta and Bartolomé de Las Casas for redeeming the cultural and spiritual potential of natives. Positions such as Santa Cruz's were part of a growing concern in Iberian intellectual circles with the progressive privatization of the conquest (Cuesta Domingo 67) and the growth of extractive, exploitive interests justified under the guise of spreading of Christianity. His literary reimagining of the region redeployed Vitoria's propositions and fashioned new territorialities that called into question the morality of Hapsburg global expansion. In a more radical move, Iberian claims to sovereignty over New World "barbarity" were discredited through direct observation of local landscapes and peoples by cosmographers and chroniclers in charge of assimilating the environments and peoples of new territories to the European imagination.[4]

Acosta's *Historia moral y natural de las Indias* (1590) enters the debate by deploying one of the first systematic theorizations of cultural evolution and the earliest system of comparative ethnology[5]—an artful and largely successful campaign of occlusion of the pernicious logics of conquest.[6] He reified social and cultural constructs (writing, culture, social regulation, Western logical reasoning based on specific parameters for confirmation of knowledge, etc.) into ostensibly inherent attributes of civilized peoples, thereby proposing techniques to "civilize"—understood as a transitive verb and an attribute of Spain as imperial power—and to regulate this new world order. Ivonne del Valle reminds us that sixteenth- and seventeenth-century Spanish colonialism is where we can first discern the bold viewpoint that all of humanity should follow a particular direction of development and that technologization was the appropriate process for reaching this goal ("From José de Acosta" 439). In his treatise, Acosta places human regency over nature's resources as part of a divine plan, where science, technology, and religion are collapsed into one, and advances a notion of knowledge that displaces textuality and favors experience ("From José de Acosta" 442).[7] Francis Bacon was familiar with the *Historia natural y moral* and, adopting several of its ideas and epistemological approaches, built on Acosta's premises to propose natural history as the basis of philosophy and science and to urge that scientific knowledge be put in the service of human progress and "empowerment" ("From José de Acosta" 441).[8]

It is in this early context that extraction is essayed and progressively imposed as a normative practice, and its concurrent toxicity—in all its expansive forms—is concealed behind its controlling logics. It is where technology transcends its early connections with the capitalism that brought it about[9] and where it becomes "part of a larger system (epistemological, religious—even in its secular manifestations) entailing a particular relationship with nature in which *anything is possible* in order to modify perceived obstacles" ("From José de Acosta" 439). Subsequent complex processes of occlusion further severed technology from its historical contexts and transformed it into what del Valle perceptively describes as the birth of the notion of technology as the end of history itself (del Valle, "*Grandeza Mexicana* and the Lakes of Mexico City" 46). These skillful amalgamations of occlusive arguments and practices that naturalize technologies—a sort of technological fetishism—launched processes of slow violence early in the sixteenth century.[10] Moreover, they have been detected in what Nobel-winning chemist Paul J. Crutzen poses as a new "Anthropocene" geological epoch based on their findings of anthropogenic emissions of carbon dioxide and methane that exceed natural behavior in the air trapped in polar ice dating back to the end of the eighteen century—a time coinciding with the design of James Watt's steam engine in 1784 ("Geology of Mankind").

Latin America faces a wide variety of environmental problems that can be described as forms of slow violence on human societies and nonhuman nature. Rapid deforestation, soil erosion and depletion, water contamination, air and water pollution in urban centers, melting glaciers, declining biodiversity, and loss of natural areas due to mining, industrialized monoculture, and tourism are some of the environmental challenges faced by the region. Mark Anderson observes that, although some of these problems are manifestations of current global environmental crises, others can be seen as the result of an intensification of the extractive practices implemented in Latin America since colonial times (x). Scholars agree that the extraction of raw materials as a central feature of Latin American economy can be traced back to Iberian colonialism.[11] The colonial enterprise of "plunder and appropriation" (Acosta, "Extractivism" 63) shaped the notion of nature as a commodity and of Latin America as a resource-rich region that was there to supply richer nations with its raw materials, a notion famously portrayed by Eduardo Galeano in 1971 through the image of Latin America as a continent of open veins.[12]

This enduring legacy is seen today in the intensification of the extractive growth model in the region, a model that in its most recent manifestation has been termed *neoextractivism*.[13] This mode of development, based on the extraction and export of raw materials and elements such as oil, minerals, and agricultural goods, differs from conventional extractivism[14] in that the state plays a bigger role in regulating extractive economies and investing the extra revenue in public infrastructure and social programs (Gudynas, "Diez," "Estado"). If neoliberal regimes during the 1990s strengthened transnational corporations and favored the local elites in the exploitation of natural resources, the progressive governments that followed gave the state a more prominent role in the extraction of raw materials and the appropriation of revenue for social expenditures while maintaining the exploitation of nature as the basis of their development strategy (Gudynas, "Diez," "Estado"). Despite the recent decline in commodity prices, which troughed in 2011, causing a reduction of foreign direct investments in extractive industries (ECLAC), the fear remains that governments of all ideological stripes still support a developmental model based on natural resource exploitation.

This consolidation and intensification of a resource-based model of development in Latin America[15] by governments from the left and the right of the political spectrum is worrisome to social and environmental scientists alike. Among its deleterious ecological effects, Burchardt and Dietz mention "global climate change, soil depletion, deforestation, loss of food sovereignty, declining biodiversity, contamination of freshwater" (469), while Acosta, focusing on the environmental impacts by large-scale mining, brings attention to the public health problems brought on by water contamination with toxic chemicals ("Extractivism" 69–70). There is also a concern regarding the depletion of natural resources that were considered "renewable" but are no longer so because the speed of extraction far surpasses the rate at which they can regenerate (Acosta, "Extractivism" 62). Finally, there is the debate surrounding the adverse economic and social effects brought about by a reliance on extracting natural resources for growth. While some scholars contend that there is a "natural resources curse" (Gudynas, "Diez" 192; Acosta 61) by which countries rich in raw materials are often doomed to poverty, constant economic crises, social inequalities, and authoritarian regimes, others refute this causal relationship by bringing up examples of countries rich in natural resources that have achieved high income levels and economic diversification while also maintaining their status as stable democracies (Andrade 115).[16]

This controversy and the growing focus of Latin American environmentalists on the destructive effects of extractivism demonstrate that this has become a central issue for the environmental movement in the region.

Alternative approaches to the commodification of nature and export-driven growth have been developed in response to these preoccupations. The most salient example is *Buen Vivir*, which, drawing on indigenous knowledges, rejects the idea of progress and well-being as dependent on material consumption and instead proposes the communion of humans and nature and the diversity of knowledge, displacing the centrality of Western thought (Gudynas, "Las disputas," 25). Moreover, not only are environmentalists and academics contesting the expansion of an extractive political economy; increasingly, local communities that are directly affected by extractive policies—and also civil society more broadly—are taking action against extraction or demanding more control over extractive processes through social mobilization in the form of protests and the use of legal and political mechanisms (McNeish 14). As John Andrew McNeish explains, the discussion surrounding the political economy of extraction in Latin America often presents a false narrative where local communities are defenseless against corporate and international dealings, while, in reality, despite the power imbalance, there are "powerful expressions of organization and political agency by communities threatened by, or wishing to participate in the control of, extractive processes" (6).

Our volume testifies both to a growing preoccupation with the rise of extractivism in Latin America and to the determination and political engagement of artists, activists, and cultural agents to bring awareness to the ecological and social costs associated with it. Most essays in this collection address the slow violence of ecological degradation caused by extractive practices: the devastating effects of gold mining in Peru and Bolivia as well as the destruction of the marine culture of western Patagonia (Day), the toxic dumping of waste in rivers and lakes by extractivist companies in Nicaragua (Price), the toxic effects of the excessive use of agrochemicals by export monoculture in Argentina (Heffes, Mutis), and the deforestation caused by logging in Mexico (Kane, Woolbright). Other chapters do not refer to specific forms of resource extraction but present the dismal consequences of modernity and capitalism on people and the environment: the ecological damage caused by urban expansion and industrialization in Puerto Rico (Rogers); the ecocide, displacement, alienation, and psychological malaise caused by the "spatiality of injustice" of globalized modernity (Barbas-Rhoden, Gardeazábal Bravo); and the socioecological degradation and exploitation brought about by neoliberalism in the Mexican border with the United States and in the Triple Frontier of Argentina, Brazil, and Paraguay (Aldrete, Pettinaroli). Even global environmental problems, such as toxic contamination by plastic waste, are presented in connection to local extractive practices such as the rubber boom in Brazil (Kressner).

All works of art discussed in this collection share a preoccupation with the damaging effects of capitalist globalization, neoliberalism, and in some cases U.S. intervention. Our volume opened with a painting describing a form of slow violence seemingly unrelated to extractivism: the aerial spraying of glyphosate over coca and poppy crops as part of the so-called war on drugs. Two chapters address this form of institutional violence (Gardeazábal Bravo, Kane) where a war waged against illegal drugs is indifferent to the deferred devastation it inflicts upon human and non-human innocent victims. It could be argued that this drug eradication program is the opposite of extractivism because it aims to prevent the extraction and export of an agricultural product. However, as with drug trafficking, it is shaped by global capitalism and displays patterns of domination and intervention akin to those seen in extractivism. Similarly, the tourism industry discussed in Pettinaroli's chapter is founded on the accumulation of global capital and often depends on the physical rearrangement of landscapes with great socioecological costs. Furthermore, tourism in the Global South has often been characterized as a form of neocolonialism wherein old structures of power and domination reemerge (Mowforth and Munt 60). Another form of neocolonialism may be seen within the scientific community, in the dynamics of domination by which scientific production from the Global South is relegated to a peripheral role (Oliveira). These neocolonial processes of domination and manipulation, with their exclusionary tactics, resemble those inscribed in the appropriation of natural resources and territories.

Ecocriticism in Latin America

Three specificities of Latin American and Latinx ecocritical thinking inform the creative and critical responses to long-term environmental violence in the region in meaningful ways.[17] First, inquiries and concerns with ecological and spatial justice are inflected by the colonial legacy of oppressive extraction and its contemporary permutations. It is at this juncture that we can find the origins of many of the logics and justifications for coloniality—the negative underside and constitutive component of modernity (Mignolo, *The Darker*). The Latin American ecological imagination is concerned with both environmental problems and unresolved social issues such as indigenous rights, sexism, poverty, and the damaging effects of neocolonial and neoliberal enterprises (Barbas-Rhoden). Accordingly, environmental justice in the region has been marked by a long trajectory of conversations with the field of political ecology and with critiques of the paradigms of development, and, as Barbas-Rhoden so aptly explains in the introduction to her chapter in this volume, Latin American ecocriticism has developed alongside a concurrent decolonial turn. Second, ecocritical thinking transcends intellectual realms and is marked by its focus on materiality and groundedness. Going beyond the realm of abstraction, the region's indigenous *saberes* (knowledges) and sensibilities underlying *Buen Vivir* (Acosta, *El buen vivir* 21) allow for encompassing notions of physical manifestations that permit designs for pluriversality—the possibility for the coexistence of multiple, concurrent cosmologies (Mignolo, "On Pluriversality"). Third, the specificity of Latin American and Latinx environmental voices testifies to a vision of art as a public, political praxis capable of reaching varied audiences and of artists as cultural agents.

Political Ecology and Spatial Justice in Latin America

The spatiality of justice is of paramount importance to the region, given that a debate central to—and perhaps even constitutive of—the configuration of what we know today as Latin America has been the polemics of possession. The clash of multiple notions of sovereignty, land ownership, governance, and self-governance has also, as Rolena Adorno proposes, transcended into debates over the attributions of human intellect and reason and over the logics of real or perceived authority (political, historical, or literary) underlying the organization of the New World (viii). These battles have generated persistent interest throughout the years and continue to inform "the conundrums of history's secrets and the potential of narrative to reinterpret the phenomena that it simultaneously conceals and reveals" (Adorno 324). Moreover, European expansionism into the New World brought with it the ecological imperialism that Alfred Crosby identified in his foundational study of the biological expansion of Europe. These two developments entailed the negotiation, appropriation, reconfiguration, and at times complete reinvention of territorial expanses (Porto-Gonçalves and Leff) and the ecologies existing in them.

Spatial (in)justice, Edward Soja argues, is found at the intersection of three levels of geographical character: the first is brought by the external creation of unjust metageographies (including boundary making and the imposition of a political organization of space, among others); the second occurs through the local, endogenously implemented, unequal distribution of resources brought by discriminatory decision making (for example, the location of toxic repositories, spatial and racial segregation, etc.); and the third subsists at regional or "mesogeographical" levels and underlies characterizations of uneven development for urban and global contexts (such as the imposition of intrametropolitan frameworks of Worldism, the transhemispheric inequities between Norths/Souths, and many other) (9). An assertive spatial perspective yields significant explanatory power in relation to the meaningful geographies of justice for today. These geographies are not solely the consequence of sociopolitical processes; rather, they also emerge from the dynamic force of human agency as it affects these processes in critical ways (103–4).

Inquiry into spatial justice in the context of environmental studies in the region has centered on two main areas: practices of violence unleashed by neoliberal policies (Salamanca Villamizar 16) and rural–urban inequities as studied in the field of environmental justice. The first area of inquiry has examined both urban settings and rural. Studies include explorations of institutional violence in working-class and poor neighborhoods as well as in marginal settlements, links between national security policy and everyday violence, and spatial injustices and their relationship to the Cold War and the global drug and weapons markets (21). Barbas-Rhoden's contribution to this volume engages an ecocritical perspective informed by such spatial justice theory to unveil the visual power of films in the representation of overlapping and sometimes contradictory regulation, corresponding to both contemporary and past structures of authority, as it affects bodies and territories—what Nixon notes as an intersection between narratives of nation formation and the annulling complexity of global systems (150–1). Similarly, several of the authors in the volume *Llubia Negra*, studied in Pettinaroli's chapter, explore the local repercussion of global neoliberal policies—that "mesogeographical" level of injustice identified by Soja. Price's exploration here of environmental poetry sheds light on the creation of a lyrical history to recount the effects of the Cold War and the fragile character of the Nicaraguan landscape, while Gardeazábal Bravo engages the notion of "spectral topographies" (Martínez) to reveal the intersections between dramatic, event-centered violence and slow violence in the contexts of Colombian guerrilla warfare and transnational drug trafficking.

The second area of investigation owes recognition to and aligns with the field of political ecology in Latin America, the study of environmental impacts, and consideration of the notion of *saberes* and *Buen Vivir* from the region. These have been thriving areas of scholarly research in the region for several decades, with particular focus on challenging social and spatial injustice (Leff, *Justicia ambiental*; *Racionalidad ambiental*; Martinez-Alier, *The Environmentalism of the Poor*; "Conflictos ecológicos y justicia ambiental"; Acselrad et al.; Carruthers; Merlinsky, among others) and on condemning extractivism, monoculture, and other forms of neoliberal modes of production (Latta and Wittman; Martinez-Alier *The Environmentalism of the Poor*; Alimonda). Scholarship and praxis in the region offer a particular perspective that informs the field of political ecology from a politics of difference embedded in the ecological and cultural conditions of its peoples and from projects to decolonize knowledge, reconfigure territories, and reenvision nature (Leff, "Political Ecology" 34). The region has witnessed the confluence of these extractive projects and the mobilization of communities to combat them. Day's study here of *The Pearl Button* and *Daughter of the Lake* explores the denunciation of extractivism in Peru and in Chile and the attendant mobilizations to fight it.[18] The reimagining of the place of literature and art in the context of contracting publishing markets reveals a new spatial consciousness where the transdisciplinary diffusion of the publishing houses known as *editoriales cartoneras* extends beyond traditional spaces into such locales as incarcerated communities and the Triple Frontier region of Paraguay-Argentina-Brazil.

A critical examination of the coloniality of knowledge (Lander; Mignolo, *Local Histories/Global Designs*; Mignolo and Escobar; Quijano), as well as an exploration of epistemologies of the South (Sousa Santos), has brought new perspectives on the configuration of the world, fostered retheorizations of the inherited frameworks that facilitated the appropriation of ecological and cultural patrimony (Leff, "Political Ecology" 35), and advanced new potentials for justice and ethics arising from pluriversal perspectives (59). The consideration of materialities informed by the concept of *Buen Vivir* opens the door to renouncing the instrumentalization of social and environmental relationships that dualize humankind and nature, and to postulating instead a more encompassing ontology of being in the world. Such an approach supports nonanthropocentric perspectives and views of life that pose alternatives to models of perpetual growth and accumulation. Native Amazonian theories of materiality offer an opportunity to reconsider the notion of being to include "things" (artifacts, songs, images, designs, names) in order to embrace a more capacious idea of agency and of living (Santos-Granero 20). The notion of place shared by indigenous communities in the Peruvian highlands makes no differentiation between humans sensing a space and said material space constituting a sentient entity—the former as subject of awareness and the latter as a cognizant object (de la Cadena 101). Consideration of forests as thinking entities among the Runa of Ecuador's Upper Amazon prompt us to consider the need to attend to a forest ecology of self (Kohn 226–7). In other words, these new perspectives call us to take seriously multiple epistemologies and ontologies.

Material Art

All the chapters in this volume look into the profound and precarious nexus of human–nonhuman relationships and the perils of ecocide from a decidedly Latin American and Hispanic/Latinx viewpoint. The topic of slow environmental destruction is examined within historical, geographical, and spatial contexts (what Tim Bristow coined as "place perception" in ecocriticism, 5–7 and 127) and related to colonial legacies and systemic inequalities (Escobar 51).[19] In addition, the studies included here engage in a critical rethinking of materialities, oftentimes in the form of a material presence of an object or artifact that contrasts with an abstract and invisible "force" (such as a globalizing or developmentalist agenda). The material nexus functions as a fundamental human–human and human–nonhuman connector and refutation of any attempt at positioning a single, essential, abstract, and autonomous identity. In a region marked by violent extractivism and the pillaging of raw materials to be processed, refined, and consumed elsewhere by absent others, this contemporary attention to matter, its shared experience, and the underlying conception of materiality is of particular significance.[20] It is intricately involved with the contemporary "turn to the material" (Iovino and Oppermann 2), an extensive environmental conversation across sciences and humanities that aims at shedding "light on the way bodily natures and discursive forces *express* their interactions whether in representations or in their concrete reality" (Iovino and Oppermann 2, emphasis in original). The Latin American/Latinx cultural agents within this broad conversation highlight the voice of the arts, in addition to traditional academic disciplines and fields.[21]

Art is particularly positioned to "materialize" literally, metonymically, and in medial variations, to make visible that which we overlook. A case in point is Ruiz's painting, where the accumulation of a suspension of white polymer particles on a canvas refers to a trail in the sky, initially taken as the exhaust from the engine of an airplane (a common reminder of the huge environmental cost of flying); however, after the viewer reads the work's title, the trail transforms into the mark of the invisible herbicide that is sprayed on ecosystems in a systematic attempt to kill, carried out at a safe distance and from within a highly technologized mobile object.

If we understand our current ecological crisis as a "crisis of modernity, to the extent that modernity has failed to enable sustainable worlds" (Escobar 51), we cannot simply read the recent focus on material as a counterpart to the traditional focus on the mind. Instead, we need to find ways to think beyond dualities (embodiment vs. thought, matter vs. meaning, body vs. identity) and become attuned to examining matter as an "ongoing process of embodiment that involves and mutually determines cognition, social constellations, scientific practices, and ethical attitudes" (Iovino and Oppermann 5). Agency, as Serenella Iovino and Serpil Oppermann remind us in their introduction to *Material Ecocriticism*, "assumes many forms all of which are characterized by one important feature: they are *material*" (2; emphasis in original). Hence, the material nexus connects diverse agents: the human hand and the piece of cardboard, once discarded as waste, once serving as the building material of houses and once becoming the material for book covers that form part of an alternative publishing venture; the water of the Puerto Rican Portuguese River that connects with an alternative, spiritual belonging in contrast to a present one; and descriptions of bodily deformations of inhabitants of the Argentinean pampa that convey a neoextractivist agricultural practice. (All of these examples are taken from essays included in *Ecofictions and Ecorealities*.) While, in the words of Francine Masiello, "neoliberalism attempts to render complete and totalizing narratives about culture under globalization, covering the unevenness of our different stories with a homogenizing gloss" (181), a focus on the tangible material reality (or second reality in a work of art) precisely highlights these fissures, stains, sutures or scars on the material and brings forth the stories that are "stored" in matter (Iovino and Oppermann 1). This focus allows us to examine the use and symbolic value of things; it also helps us to explore the impact of those values for our environment.[22] A case in point is the significance of timber for an entrepreneur in early twentieth-century Mexico (Héctor Aguilar Camín's *El resplandor de la madera*) and its vivid contrast with the meaning in the life of a woman of color in mid-twentieth-century Puerto Rico connected to a single balata tree that is being carved to become stools, toys, or cooking spoons (Mayra Santos Febres's *Nuestra Señora de la Noche*).

In addition to the connecting function of objects, the focus on materialities in today's art from Latin America and the Latinx world also emphasizes how objects may "compete with humans for the scarce resources in the same ecosystem" (Csikszentmihalyi 20–1). To give two examples from works of art discussed in the present volume: plastic takes over the space of the dancers in *Agwa*, and the mushrooms (powerful agents in their own right) in the video game *Mulaka* poison the air, hence threatening the hero's survival. A focus on material things unveils our addiction to materialism as a result of our "paradoxical need to transform the precariousness of consciousness into the solidity of things. The body is not large, beautiful, and permanent enough to satisfy our sense of self. We need objects to magnify our power, enhance our beauty, and extend our memory into the future," describes Csikszentmihalyi in a rather ironical tone (28). What is needed and what is suggested in artworks that precisely focus on materialities is a critique of this obsessive "need" to accumulate and transform the precarious into solidity and permanence.[23] Instead of such a solidifying impetus, materialist art insists on the reality of our shared precariousness, based on our (this "our" including humans and nonhumans alike) materiality in an era of environmental vulnerability.

The Work of Art for and in Latin America: A Participatory Project

In her book *The Work of Art in the World. Civic Agency and Public Humanities* and recent essay "Lessons Learned from Latin America," Doris Sommer describes as the salient characteristic of artistic praxis in Latin America the deep structural link between art, pedagogy, and politics ("Lessons" 189). The artists whose works she examines are both thinkers and creators; they conceive of their works as means of interruption and questioning of established political and cultural discourses, similar to the impetus of the historical avant-gardes. The films, novels, poems, video games, and performances discussed in this volume, too, all aim at, returning to Sommer, "good social change [that] begins with incremental work to enchant hearts and minds" (189); artistic expression in and from

Latin America that may spark audiences' interests is oftentimes activist and participatory in its orientation. In our view, this is even more the case for artworks that are explicitly dedicated to environmental issues. Artists focusing on environmentalism highlight the dense web of material relations into which all is enmeshed and argue for a concept of agency that undermines and transcends the anthropocentric toward a cognition that is envisioned to be embodied, relational, and collective.

In our volume, this participatory orientation and invitation to focus on the more-than-human world[24] has different guises: it includes focalizations of minor contexts within event-driven narratives, metanarrative playing with audiences' expectations, invitations to actively connect metaphorical "dots" dispersed among narratives, and sudden changes of and oscillations between contrasting ideological viewpoints. Audiences find themselves on slippery conceptual ground and need to read actively and carefully, at times even against the grain of present enunciations; to refocus in order to examine minor or marginal contexts beyond central narrative arcs; and to take part in the interpretative process of a given work. At times, we are confronted with radical conceptual openness that we need to make sense of—actively, creatively, and emotionally.

In her reviewing of *Y tu mamá también*, Laura Barbas-Rhoden strolls away from the alluring main narrative thread and carefully observes marginal scenes and situations that allow for original, ecologically informed insights. Diana Aldrete, Charlotte Rogers, and Carlos Gardeazábal Bravo, all in different ways, embark on innovative readings of violence inflicted upon the environment in novels that have not previously been approached from an ecological vantage point: Roberto Bolaño's *2666*, Mayra Santos-Febres's *Nuestra Señora de la Noche*, and Evelio Rosero's *Los ejércitos*, respectively. The critics ask the readers of the novels to refocus their gaze beyond the immediately visible and event-centered narratives toward the seeming backgrounds that nevertheless hold the potential to perceive contexts in much broader terms and with a much less limiting, human-centered emphasis. In her study of the two films *Daughter of the Lake* and *The Pearl Button*, Ida Day reexamines the figure of the metaphor of water and flow toward a less exclusively intellectual (in the Western sense) and instead more associative awareness of the element. The short literary form of poetry, as Jacob G. Price proposes, is a medium of many appearances: it may flirt with political discourse or be a visionary voice, but it may also act as an instrument of campaign promises. Price studies such oscillations of discourses in Pablo Antonio Cuadra's *Cantos de Cifar y del mar dulce* and Ernesto Cardenal's "Nueva ecología."

Art is playful, and precisely because of its conceptual openness, it is also an ethical endeavor for many working in and reflecting on precarious times and spaces marked by forms of abuse and neoliberal domination. Artists, and we engaged humanists as well, are creative when we think beyond the present (Strupples 4), as this is actually a form of deep examination of our present. This is all the more urgent today as many neoliberal discourses use the language of democracy, minority rights, and spatial justice by co opting a counter discursive position. All authors of this volume engage in patient, careful, and contextual readings to detect pernicious euphemisms and differentiate new nuances within artistic enunciations; for example, Adrian Taylor Kane's essay here elucidates this in his examination of the language of warfare, in particular the notion of collateral damage within two contemporary narratives from Mexico and the Latinx world: Héctor Aguilar Camín's *El resplandor de la madera* and Jennifer Clement's *Prayers for the Stolen*.

In the literary works included here, the realist mode is less prominent than various forms of metaphorical, meta-narrative, and allegorical expressions. The purpose is clear: artists-activists invite us to challenge our established habits of reading and expectations related to storytelling in order to understand topics and realities that oftentimes resist being described in a straightforward, chronological, and realist manner. Thus, for instance, Gisela Heffes, in order to elucidate the invisible agrochemical pollution (among other types) that extends beyond commonly imaginable space-times, explores corporeal crises of fictional bodies described in a selection of contemporary novels and poetry from Argentina, while Ana María Mutis examines the figure of the monstrous child and the figural speech of the Gothic genre in Argentinean Samantha Schweblin's *Distancia de rescate*. The videogame *Mulaka* might be the most radical artistic challenge to conventional narrative arcs included in this volume: Lauren Woolbright describes how the game at times has its heroes (and players) fail in the quest to save the world and interprets this video-narrative refusal to deliver a positive conclusion as a direct environmental cautionary tale.

In her examination of Colombian sculptor Doris Salcedo's work, *Of What One Cannot Speak: Doris Salcedo's Political Art*, critic Mieke Bal describes political art as a work that "emphatically endorses the inescapable fact that it is part of the world in which it occurs . . . and as part of the world, the work labors for its transformation" (9). In all artworks included in this volume, the two elements of the concept of *political art* cannot be separated (Bal 1); moreover, they condition each other. An artwork "in situ, in process, . . . inspires thoughts that pertain to the social collective that in turn inspired it" (7). A case in point is Elizabeth M. Pettinaroli's study of the artistic repercussions of the experience of a mysterious black rain that fell in Asunción in Paraguay in 2009, which she examines in three short stories included in a *cartonera* publishing project that, through new eco- and carto-poesis, offer theorizations of the dialectic between environmental slow violence and the historical process, within new cultural ecosystems that provide a framework to rethink form, matter, and historical agency.

Bal asks how art's collective, political potential can be deployed and performed in the singular, by a sentence written by one single author, a gesture performed by one actress, or a motion performed by an individual dancer. The response from the artists and critics assembled here is twofold: one avenue is the insistence in the participatory valence of an artwork and its attendant need for an active reader/viewer/listener/member of the audience. Artists create scenes and images that instead of merely transmitting meaning, evoke and connote it and even puzzle us, thereby creating the need to wrestle with their meaning (Bal 14). Thus, art functions as an invitation to have a dialogue and expand beyond the singular (that might extend over time and space but is nevertheless conceived as a dialogue of at least two participants). The second avenue to underscore art's collective potential involves the direct creation of art by various participants. The last part of this volume is therefore dedicated to plural voices, conversations, and artists' collectives. Pettinaroli's study of the *cartonera* project goes beyond the bourgeois notions of the individual author, reader, and professional publisher and examines recycling as a creative practice of the literary project itself. Ilka Kressner's exploration of the dance performances *Agwa* and *Pixel* studies the creation of a political space composed of human and nonhuman interaction that binds existential and performative claims toward a creation of a political space in and of art. Returning to Sommer's "Lessons Learned from Latin America," in "acknowledging art's work" (189), we become "cultural agents" (189), connect with others, establish alliances, and become participants in a conversation of creative and productive dissent of the human-centered hubris. During our era of environmental vulnerability and ecocide, such a creative—which is also a critical and vice versa—practice of collaborative ventures may be one of the fruitful responses to rethink and change our relation to our environment.

Organization of This Volume

This volume brings together 13 original, critical studies of Latin American and Latinx writing, film, video games, and visual and performing arts addressing the topic of ongoing ecological violence. It takes seriously Nixon's call for "more than simply diversifying the canon: we need to reimagine the prevailing paradigms" (257). It proposes both a diversification of the corpus of environmental discourse and a recasting of the ecocritical tool kit from a Latin American and Latinx vantage point. The chapters of this volume are divided into four parts. The title of the first part, "Bad Living: Mutations, Monsters, and Phantoms," refers to the dark side of the concepts of *Buen Vivir*, and its essays explore bodily transformations as visible effects of slow violence's invisible harm. Ana María Mutis's "Monsters and Agritoxins: The Environmental Gothic in Samanta Schweblin's *Distancia de rescate*" examines Schweblin's engagement with the Gothic genre in her recent novella to represent the slow violence of agrochemical pollution and to criticize, at the same time, our tendency toward privileging the visible and the immediate. The novella's protagonist, a poisoned child whose soul has transmigrated to another body, incarnates the figure of the monster that in Gothic fiction fulfills the double function of showing and warning about hidden and invisible threats. For Mutis, this child-monster shares significant characteristics with the zombie, a figure linked to human and ecological exploitation within the plantation economy, and to the capitalist expansion that emerged from the implementation of this agricultural-based colonial economy. These resonances, she argues, suggest a critique of the model of agricultural production in Argentina, as the novella's combination of elements from Gothic fiction and its exploration of maternity give visibility and imminence to the gradual poisoning of the world through the biotechnological manipulation of crops and the excessive use of agrochemicals. Gisela Heffes's "Toxic Nature in Contemporary Argentine Narratives: Contaminated Bodies and Ecomutations" then analyzes the portrayals, transformation, and mutation of bodies in three novels, *Las estrellas federales* by Juan Diego Incardona, *La vi mutar* by Natalia Rodríguez, and *Distancia de rescate* by Samanta Schweblin, as well as the poetry collection *Un pequeño enfermo* by Julián Joven. Heffes asks how the degradations and mutations of these bodies become the last frontier of environmental pollution and destruction, with her analysis being informed by the concept of hyperobjects (Tim Morton) as objects that emerge in a moment of ecological crisis, that blur the material, spatial, and temporary borders. She concludes that if modernity brought about the idea of an objectified nature, the immediate present has added to this landscape by becoming an era of objectified bodies. Diana Aldrete's "The Ruins of Modernity: Synecdoche of Neoliberal Mexico in Roberto Bolaño's *2666*" then concludes this section by proposing an ecocritical and environmental justice approach that engages Rob Nixon's concept of slow violence and focuses on the portrayal of the clandestine garbage dump "El Chile" as a metonym of the fictional city of Santa Teresa, where criminal violence and industrial neglect overlap. In this abominable dust heap, the materiality of bodies is degraded to the status of corpses, creating a place where humans and industrial materials share the same space. Aldrete argues that literature allows for a deeper analysis of the systems, such as capitalism and globalization, that often fragment and obscure realities. The literary praxis counteracts this tendency by materializing the stories of the poor and making visible the harms of the environment. Situated in the fictional border town of Santa Teresa, the literary representation of Ciudad Juarez, *2666* displays the tensions lived at the border with the effects of legal and illegal production as it pertains to the maquiladoras, drug trafficking, snuff movies, and the murders of hundreds of women.

The second portion of the volume, "Econarratives and Ecopoetics of Slow Violence," includes three studies dedicated to portrayals of slow violence in filmic and literary narratives and poetics from Mexico, Peru, Bolivia, Chile, and Nicaragua. Laura Barbas-Rhoden's "The Representation of Slow Violence and the Spatiality of Injustice in *Y tu mamá también* and *Temporada de patos*" offers an ecocritical reading of the films *Y tu mamá también* by Alfonso Cuarón and *Temporada de patos* by Fernando Eimbcke through the lenses of spatial justice theory and decolonial theory. She argues that both works, though not explicitly environmental in focus, actively engage with the representation of slow violence and its consequences at different levels of scale, from the scope of the human body and psyche to the macro level of nation-scapes. She proposes that the poetic sensibility and narrative arcs of each film, analyzed with attention to the spatial dialectic and its ecopsychosocial dimension, may be read as ethical and ideological interventions to foreground processes and places that are "normalized" in globalization. Barbas-Rhoden's reading considers the commentary each film makes about agency, space, and place—and the possibility of addressing the malaise of modernity. Next, Ida Day's study, "The Voice of Water: Spiritual Ecology, Memory, and Violence in *Daughter of the Lake* and *The Pearl Button*," examines the deep spiritual connection between indigenous populations and water in two poetic documentaries by Ernesto Cabellos Damián and Patricio Guzmán. These films focus on various forms of violence against local people and the environment that are intimately linked to extractive practices such as mining and industrialized agriculture. The sea and lakes are presented as living beings who have a spirit, memory, and voice. Day elaborates how this attitude toward nature, advocated by spiritual ecology, may serve as a model of resistance against unlimited economic growth and modern forms of economic colonialism. Her chapter furthermore examines how the poetic mode expresses forms of slow violence and adequately communicates the objectives of spiritual ecology. Jacob G. Price's chapter, "From Polluted Swan Song to Happy Armadillos: The Cold War's Slow Violence in Nicaragua," then studies how Pablo Antonio Cuadra's *Cantos de Cifar y del mar dulce* and Ernesto Cardenal's *Vuelos de victoria* outline the history of slow violence in Somocista Nicaragua through the Revolution. He explores how the lyrical enunciations in both volumes elucidate how contamination incrementally damaged the environment as well as how the Sandinista government used their political victory as a moment to renegotiate human and nonhuman relationships. Both collections convey environmentalist sensibilities through the portrayals of mysterious deaths and environmental damage; by so doing, they demystify Nicaragua's environmental history. *Cantos de Cifar y del mar dulce* expresses a latent worry for irregular changes in nature that do not coincide with a literary sense of place. Similar to Barbas-Rhoden's and

Day's elaborations, Price's chapter examines how nonhumans emerge as actors, now in the context of Nicaragua's political history, first as victims of slow violence during Somocismo and later as corevolutionaries during Sandinismo. In this way, Cuadra's and Cardenal's works underscore the ecological trajectory of Nicaraguan politics throughout the Cold War.

The third part of this volume is dedicated to an exploration of "Protracted Degradation and the Slow Violence of Toxicity" in today's Latin American and Latinx Worlds. In his "Collateral Damage: Nature and the Accumulation of Capital in Héctor Aguilar Camín's *El resplandor de la madera* and Jennifer Clement's *Prayers for the Stolen*," Adrian Kane analyzes the relationship between nature and culture in both novels. He examines how *El resplandor* traces the feats, misfortunes, and scandals of one family over five generations, providing a broad historical scope for the portrayal of business practices from the colonial era to the present and for the depiction of humans' desire for profit and their indifference to the environmental implications. Jennifer Clement's *Prayers for the Stolen* offers similar insight into nature-culture relations in Mexico, now within the context of the trafficking of women, the violence of Mexican drug cartels, and ecocide. Kane's study thus portrays the systematic forms of slow violence through the representation of nature as collateral damage in the battle for the accumulation of capital. The section then continues with "Violence, Slow and Explosive: Spectrality, Landscape, and Trauma in Evelio Rosero's *Los ejércitos*" by Carlos Gardeazábal Bravo, who examines different forms of violence in Rosero's novel that is set amidst the most violent period of the Colombian civil war. He argues that, although the narrative is ostensibly focused on event-centered violence, it also provides literary insight into other forms of attritional or systemic violence. This chapter appeals to Rob Nixon's concept of slow violence and Slajov Žižek's notion of objective violence to analyze how the long-term consequences of the conflict are represented in the novel. Moreover, it connects these forms of violence with the spectral topographies present there as well and explores how *Los ejércitos* addresses collective trauma and also challenges the binaries implied by the concept of slow violence. Charlotte Rogers's chapter, "The Environmentalism of Poor Women of Color in Mayra Santos-Febres's *Nuestra Señora de la Noche*," then adds an ecocritical interpretation of Santos-Febres's novel. Drawing on the work of Rob Nixon and Greta Gaard, Rogers argues that the novel's protagonist, a woman of color named Isabel Luberza Oppenheimer, exemplifies the environmentalism of poor women of color in twentieth-century Puerto Rico. Luberza impedes the slow violence of industrial development by refusing to sell land she owns on the banks of the Portuguese River in the impoverished neighborhood of San Antón in Ponce. Santos-Febres's protagonist uses her Afro-diasporic spirituality, ecological sensitivity, and sense of community to resist the exploitation of the river and the displacement of vulnerable local residents.

The final section of this work is then dedicated to "Materialities, Performances, and Ecologies of Praxis." It begins with Lauren Woolbright's "Slow Violence in a Digital World: Tarahumara Apocalypse and Endogenous Meaning in *Mulaka*," which explores concerns among the Tarahumara about environmental degradation through an ecocritical reading of the video game *Mulaka*, created by the Chihuahua-based game development team Lienzo. In the virtual world of *Mulaka*, gamers seeks to raise awareness of Tarahumara culture and folklore. The Tarahumara people are no strangers to environmental degradation; as recently as 2017, they witnessed the assassination of environmental activist leaders in their community who protested illegal logging, and they continue to endure a decade-long drought that exemplifies the environmental degradation brought on by climate change. Woolbright describes how through the deployment of industry-standard mechanics reifying certain forms of violence, the game's endogenous meaning nonetheless conveys an ecocritical message about the outcome of slow violence, should we choose inaction. Indigenous games such as *Mulaka* demonstrate that, despite being plagued by many forms of toxicity, our deeply interconnected lands and peoples are worth fighting for. Thaiane Oliveira's "Slow Violence in the Scientific Ecosystem: Decolonial Ecocriticism on Science in the Global South" then sounds a note of caution about the dynamics of power and the epistemic violence within the scientific ecosystem from a decolonial ecocritical perspective, thus unveiling the strategies of domination imposed on scientific production from the Global South. Understanding the scientific community as an ecosystem, she seeks to bring concepts of ecology such as resilience, resistance, revolution, and networks to advance a declassification of knowledge that overcomes the abyssal divisions between North and South. Through the specific example of Brazilian scientific production on climate change, Oliveira observes new forms of colonization manifesting within science, with true knowledge becoming increasingly discredited. Next, "Bodies, Transparent Matter, and Immateriality: Compagnie Käfig's Eco-Dance Performances," by Ilka Kressner, examines dance as an art form used to engage environmental issues and communicate unsettling slow violence in two contemporary performances by Mourad Merzouki's dance troupe Compagnie Käfig: *Agwa* and *Pixel*. Given the copious and constant presence of plastic cups and pixels in both performances, this chapter explores the connections and conversions between the accentuated physicality of the human body and the transparency of plastic cups and immateriality of pixels. By doing so, it invites us to rethink notions of presence and materiality from an ecocritical perspective grounded in performance studies. Moreover, it examines the challenges of creating and referencing metaphors of intrusion and interference in the context of dance. Based on dance's nonchronological form and its emphasis on collaboration, both performances become countermodern reflections on the fragility of ecosystems, conceived as sites of dynamic and synchronous interactions of human and nonhuman agents. It is precisely the dancers' and their audiences' shared bodily perception that alerts us to the interferences related to the more-than-material spaces we inhabit and carves out a potential of agency that goes beyond the mere individual. Finally, Elizabeth M. Pettinaroli's "*Llubia Negra:* Fetishism of Form, Temporalities of Waste, and Slow Violence in Cartonera Publishing of the Triple Frontier (Paraguay, Brazil, Argentina)" explores the mysterious *lluvia negra* (black rain) that fell over Paraguay's capital city Asunción on the afternoon of April 4, 2009 and the response to the event by writers in the cooperative *cartonera* publishing house Yiyi Yambo. The manifestly inadequate scientific assessment of the phenomenon by governmental entities charged with its investigation contributed to massive public distrust. In true ecocritical practice, the authors in *Llubia Negra* collaboratively created and published a volume in which they innovatively envisioned the black rain of 2009 as an utterly extraordinary phenomenon: temporally and chronologically diffuse, unfolding on a scale unimaginably vast and utterly disproportionate to human experience, "massively distributed in time and space relative to humans"—that is, a hyperobject (Morton 1). Pettinaroli examines how the writer-activists working with *cartonera* publishers explore the dialectical relationship between environment and historic processes and expose the occlusions that complicate the representation of slow violence (Nixon). The stories' rainy eco- and cartopoesis offer liquid theorizations centered on the reconsideration of the dialectic between environmental slow violence and historical processes, provide a framework for rethinking form and matter, and foster collective communities that reimagine worlds with an ecocritical foundation and purpose at their core.

Total Eclipse of the Heart (Replay)

Returning to Ruiz's painting that opened this introduction, we look again at its long, white line that by now has come to dominate the entire painting. We read William Ospina's lucid interpretation that gives motion and context to its abstract stillness: "La parte del diablo es ese trazo blanco, al comienzo sólo una línea, más lejos una franja liviana que permanece en el aire, y al final una suerte de bruma que se va disolviendo sobre los campos" ("Pedro Ruiz y la caricia del veneno"; "The devil is in the white line, at the beginning only one line, then a light strip that remains in the air, and at the end a kind of mist that dissolves over the fields"). Art selects, singles out certain images, situations, even lines; it distorts, provides context, connects, and communicates among fields of knowledge that are usually kept separate. Its impact may be immediate, or, as in the case of Ruiz's painting and many of the works of different media analyzed in this volume, it may necessitate careful contemplation, brooding over, even lingering and revisiting, to be perceived and in order to become the inspiration to change attitudes—among those, our view of the human–nonhuman relationship within a space now perceived as an ecozone.

Bonnie Tyler's hit song that might have blasted in the cockpits of the pilots while they were spraying their horror builds on a pop cliché: "Once upon a time I was falling in love/But now I'm only falling apart/There's nothing I can do/A total eclipse of the heart" (lines 24–7). Put as a quote in the paratext of a painting about ecological poisoning, the self-absorbed, first-world lament expands, becomes part of a bigger picture (literally), finds a "we" beyond a mere "I," and starts to reassemble those pieces of the lonely heart. At the end of the song, Tyler's voice rather vaguely calls for a collective action: "together we can take it to the end of this line" (line 17). The line might, of course, be taken as referring to a line within the song (perhaps as an invitation to sing along). But looking at Ruiz's painting, we read Ospina's interpretation, according to which:

> El arte no resuelve los temas con argumentos sino con eficientes metáforas, con esos trazos inspirados, signos que evitan largas disquisiciones. . . . La flor de la sangre [la amapola] revienta bajo el vuelo de una blanca línea mortal, la diversidad del mundo desaparece bajo la monotonía de los cultivos que quieren convertir la realidad en un solo tema persistente y sangriento.
>
> ("Las sutiles transgresiones de Pedro Ruiz")

> art does not resolve issues with arguments but with efficient metaphors, with those inspired lines [brushstrokes], with signs that avoid long disquisitions. . . . The flower of blood [poppy] bursts below the flight of the white, lethal line, the world's diversity disappears under the monotony of the cultivations that aim at converting reality in a sole topic, persistent and sanguinary.

The single white line, connecting Tyler with Ruiz with Ospina with us, becomes a catalyst to think through abstraction and technological formalism and make visible and palpable their pernicious effects for us, humans and nonhumans. The chapters that follow, in all their diversity, join forces in this endeavor to appreciate ecofictions as a powerful means to think about and ultimately to help change our shared ecorealities.

Notes

1. Corporate social responsibility (CSR) has been upheld as a valid principle to reduce mining-related violence in rural communities facing the extractive operations of the industry.
2. A newly surging interest in descriptive geography within humanism, aligned with the renewal of Ptolemaic cartography and the Copernican challenge to the inherited planetary configuration, brought about new theorizations that served as the foundation for philosophical, empirical, and political interpretations

of the notion of "Nature." Displacing the knowledge received from the ancients, many scholars discarded Aristotle's abstract ideas about nature as matter and privation, and they therefore searched for more concrete explanations of natural phenomena.

3. From early on, eyewitness accounts reported that the old Ptolemaic model that tied civility to the location of peoples within the tripartite division of the globe—with a torrid belt of uninhabitable hot climates and environments in the Equinox region surrounded by two temperate regions that extended to the poles and provided the only areas where a *polis* and civility were possible—did not fully match the reality of the Indies. Bartolomé de Las Casas is first to articulate a full challenge to the spatial principles sustaining the geopolitical model of expansion, in particular in his *Historia de las Indias* and the accompanying *Apologética.* Pagden studies the ideologies behind Hapsburg imperial aspirations. Lupher traces the adoption of this model of imperialism in the debates on New World conquest.
4. Vitoria reinstated the legitimacy of the Roman model, upholding Domingo de Soto's challenge to divine legitimacy but playing down arguments of geographic limitation and exercise of good rule. Grounding legitimacy in human justice, he affirmed the virtues of the ancient imperial model towards a positive application of the model of Roman imperialism to the case of Spanish dominion in the Indies. For a full discussion of the argumentative maneuvers in this debate, see Bentancor and Pettinaroli.
5. For a longer discussion, see Pagden.
6. As Del Valle notes (442), this theorization came to sustain what Carl Schmitt labels a Christian-based "Nomos of the Earth," understood as a territorial appropriation where the sovereign resides solely in Christian European aspirations to maintain absolute power over the Earth's peoples.
7. Emiliano Aguirre notes that Acosta is the first scholar to posit the possibility of biological evolution and essays a theorization about the origins of new territories (cited in del Valle's "From José de Acosta" 440).
8. This work became a sixteenth-century best seller and greatly influenced such later thinkers as John Locke, Adam Smith, Alexander von Humboldt, and possibly Carl Linnaeus (del Valle, "From José de Acosta" 440).
9. See Charles Gibson's discussion of how the drainage of the lakes surrounding the basin of Mexico brought about the destruction of the pre-Hispanic ecosystem as well as the displacement of indigenous communities whose lives were supported by this complex precolonial aquatic ecological system. Also see del Valle's "*Grandeza Mexicana* and the Lakes of Mexico City" (46–50).
10. Del Valle characterizes these invasive, violent processes as "countless daily catastrophes [lower-case *c*] that since the early sixteenth century have destroyed the world and lives of millions of people through invasive and violent technological practices" (438).
11. Alberto Acosta explains that the long history of extractivism was established 500 years ago (62), Burchardt and Dietz describe the history of the region as a history of extractivism starting in colonial times and continuing during the nineteenth century up through today (481), and Anderson briefly traces the legacy of colonial extractivism from the sixteenth century until the present (x–xi). Furthermore, Anderson identifies the origins of current global extractive technologies and land practices in colonial and contemporary Latin America (x).
12. Latin America is indeed a resource-rich region. As Fábio de Castro, Barbara Hogenboom, and Michiel Baud explain: "Nearly half of the world's tropical forests are found in the region, next to several other natural biomes, which together carry a wealth of biodiversity. It holds one-third of the world's

freshwater reserves and one-quarter of the potential arable land. And despite five centuries of extractive activities to serve global markets, the region still holds large volumes of important mineral reserves, including oil, gas, iron, copper and gold" (1).

13. Eduardo Gudynas introduced the term in 2009 in "Diez tesis urgentes sobre el nuevo extractivismo. Contextos y demandas bajo el progresismo sudamericano actual." Studies built on this definition include, but are not limited to, those of Acosta, Burchardt and Dietz and of Castro et al.
14. Alberto Acosta defines extractivism as "those activities which remove large quantities of natural resources that are not processed (or processed only to a limited degree), especially for export. Extractivism is not limited to minerals or oil. Extractivism is also present in farming, forestry and even fishing" (62).
15. The study by Burchardt and Dietz relies on empirical evidence to demonstrate the consolidation of the extractive growth model in Latin America. By evaluating the increase in exports of primary goods and the growing share of the primary sector (agriculture, forestry, mining, and hunting) in GDPs from 2000 to 2011, the authors show a rise of extraction economies in the region.
16. Pablo Andrade includes in this list developed countries (Canada, the United States, the UK, Australia, and Norway), emerging countries (Brazil, Chile, South Africa, and Indonesia), and developing countries (Botswana, Bolivia, and Ecuador) (115). For a more detailed discussion of the debate between critics and the proponents of the extractive model, see Anthony Bebbington (5–8).
17. In her 2014 article, Gisela Heffes provides an extensive treatment of Latin American ecocritical scholarship.
18. Marcone notes that the representations of these conflicts are insufficient, for they elide the political ontologies that inform them, thus challenging environmentalism and ecological thinking as effective critical perspectives for addressing social exclusion and poverty (209).
19. Arturo Escobar summarizes this poignantly as follows: "Ecology and environmentalism imply different ways of thinking (necessarily relational, situated and historical)" (51).
20. Examples of works of art from Latin America, in addition to the ones discussed in the present volume, are the works explored by sculptural artist Doris Salcedo, such as *A flor de piel* (2012; literally "flowers touching the skin," figuratively "nerves on edge, hypersensitive"), a shroud sutured of rose petals to commemorate the disappeared victims of the civil war; or Brazilian visual artist Maria Thereza Alves's (who was also the cofounder of the Brazilian green party) project *Seeds of Change*, based on her research into old shipping routes. Between the sixteenth and nineteenth centuries, English captains returning from the American colonies had their hulls filled with earth and stones to weigh down the ships. "This ballast contained the seeds of plants from wherever the ship had sailed. Once back in Britain, the earth was offloaded into the rivers. Alves discovered that these ballast seeds can lay dormant for hundreds of years but by excavating the river bed, it is possible to germinate and grow them into flourishing plants" (Brown 136). In her project *Seeds of Change* (1999–ongoing), Alves grows those seeds into flourishing plants and living art installations and outdoor sculptures at various places in the world, one of which we visited (*Seeds of Change—A Botany of Colonization*, New York City, 2017).
21. Moreover, materialities point to an inclusive conception of art that goes beyond the written word and its colonial legacies. Tara Daly describes the Latin American context as follows: "written language, as distinguished from spoken, and Spanish and Portuguese, as relative to indigenous languages, carry colonial legacies because as imperial tools, even as they have evolved,

they continue to serve as privileged modes of telling Latin American's history" (122).

22. Our relation with material objects, in particular wood, has been analyzed by philosopher Hannah Arendt in a frequently quoted section of *The Human Condition*. Arendt specifies that "[t]he durability of the human artifice is not absolute; the use we make of it, even though we do not consume it, uses it up. The life process which permeates our whole being invades it, too, and if we do not use the things of the world they also will eventually decay, return into the over-all natural process from which they were drawn and against which they were erected. If left to itself or discarded from the human world, the chair will again become wood, and the wood will decay and return to the soil from which the tree sprang before it was cut off to become the material upon which to work and with which to build" (136–7).
23. Walter Mignolo's concept of "delinking" as a changing "of the terms of conversation, and above all, of the hegemonic ideas of what knowledge and understanding are" (313) is useful in this context.
24. Tom Bristow gives a lucid definition of the position of the human being in the context of the more-than-human world. The framework "asks us to think of the human as one part of the More-than-human world, which is to think of us not within the world but of the world; this position not only turns away from human instrumentalism but it shifts focus from the significance of human species to transcorporeality and personhood" (2). The term furthermore "remind[s] us that the nonhuman world (on which humans are absolutely dependent) has agencies of its own" (126). Bristow follows David Abram's argumentation from *The Spell of the Sensuous: Perception and Language in a More-than-Human World*, according to which we are only human when in contact with the nonhuman (126). Referencing cultural geographer Sarah Whatmore's "Materialist Returns: Practising Cultural Geography in and for a More-Than-Human World," Bristol describes the need to shift "disciplinary registers from material concerns (which speak of an external nature) to the fabric of corporeality. This shift entails a literacy of intimacy by which the human is redistributed in relational space" (126–7).

Works Cited

Abram, David. *The Spell of the Sensuous: Perception and Language in a More-Than-Human World*. Vintage, 1996.

Acosta, Alberto. *El Buen Vivir*. Icaria, 2013.

———. "Extractivism and Neoextractivism: Two Sides of The Same Curse." *Beyond Development. Alternative Visions from Latin America*, edited by Miriam Lang and Dunia Mokrani, Amsterdam, Transnational Institute, 2013, pp. 61–86.

Acselrad, Henri, et al. *Justiça ambiental e cidadania*. Relume Dumará, 2004.

Adorno, Rolena. *The Polemics of Possession in Spanish American Narrative*. Yale UP, 2008.

Ahmann, Chloe. "'It's Exhausting to Create an Event Out of Nothing': Slow Violence and the Manipulation of Time." *Cultural Anthropology*, vol. 33, no. 1, 2018, pp. 142–71.

Alimonda, Héctor. *La naturaleza colonizada: Ecología política y minería en América Latina*. CLACSO, 2011.

Alves, Maria Thereza. *Seeds of Change: A Botany of Colonization*. Installations at Different Sites in New York City, 2017.

Anderson, Mark. "Introduction: The Dimensions of Crisis." *Ecological Crisis and Cultural Representation in Latin America: Ecocritical Perspectives on Art, Film, and Literature*, edited by Mark Anderson and Zélia Bora, Lexington Books, 2017, pp. ix–xxxii.

Andrade, Pablo A. "The Government of Nature: Post-Neoliberal Environmental Governance in Bolivia and Ecuador." *Environmental Governance in Latin America*, edited by Fabio de Castro et al., Palgrave Macmillan, 2016, pp. 113–36.

Arendt, Hannah. *The Human Condition*. U of Chicago P, 1958.

Bal, Mieke. *Of What One Cannot Speak: Doris Salcedo's Political Art*. Chicago, U of Chicago P, 2010.

Barbas-Rhoden, Laura. *Ecological Imaginations in Latin American Fiction*. UP of Florida, 2011.

Barnwell, Ashley. "Family Secrets and the Slow Violence of Social Stigma." *Sociology*, 8 May–June, 2019, doi:10.1177/0038038519846443.

Bebbington, Anthony. "Extractive Industries, Socio-Environmental Conflicts, and Political Economic Transformations in Andean America." *Social Conflict, Economic Development and the Extractive Industry: Evidence from South America*, edited by Anthony Bebbington, Routledge, 2011, pp. 3–26.

Bentancor, Orlando. *The Matter of Empire: Metaphysics and Mining in Colonial Peru*. U of Pittsburgh P, 2017.

Bristow, Tom. *The Anthropocene Lyric: An Affective Geography of Poetry, Person, Place*. Palgrave Macmillan, 2015.

Brown, Andrew. *Art and Ecology Now*. Thames and Hudson, 2014.

Burchardt, Hans-Jürgen, and Kristina Dietz. "(Neo-)Extractivism: A New Challenge for Development Theory from Latin America." *Third World Quarterly*, vol. 35, no. 3, 16 Mar. 2014, pp. 468–86.

Caña Jiménez, María Del Carmen. "Violencia, necropolítica y capitalocene en *Cromo*." *Revista Canadiense de Estudios Hispánicos*, vol. 42, no. 1, 2017, pp. 1–23.

Carruthers, David V. *Environmental Justice in Latin America: Problems, Promise, and Practice*. MIT Press, 2008.

Castro, Fabio de, et al. "Introduction: Environment and Society in Contemporary Latin America." *Environmental Governance in Latin America*. Springer, 2016, pp. 1–25.

Cecire, Natalia. "Environmental Innocence and Slow Violence." *WSQ: Women's Studies Quarterly*, vol. 43, no. 1, 2015, pp. 164–80.

Coddington, Kate. "The Slow Violence of Life without Cash: Borders, State Restrictions, and Exclusion in the U. K. and Australia." *Geographical Review*, 27 Dec. 2018, doi:10.1111/gere.12332.

Crosby, Alfred W. *Ecological Imperialism: The Biological Expansion of Europe, 900–1900*. Cambridge UP, 2015.

Crutzen, Paul J. "Geology of Mankind." *Nature*, vol. 415, no. 6867, Jan. 2002, p. 23, doi:10.1038/415023a, www.nature.com.

Csikszentmihalyi, Mihaly. "Why We Need Things?" *History from Things: Essays on Material Culture*, edited by Steven Luba and W. David Kingery, Smithsonian, 1995, pp. 20–9.

Cuesta Domingo, Mariano, ed. *Alonso de Santa Cruz y su obra cosmográfica*. Consejo Superior de Investigaciones Científicas, Instituto "Gonzalo Fernández de Oviedo," 1983.

Daly, Tara. "The Air as Decolonial Critique of Being in César Calvo's *Las tres mitades de Ino Moxo y otros brujos de la Amazonía*." *Decolonial Approaches to Latin American Literatures and Cultures*, edited by Juan G. Ramos and Tara Daly, Palgrave Macmillan, 2016, pp. 121–39.

De la Cadena, Marisol. *Earth Beings: Ecologies of Practice Across Andean Worlds*. Duke UP, 2015.

Del Valle, Ivonne. "From José de Acosta to the Enlightenment: Barbarians, Climate Change, and (Colonial) Technology as the End of History." *The Eighteenth Century*, vol. 54, no. 4, 2013, pp. 435–59.

———. "*Grandeza Mexicana* and the Lakes of Mexico City: Economy and Ontology in Colonial Technological Development." *Troubled Waters: Rivers in Latin American Imagination*, edited by Elizabeth M. Pettinaroli and Ana Maria Mutis, Hispanic Issues Online 12, 2013, pp. 38–54.

ECLAC, Economic Commission for Latin America and the Caribbean. *Foreign Direct Investment in Latin America and the Caribbean 2018*. CEPAL, 2018. www.cepal.org/en/publications/43690-foreign-direct-investment-latin-america-and-caribbean-2018.

Erler, Carolyn. "Targeting 'Plan Colombia': A Critical Analysis of Ideological and Political Visual Narratives by the Beehive Collective and the Drug Enforcement Administration Museum." *Studies in Art Education*, vol. 50, no. 1, Fall, 2008, pp. 83–97.

Escobar, Arturo. "Worlds and Knowledges Otherwise: The Latin American Modernity/Coloniality Research Program." *Globalization and the Decolonial Option*, edited by Walter D. Mignolo and Arturo Escobar, Routledge, 2010, pp. 33–64.

Galeano, Eduardo. *Las venas abiertas de América Latina*. Siglo XXI, 1971.

Gamu, Jonathan Kishen, and Peter Dauvergne. "The Slow Violence of Corporate Social Responsibility: The Case of Mining in Peru." *Third World Quarterly*, vol. 39, no. 5, 2018, pp. 959–75.

Gibson, Charles. *The Aztecs Under Spanish Rule: A History of the Indians of the Valley of Mexico, 1519–1810*. Stanford UP, 1964.

Giroux, Henry, and Laura Proasi. "Los maestros se levantan para resistir los ataques neoliberales contra la educación/Teachers Are Rising up to Resist Neoliberal Attacks on Education." *Revista de Educación*, vol. 16, 2019, pp. 13–38.

Gudynas, Eduardo. "Diez tesis urgentes sobre el nuevo extractivismo. Contextos y demandas bajo el progresismo sudamericano actual." *Extractivismo, política y sociedad*, edited by Jürgen Schuldt et al., CAAP/CLAES, 2009, pp. 187–225.

———. "Estado compensador y nuevos extractivismos. Las ambivalencias del progresismo sudamericano." *Nueva Sociedad*, 1 Jan. 2012, pp. 128–46.

———. "Las disputas sobre el desarrollo y los sentidos de las alternativas." *Revista Kavilando*, vol. 6, no. 1, 2015, pp. 15–29.

Heffes, Gisela. "Introducción: Para una ecocrítica latinoamericana: entre la postulación de un ecocentrismo crítico y la crítica a un antropocentrismo hegemónico." *Revista de Crítica Literaria Latinoamericana*, vol. 40, no. 79, 2014, pp. 11–34.

Hesse, Isabelle. "Sensory Siege: Dromocolonisation, Slow Violence, and Poetic Realism in the Twenty-First Century Short Story from Gaza." *Journal for Cultural Research*, vol. 21, no. 2, 2017, pp. 190–203.

Holterman, Devin. "Slow Violence, Extraction and Human Rights Defence in Tanzania: Notes from the Field." *JRPO Resources Policy*, vol. 40, 2014, pp. 59–65.

Hutchings, Rich, and Marina La Salle. "Arqueologia como Capitalismo do Desastre." *Revista de Arqueologia*, vol. 28, no. 2, 2015, pp. 20–44.

Ingegno, Alfonso. "The New Philosophy of Nature." *The Cambridge History of Renaissance Philosophy*, edited by Charles Schmitt et al., Cambridge UP, 1988, pp. 236–63.

Ingold, Timothy. "Globes and Spheres: The Topology of Environmentalism." *The Perception of the Environment*. Routledge, 2000, pp. 209–18.

Iom, Programa RPR. "Spotlight: Reclutamiento de niños, niñas y adolescentes: la retención como violencia lenta." Repositorio de Información de la Organización Internacional para las Migraciones OIM, 2019.

Iovino, Serenella, and Serpil Oppermann. "Introduction: Stories Come to Matter." *Material Ecocriticism*, edited by Serenella Iovino and Serpil Oppermann, Indiana UP, 2014, pp. 1–17.

Kohn, Eduardo. *How Forests Think: Toward an Anthropology Beyond the Human*. U of California P, 2013.

Lander, Edgardo. *La colonialidad del saber*. CLACSO/UNESCO, 2000.

Latta, Alex, and Hannah Wittman, eds. *Environment and Citizenship in Latin America: Natures, Subjects and Struggles*. Berghahn Books, 2012.

LeBel, Sabine. "Fast Machines, Slow Violence: ICTs, Planned Obsolescence, and E-Waste." *Globalizations*, vol. 13, no. 3, 2016, pp. 300–9.

Leff, Enrique. *Justicia ambiental: construcción y defensa de los nuevos derechos ambientales culturales y colectivos en América latina*. PNUMA Red de Formación Ambiental, 2001.

Leff, Enrique. "Political Ecology: A Latin American Perspective." *Desenvolvimiento e Meio Ambiente*, vol. 35, Dec. 2015, pp. 29–64.

———. *Racionalidad ambiental: la reapropiación social de la naturaleza*. Siglo XXI, 2004.

Liu, Xinmin, and Hua Li. "The Environment, Humankind, and Slow Violence in Chinese Science Fiction." *Communication and the Public*, vol. 3, no. 4, 2018, pp. 270–82.

Llamas-Rodriguez, Juan. "Toward a Cinema of Slow Violence." *Film Quarterly*, vol. 71, no. 3, 2018, pp. 27–36.

López, Ronald Campos. "De lagos y lagunas en la ecopoesía hispánica contemporánea." *Revista Estudios*, 2018, https://revistas.ucr.ac.cr/index.php/estudios/article/view/35001/34563.

Lupher, David A. *Romans in a New World: Classical Models in Sixteenth-Century Spanish America*. U of Michigan P, 2003.

Marcone, Jorge. "Filming the Emergence of Popular Environmentalism in Latin America: Postcolonialism and Buen Vivir." *Global Ecologies and Environmental Humanities: Postcolonial Approaches*, edited by Elizabeth DeLoughrey and Anthony Carrigan, Routledge, 2015, pp. 207–25.

Martínez, Juliana. "Fog Instead of Land, Spectral Topographies of Disappearance in Colombia's Recent Literature and Film." *Espectros, Ghostly Hauntings in Contemporary Transhispanic Narratives*, edited by Alberto Ribas-Casayas and Amanda L. Petersen, Bucknell UP, 2016, pp. 117–31.

Martinez-Alier, Joan. "Conflictos ecológicos y justicia ambiental." *Papeles de relaciones ecosociales y cambio global*, vol. 103, 2008, pp. 11–27.

———. *The Environmentalism of the Poor: A Study of Ecological Conflicts and Valuation*. Edward Elgar Publishing, 2003.

Masiello, Francine. *The Art of Transition: Latin American Culture and Neoliberal Crisis*. Duke UP, 2001.

Massey, Rachel. "The 'Drug War' in Colombia: Echoes of Vietnam." *Journal of Public Health Policy*, vol. 22, no. 3, 2001, pp. 280–5.

McNeish, John Andrew. "Resource Extraction and Conflict in Latin America." *Colombia Internacional*, 25 Jan. 2018, pp. 3–16, https://revistas.uniandes.edu.co/doi/abs/10.7440/colombiaint93.2018.01.

Merlinsky, Iliana. *Política, derechos y justicia ambiental. El conflicto del Riachuelo*. Fondo de Cultura Económica, 2012.

Mignolo, Walter D. "Delinking: The Rhetoric of Modernity, the Logic of Coloniality, and the Grammar of De-Coloniality." *Globalization and the Decolonial Option*, edited by Walter D. Mignolo and Arturo Escobar, Routledge, 2010, pp. 303–68.

———. *La idea de América Latina: La herida colonial y la óptica decolonial*. Gedisa, 2005.

———. *Local Histories/Global Designs*. Princeton UP, 2000.

———. "On Pluriversality and Multipolarity." *Constructing the Pluriverse: The Geopolitics of Knowledge*, edited by Bernd Reiter, Duke UP, 2018, pp. 90–116.

———. *The Darker Side of Western Modernity: Global Futures, Decolonial Options*. Duke UP, 2011.

Mignolo, Walter D., and Arturo Escobar, eds. *Globalization and the Decolonial Option*. Routledge, 2009.

Morton, Timothy. *Hyperobjects: Philosophy and Ecology After the End of the World*. U of Minnesota P, 2013.

Mowforth, Martin, and Ian Munt. *Tourism and Sustainability: Development, Globalisation and New Tourism in the Third World*. Routledge, 2015.

Nixon, Rob. *Slow Violence and the Environmentalism of the Poor*. Harvard UP, 2011.

O'Lear, Shannon. "Climate Science and Slow Violence: A View from Political Geography and STS on Mobilizing Technoscientific Ontologies of Climate Change." *Political Geography*, vol. 52, May 2016, pp. 4–13.

Ospina, William. "Las sutiles transgresiones de Pedro Ruiz." *Pedro Ruiz*, www.pedroruiz.co/textos.

———. "Pedro Ruiz y las caricias del veneno." *Pedro Ruiz*, www.pedroruiz.co/love-is-in-the-air.

Pagden, Anthony. *Lords of All the World: Ideologies of Empire in Spain, Britain and France c.1500–c.1800*. Yale UP, 1995.

Pérez, Moira. "Violencia epistémica: Reflexiones entre lo invisible y lo ignorable." *El lugar sin límites: Revista de Estudios y Políticas de Género*, vol. 1, no. 1, 2019, pp. 81–98.

Pérez-Melgosa, Adrián. "Low-Intensity Necropolitics: Slow Violence and Migrant Bodies in Latin American Films." *Arizona Journal of Hispanic Cultural Studies*, vol. 20, 2016, pp. 217–36.

Pettinaroli, Elizabeth M. "Watershed of Sorrows: The Epic of Impossibility and New Theorizations of Tropicalia in *El epítome de las conquistas del Nuevo*

Reino de Granada." *Troubled Waters: Rivers in Latin American Imagination*, edited by Ana María Mutis and Elizabeth M. Pettinaroli, *Hispanic Issues Online* 12, 2013, pp. 19–36.

Porto-Gonçalves, Carlos Walter, and Enrique Leff. "Political Ecology in Latin America: The Social Re-Appropriation of Nature, the Reinvention of Territories and the Construction of an Environmental Rationality." *Desenvolvimento e Meio Ambiente*, vol. 35, 2015. Research Gate, doi:10.5380/dma.v35i0.43543.

Quijano, Alonso. "Coloniality of Power, Eurocentrism, and Social Classification." *Coloniality at Large: Latin America and the Postcolonial Debate*. Duke UP, 2008, pp. 181–224.

Ruiz, Pedro. "Total Eclipse of the Heart." 2006, www.pedroruiz.co/love-is-in-the-air?lightbox=image_234p.

Salamanca Villamizar, Carlos, et al. "Trayectorias de las (in)justicias en América Latina: Un estudio introductorio." *Justicias e injusticias espaciales*, editado por Bernard Bret et al., Editorial de la Universidad de Rosario, 2016, pp. 11–65.

Salcedo, Doris. *A flor de piel*. Cambridge, Harvard Art Museum, 2012.

Santos-Granero, Fernando. "Introduction." *The Occult Life of Things: Native Amazonian Theories of Materiality and Personhood*. U of Arizona P, 2009, pp. 1–29.

Schmitt, Carl. *The Nomos of the Earth in the International Law of Jus Publicum Europaeum*. Translated by G. L. Ulmen. Telos Press Publishing, 2006.

Skotnicki, Tad. "Unseen Suffering: Slow Violence and the Phenomenological Structure of Social Problems." *Theory and Society: Renewal and Critique in Social Theory*, vol. 48, no. 2, 2019, pp. 299–323.

Soja, Edward W. *Seeking Spatial Justice*. U of Minnesota P, 2010.

Sommer, Doris. "Lessons Learned from Latin America." *Modern Language Quarterly*, vol. 77, no. 2, June 2016, pp. 175–91, doi:10.1215/00267929-3464850.

———. *The Work of Art in the World*. Duke UP, 2014.

Sousa Santos, Boaventura de. *Epistemologies of the South: Justice Against Epistemicide*. Routledge, 2014.

Strupples, Peter. "Introduction." *Art and Future: Energy, Climate, Cultures*, edited by Peter Strupples, Cambridge Scholars, 2018, pp. 1–5.

Ureta, Sebastián, et al. "Sujetos de desecho: violencia lenta e inacción ambiental en un botadero minero abandonado de Chile." *Canadian Journal of Latin American and Caribbean Studies/Revue canadienne des études latino-américaines et caraïbes*, vol. 43, 2018, pp. 337–55.

Whatmore, Sarah. "Materialist Returns: Practising Cultural Geography in and for the More-Than-Human World." *Cultural Geographies*, vol. 13, no. 4, pp. 600–9.

Part I

Bad Living

Mutations, Monsters and Phantoms

1 Monsters and Agritoxins

The Environmental Gothic in Samanta Schweblin's *Distancia de Rescate*[1]

Ana María Mutis

In Samanta Schweblin's short novel *Distancia de rescate* (2015) (*Fever Dream*, 2017), terror emanates from the first line: "Son como gusanos" (7) ("They're like worms" 1), says David, a nine-year-old boy to Amanda, a woman dying in a hospital.[2] Worms, as we later find out, is the simile David uses to describe the physical sensation of agrochemical poisoning. This frightening image opens the sinister dialogue between these characters around which Schweblin constructs a fable of the destruction of the Argentine countryside through toxic agriculture. The nightmare of ecocide that the novel depicts can be traced back to 1996, the year Monsanto's genetically modified soy was approved in Argentina as an integral part of a new model of agriculture: no-till farming. This technique, which does not require ploughing and which depends primarily on agrochemicals to prepare the soil and eliminate weeds, depends on great quantities of glyphosate-based herbicides and genetically modified (GM) seeds that resist these herbicides.[3] The widespread adoption of no-till farming in Argentina spiked the production of transgenic soy and, consequently, the use of glyphosate.[4]

Among the environmental costs of the increase in the direct-seeding of transgenic soy is the degradation of soil, the appearance of new pests and weeds resistant to glyphosate, and the deforestation of ecoregions through intensive agriculture (Pengue). Though the health impacts of glyphosate have been the subject of much debate, in 2015 the International Agency for Research on Cancer (IARC), an agency of the World Health Organization, declared that this pesticide could have carcinogenic effects. This announcement lent credence to studies demonstrating the toxicity of the herbicide and its relationship to the incremental surge in cancer rates and genetic malformations that had been ignored by the scientific community and the government until then.[5]

Agrochemical aggression toward the population and the environment adheres to the definition of slow violence that Rob Nixon offers in his book *Slow Violence and the Environmentalism of the Poor*, since it constitutes a form of violence that occurs slowly and almost imperceptibly. Because the devastating effects of slow violence are dispersed over time and space, they do not capture the public's attention the way other more spectacular catastrophes, such as natural disasters, do. Nor do they provoke the same repudiation as other forms of more visible and immediate violence, which conform to the traditional notion of violence as "a highly visible act that is newsworthy because it is event focused, time focused, and body bound" (3). Nixon explains that in order to confront slow violence, we must overcome the difficulties of representation caused by its dilated temporality and limited visibility. The challenge, says Nixon, is in "how to devise arresting stories, images, and symbols adequate to the pervasive but elusive violence of delayed effects" (3). Yet in order to achieve a deeper awareness of the catastrophic consequences of slow violence, we must also question our partiality to the visible. Following this concern, the critic asks, "How do we both make slow violence visible yet also challenge the privileging of the visible?" (15).

This chapter explores Schweblin's engagement with Gothic fiction in *Distancia de rescate* to resolve the difficulties of representing agrochemical pollution and, at the same time, to criticize the tendency to privilege the visible and the immediate, which makes the silent and hidden damage of slow violence easier to ignore. The catastrophic consequences of the abuse of pesticides become obvious and urgent in *Distancia de rescate*, thanks to Gothic fiction's ability to give form to invisible threats (Haber 2). In its origins in the eighteenth century,[6] Gothic literature deployed horror, the supernatural, the irrational, and especially the ineffable to manifest a lack of confidence in the pretensions of the Enlightenment's ability to explain the world through reason and logic. In the nineteenth and twentieth centuries, the Gothic aesthetic gave voice to societal fears about industrial and technological changes (Del Principe 2). As such, its application in contemporary literature to anxieties around environmental problems can be seen as a logical evolution. As Kelly Hurley notes, the Gothic is "a cyclical genre that reemerges in times of cultural stress in order to negotiate anxieties for its readership by working through them in displaced (sometimes supernaturalized) form" (194). This is demonstrated by its growing use in contemporary narratives that deal with environmental destruction and climate change.

This is the case with *Distancia de rescate*, a novel with an environmental theme that adopts various conventions from Gothic literature to represent the dangers of the pesticide-filled countryside. As previously mentioned, the work opts to narrate through a dialogue between little David and Amanda, and from this conversation, two interconnected stories are revealed. On the one hand, there is Amanda's story. She has come to the countryside on holiday with her young daughter, Nina, and suddenly finds herself alone in a hospital on the verge of death, without understanding why. The events that led her to this place and the whereabouts of her daughter are discovered thanks to David's insistence that she search her hazy remembrances until she finds the exact moment that her tragedy began. In this reconstruction, Amanda reveals to us a second, more remote story: David's, the child with whom she is speaking. Carla, David's mother, told Amanda that when the boy was only three, he suffered from poisoning after drinking water from the river. His mother brought him to a healer who performed a healing ritual whereby his soul transmigrated to another unknown body, taking part of the poisoning with it. Another soul then occupied the boy's body and waged a battle against the remaining toxins. Thus, David survived the poisoning, but he is no longer himself; another soul occupies his body.

In her book *The Coherence of Gothic Conventions*, Eve Kosofsky Sedgwick offers a list of Gothic elements, among them the use of narrative discontinuity that seeks to reconstruct a hidden past. The presence of multiple narrators, in this case Amanda and David, who, together, spin two stories and three different time periods—recent past, present, and remote past—is a frequent device in Gothic narrative to portray the difficulty of saying the unsayable and to reveal a secret story (Sedgwick 8). Similarly, semiconscious or dreaming states are another Gothic element we find in Schweblin's novel, in her depiction of Amanda, who confuses her memories with dreams and even wonders if her conversation with David is actually happening. Amanda has trouble remembering and narrating what happened because, according to her, she is "anclada en este relato" (13) (anchored in this tale, my translation).[7] Confronted with David's insistence that she put the events in order and narrate them, Amanda responds anxiously: "lo veo perfectamente, pero a veces me cuesta avanzar" (13) ("I can see the story perfectly, but at times it's hard to move forward" 5). Amanda's torpor and confusion are effects of her poisoning and the reason why she has told David the same story four times without ever completing it or even realizing that she has recounted it before. The repetition of Amanda's story has a counterpoint in the constant reiteration by David of sentences such as "eso no es importante" ("that is not important") and "estamos perdiendo el tiempo" ("we're wasting time"), which frame the narrative as circular and reinforce the idea of enclosure. If on a communicative level Amanda is trapped in her story, on a physical level she is stuck in a hospital that she won't escape alive.

Situating the stories in closed, often subterranean spaces is another Gothic element according to Sedgwick (8). The doors of the hospital in which Amanda and David find themselves can't be opened from within, and this physical enclosure is replicated in a suffocating narrative, in turn itself a metaphor for the reclusion of illness. Amanda is trapped in her poisoned body, confined to a place she won't leave alive and tied to a narrative that reaches no conclusion. Just as Amanda's captivity in the emergency room redefines the hospital as prison, the countryside is presented as a malignant space that poisons and destroys and that is also impossible to escape, as Amanda and later her husband will find out. Soy comes up a few times in the narration, and its description as green and perfumed constitutes a seductive but threatening presence. Interior and exterior spaces are similarly toxic and claustrophobic, relaying the imprisonment of poisoning. Sharon Rose Yang and Kathleen Healy explain that Gothic landscapes, whether natural or human-made, are more than the background in which action takes place; they are the principle vehicle to create "[an] ambience of uncertainty, delusion, fluidity, isolation, and instability" (5). *Distancia de rescate*'s rural landscape inverts the traditional association of the countryside retreat as a space of leisure and recreation to one of oppression and mortality, and the hospital from a place of healing and recovery to a prison cell where one meets certain death. The Gothic register under which these sinister mutations of space and nature are effected point to agrochemical pollution as a violence that asphyxiates, traps, and kills.

Another convention that Sedgwick mentions is the presence of duplication (8). This might be the most revealing Gothic element in Schweblin's novel because it serves as a metaphor for pollution. *Distancia de rescate* is constructed on the number two: two mothers (Carla and Amanda), two children (Nina and David), and two husbands (Omar and Amanda's) in almost identical circumstances, since in both families the children suffer from agritoxin poisoning and are subject to the transmigration of their souls. Moreover, David is Nina's double: he presages the girl's fate, and in the end he will become her. This is evident toward the end of the novel when Nina's father gets into his car and finds David sitting in the back seat, with Nina's stuffed animal, the seat belt tied and his legs crossed, just as the girl used to sit in the car. David's desperate pleading look to leave with Nina's father confirms the suspicion that Nina's soul has transmigrated to David's body and that she also finds herself trapped in a strange place. Nina had announced it before in her preference for the first person plural and when in Amanda's dream she had said to her mother: "Soy David" (56) ("I'm David" 74).

In this way, the uncontrolled propagation of agrochemical poisons is paralleled in the Gothic code to the dispersion of souls invading foreign bodies. The transmigration of souls is in a way a form of duplication, since each child becomes two: the body of one person and the soul of another. Although, as the healer explains, each body harbors only one spirit, in the transmigration process "Algo de cada uno quedaría en el otro" (28) ("Something of each of them would be left in the other" 29–30). Duplication is, as Freud contends, directly related to the *unheimlich*, or uncanny, that frightening feeling produced by something that is both unknown and familiar, in this case a beloved child possessed by an alien spirit. Schweblin heightens the horror of splitting and altering a child's identity by showing it through the lens of maternity. Having a stranger inhabit the body of one's child is to disturb what is closest and most loved, bringing the Gothic mechanism of transforming the familiar into the uncanny to an intimate extreme.[8]

The theme of maternity is central in *Distancia de rescate* and invites explorations from multiple theoretical approaches. Among them and of particular relevance to the present study is the relationship between maternal narrative and ecological concerns and how this is connected to the Gothic genre in the novel. At first glance, we can see that this novel rests on the traditional identification in Western culture of nature with motherhood by presenting both as victims of agrochemical pollution. Moreover, the work adopts an ecofeminist rhetoric that Sherilyn McGregor has termed "Ecomaternalism," which ties motherhood to a greater concern for the environment. According to McGregor, this rhetoric, prevalent in contemporary ecofeminism, explains the environmental activism of women through their maternal vocation and work, since caring for children and a concern for their future well-being leads to a greater engagement with environmental issues.[9] These ideas are present in *Distancia de rescate*, which calls on the maternal instinct and the protector function of the mother to give us the drama of environmental destruction within a maternal discourse.

But it's through the intersection of the maternal discourse with the Gothic genre that *Distancia de rescate* most powerfully politicizes motherhood in order to advance an ecological message.[10] By way of the Gothic, Schweblin invokes the figure of the missing child and the mother as a political subject to make the slow violence of agrochemical pollution visible and urgent. From the title—*Distancia de rescate* (Rescue Distance) is the name Amanda gives to the invisible thread that links her to her daughter Nina—and throughout the novel we see that the work centers on the maternal anxiety around the loss of a child. This anxiety is inserted into a Gothic register, made clear in David's soul transmigration and Carla's persistent fear that her son shelters an alien spirit while his soul inhabits an unknown body. But it is her search for her son's soul, checking all the kids his age, talking to them, and looking them in the eyes to see whether she can see David in them, that shows the desperation of a mother seeking to find her beloved child who is not dead but has disappeared. The horror of her child's vanishing haunts Amanda as well, who in her conversations with David asks, unceasingly and with ever greater dread and anguish: "¿Dónde está Nina?" ("Where is Nina?").

The Gothic genre also shapes maternal anxieties in the novel by associating the invisible thread with which Amanda defines the mother–child bond with the eerie presence of numerous sisal ropes in the narrative. The most chilling of these ropes is the one used by the healer to tie David down to prevent his body from escaping during the transmigration. The boy also uses a sisal thread to tie down different objects in an attempt to put order in his house after his mother abandons him. It's notable that David uses this same yarn to tie, from a single nail, a photo of his father and one of his mother, and, below them, photos of his father with horses, as if tracing a family tree to which he no longer belongs. In this ancestral composition, the thread that connects the boy to his mother and father is absent. Threads that tense, break, shorten, or disappear lurk in the story as ominous symbols of the fragile bond that ties a mother to her child.

Embedded in the fragility of this bond is the idea that the maternal instinct to protect the offspring fails in the presence of the imperceptible threat of agrochemical poisoning. Amanda's attempt to protect Nina from harm through her obsessive attention to the imaginary thread that binds them together—the rescue distance—is as ineffective as it is misguided. This suggests that the harm done by a toxic environment is so strong as to render our natural mechanisms for survival useless and is so intangible that any attempt to measure its risks—hence the use of the word "*distancia*"—is deemed futile.[11]

Inflected by the Gothic, motherhood is presented in the novel through a loss, through the violent and mysterious disappearance of a child. Carla's anxiety about the uncertain location of her son's soul, her unfruitful search for the real David, as well as Amanda's desperation at not knowing Nina's whereabouts while she talks to David, evokes the figure of the mother in search of a child disappeared by the military dictatorship in Argentina from 1976 to 1983. As Nora Domínguez explains, the emergence of the political group Mothers of the Plaza de Mayo inserts the figure of the mother in the Argentine political sphere while modifying the place of enunciation of maternal discourse, in the voices of mothers who call on the State to explain the disappearance of their children (21). For Domínguez, the discursive position of the Mothers and Grandmothers of the Plaza de Mayo and the subsequent politicizing of motherhood have an impact in Argentine literature and culture. As Domínguez explains, until this moment the hegemonic maternal myth in Argentine culture was told by the adult child. With the arrival of the Mothers and Grandmothers of the Plaza de Mayo, a new story of motherhood is formulated, one in which the mother acquires a political identity and takes control of the discourse.

A recent example of the politicization of motherhood that specifically addresses the noxious effects of glyphosate-based herbicides in the population is the Mothers of Ituzaingó, a group of mothers from a neighborhood in the outskirts of the city of Córdoba that borders soybean farms. As Florencia Arancibia describes in her study on collective action against the growth of bioeconomy in Argentina, this group of mothers started working in the early 2000s to change regulations on the use of glyphosate and ban the spraying of this chemical near their homes. With the help of local physicians, they conducted their own scientific survey that documented 200 cases of cancer in a population of 5,000 inhabitants. In their report, published in 2005, the mothers denounced the "subtle dynamics of concealment and invisibility" of the soybean complex, and they appropriately described their efforts as "making the invisible visible" (quoted in Arancibia 84).

A fictional politicization of motherhood can be seen in *Distancia de rescate* in the character of Amanda, who is given the task of "making the invisible visible" by narrating the destruction of the Argentine countryside because, as David insists, "es importante, es muy importante para todos" (11) ("it's important, very important for us all" 2). Although there is no thematic connection to the historical violence of state terrorism in the novel, it is possible to see some allusions to its horror in the urgent search of these mothers for their missing children. Domínguez reminds us that by assassinating mothers and appropriating their children, military leaders wrought a violence of "desmaternalización y des-filiación" ("de-mothering and de-filiation") in which "se cortan los lazos maternos y familiares de transmisión, se fabrican adopciones aberrantes y espurias" (291) ("maternal and family ties are cut, spurious and aberrant adoptions are fabricated"). Similarly, Schweblin depicts the violence of agrochemical pollution as one that destroys family ties, tears children from their mothers, and in a supernatural way creates "spurious and aberrant adoptions." Moreover, the child whose soul is elsewhere but refuses to vanish completely shares with the *desaparecidos* its present absence and the uncertainty of its fate. The echoes of the historical violence of state terrorism in Schweblin's novel make the violence of agrochemical pollution concrete and gives a sense of urgency to the situation, by associating it with a past that remains painful. However, by turning the missing child into one that can be seen but is no longer there, the novel questions our ability to see and recognize the horrific effects of agrochemical violence.

Thus the Gothic elements in *Distancia de rescate* give visible form to the violence of agrochemical pollution, while at the same time showing a profound distrust of visibility. By using the supernatural to describe a silent and invisible enemy and the Gothic register to imbue the maternal fable with allusions to the violence of state terrorism, *Distancia de rescate* signals and resolves one of the problems of representation of slow violence: its scarce visibility. The Gothic lends itself to illustrating the surreptitious and prolonged violence of agrochemical toxins precisely because, as Alison Rudd puts it, "the Gothic has a particular relation to the unseen, the unsaid and the silencing in history and through language" (5). Of all the Gothic elements mentioned so far, none captures quite as vividly the horrors of toxic agriculture or gives it visibility quite like the monster.

Given its ontological richness, the monster in literature has been the object of many studies. Its name derives from the Latin *monere*, meaning "to warn," and from *monstrare*, meaning "to show." The monster is a construction designed to reveal and warn, and its marginal, transgressive, and anomalous nature has been used to interrogate and perturb the status quo (Moraña 41), question our prejudices toward difference (Cohen, "Monster Culture [Seven Theses]" 20), prevent the crossing of certain boundaries (Rudd 21), uncover what has been erased or repressed

(Moraña 42), and reveal future calamities (Moraña 50). All of these functions point toward the monster as "the ultimate incorporation of our anxieties—about history, about identity, about our very humanity" (Cohen, "Preface" xii). Its typology is varied, but the common characteristics of all monsters are their hybrid nature—a monster is a combination of elements or properties of different living beings or the amalgam of human and animal or mechanical parts—and their difference.[12]

From this brief description we can see how David incarnates the figure of the monster. David is a hybrid being who, as the product of combining a body with an alien spirit, fits Mabel Moraña's typology: "Para construir lo monstruoso debe quebrarse la unidad orgánica, la fluidez entre cuerpo y alma, la armonía psicosomática, sustituyéndola por el *pastiche* que hace de todo monstruo un simulacro de humanidad" (36; "to construct the monstrous, organic unity, the fluidity between body and soul, the psychosomatic harmony must be broken, substituting in its place a *pastiche* that makes all monsters a simulacra of humanity"). Likewise, David is a dislocated being in whom presence and absence converge since, as mentioned, he is there but his soul is elsewhere. David's present/absent state is also a liminal state between life and death that manifests itself in the close relationship between the boy and dying animals. Before his mother's frightened eyes, David attracts poisoned animals. They seek him out, staggering toward him, before they die. The boy, as though hypnotized, looks them in the eyes, urges them on, watches them die, mourns and buries them. David's portrayal as a being that bridges the dead and the living is intensified through the simile of worms to describe his poisoning. Alluding to a state of postmortem decomposition, this image relates David to the living dead. Because of his behavior, his hybrid nature, and because he is no longer himself but another, his mother, Carla, calls him a monster (34).

It's revealing that, though David incarnates various monstrous attributes, the principal among them is actually the least noticeable. Critics agree that a monster is a monster above all because of a physical anomaly. Moraña affirms, "La monstruosidad se construye sobre la base irrebatible de la corporeidad" (12; "monstrosity is constructed on the irrefutable basis of corporality"), but in David's case his monstrosity is not glaringly visible. Though it is true that poisoning—or transmigration—have left white splotches on his skin, they are not monstrous, and even Amanda describes them as subtle (52).[13] It's only during her conversation with David that Amanda starts to pay attention to the boy's skin blemishes, as though it is in the process of remembering and narrating her poisoning that the physical markers of David's misfortune become visible. Even so, David's monstrosity is less physical than spiritual, so much so that Amanda tells him that, compared to other poisoned children, he seems normal (108). His own mother says that after the cure what bothered her most about the new David were the splotches, but soon she realized that what was really disturbing was not his body but his strange behavior: how he talked, the phrases he repeated, his nightly wanderings, his ubiquity, and his troubling relationship to the animals that he helps die and buries near the house. His body, though marked by tragedy, doesn't show the monster he carries within.

In *Distancia de rescate*, the monster serves to show and warn about the horrors of agrochemical poisoning, but by relegating the monstrous body to a secondary realm, Schweblin signals the invisibility of this type of violence. This isn't to say that the novel underestimates the physical harm that agrochemical violence causes to the health of its victims. The other poisoned children "ya no controlan bien los brazos, o ya no controlan su propia cabeza, o tienen la piel tan fina que si apretan demasiado los lápices, terminan sangrándoles los dedos" (86) ("can't control their arms anymore, or they can't control their own heads, or they have such thin skin that if they squeeze the markers too much their fingers end up bleeding" 122). Abigaíl, the daughter of the woman in the store, suffers from multiple physical deformities, and, as Carla explains to Amanda, "estamos en un campo rodeado de sembrados. Cada dos por tres alguno cae, y si se salva igual queda raro" (70) ("We're in the country, there are sown fields all around us. People come down with things all the time, and even if they survive they end up strange" 95–6). Despite this, David's monstrification privileges the spirit over the body, since horror emerges not from his physical disfigurement but from the invisible deformation of his identity.

In this way, the warning that this monster gives registers not in physical appearance but in the interior fragmentation of his being, in his otherness, and in his proximity to death. These characteristics, along with his displaced soul, his alienation, his divided subjectivity, his unconscious and repetitive acts, his nightly wanderings, and his strange ubiquity remind us of a particular type of monster: the zombie. Although David does not conform completely to the figure of the zombie we know from popular culture—he is not a resurrected corpse, a mindless slave or the cannibal popularized by Hollywood—he does bear a significant resemblance to this undead being. Following Gudrun Rath's assertion that the zombie is a multiple figure in continuous transformation that adopts various forms and meanings (394), it is important to see beyond a fixed taxonomy of the zombie and focus on the suggestive resonances of this monster in David.[14] From here, a plausible argument can be made that David is a new iteration of the zombie or is at least related to this undead being.

As mentioned, the description of his poisoning as worms infesting his body hints at postmortem decomposition, suggesting that David is a half-living/ half-dead creature, an idea that is reinforced through the boy's close relationship to dying animals. Another important trait that David has in common with the zombie is his split identity that in the case of the zombie derives from Haitian vodou's concept of the multiple soul

(Rath 387, Ackerman and Gauthier 467). His nightly walks, his emotionally dead demeanor, his unresponsive behavior are all salient characteristics of the zombie. But perhaps the most striking similarity between David and the zombie is that they are both the victims of a magical ritual by a sorcerer who has stolen their soul. As Kerstin Oloff explains, the original Haitian zombie is a soulless body or a bodiless soul, having been bewitched by a vodou priest (33). Under the control of this sorcerer, zombies are deprived of will or memory and wander staring at nothing, moving with automatic gestures, without control or consciousness of their actions. Even though David is not under the healer's control, he has undergone a similar process of zombification in which his soul has been stolen by a sorcerer, and as a result he wanders in a zombie-like state.

The other sick children, some of whom were born with the toxins and others of whom were poisoned later, also resemble zombies in that they walk in groups with no control of their limbs or their routines. There is a key scene in which Amanda is in her car with Nina when they see a group of children cross the road, guided by nurses. Amanda is surprised that there are so many (David says there are 33), that their physical deformities are conspicuous, and that "la nena de la cabeza gigante" ("the girl with the giant head") stops in front of the car and stares at Amanda until David pushes her to keep walking. As David explains to Amanda, he pushed her because "Siempre hay que empujarla" (108) ("She always needs a push" 157). The child's automatic gestures and her lack of will, her inability to speak, and her fixed gaze are all characteristics of the zombie. The fact that she is part of a group of children in similar conditions confirms their association with zombies since they "suele[n] presentarse en grupos amorfos y acumulativos, que, aunque no tienen voz ni conciencia ni admiten liderazgo . . . tienen una *presencia* proliferante e inorgánica" (Moraña 175; "appear in amorphous and cumulative groups, that, although they have neither consciousness nor leader, have a proliferating inorganic *presence*").

In order to reflect on the resonances of the zombie figure in *Distancia de rescate* in relationship to the slow violence of agrochemical pollution, it's important to consider the origins of the zombie, its main characteristics, and the logic that governs the construction of this monster. Though it has its roots in African beliefs, the Caribbean version of the zombie that has evolved into the figure we know today goes back to the experience of slaves in the Haitian plantations during the Colonial period. As Oloff explains, the original Haitian zombie is controlled by the vodou sorcerer who has stolen its soul, and as a result the zombie has become a victim of exploitation (33).[15] Characterized by its effaced consciousness and lack of will, the zombie serves to illustrate the alienation, human deprivation, and exploitation that originally referred to slavery and colonialism in the sixteenth and seventeenth centuries, and that later was coded to mark the exploitation of workers by capitalism (Oloff 42).[16]

The Haitian zombie's links with human and ecological exploitation in the plantation economy and, with the capitalist expansion that emerged from the implementation of this agricultural-based colonial economy, make the zombie an appropriate instrument with which to represent the anxieties around environmental destruction, especially that associated to new forms of agricultural production fostered by neoliberalism. Proof of this can be found in the most recent transmutation of the zombie, which Sarah Juliet Lauro has called the eco-zombie. The eco-zombie is, in Lauro's words, "an undead that results from some gross mistreatment of nature—usually by a corporation, or by someone seeking to make a profit, who thinks little of protecting the natural environment" (62). The eco-zombie, Lauro explains, differs from its predecessors in that it is not truly controlled by man and instead points simultaneously to humans' inability to control nature and our ability to destroy it. This explains why the eco-zombie is not a participant in the cycle of capitalist production as the Haitian zombie was, but rather a result.[17]

It's easy to see how the resemblance of poisoned children to the eco-zombie in *Distancia de rescate* suggests a critique of the model of agricultural production in Argentina. A criticism of monoculture, biotechnological manipulation of soy, and the excessive use of agrochemicals is implied in the representation of these zombified children who, in ever greater numbers, inhabit the region described in Schweblin's novel. Walking proof of the out-of-control spread of agrochemical poisons, they testify to the destructive power of science in the service of capitalism. As Lauro explains, the eco-zombie reveals "that the powers of man may be more terrifying than the powers of nature" (63), a thesis that is underlined in the novel by the monstrous zombification of David, due not only to agritoxins but also to the magic of the healer attempting to save him from death. David's transmigration emphasizes the human responsibility in the monstrification of human and nonhuman nature. Magic and science work together in the monstrification of the child, and this collaboration points to what they have in common: human agency. Thus Schweblin's monster, a being on the borderline between life and death, the product of the toxic manipulation of human nature and of the environment, gives form to the hidden evil of the agrochemical degradation of the countryside and its inhabitants. The form of this destruction, in the cloak of an eco-zombie, is an agonizing, unrecognizable, and toxic existence.

Thus far we have seen how *Distancia de rescate* uses the eco-Gothic to render the slow violence of agrochemical toxins visible while, at the same time, emphasizing its lack of visibility. As mentioned previously, the other essential aspect of slow violence is its prolonged duration. As Nixon points out, we live in a period in which people suffer from short attention spans, which requires accelerating the speed of slow violence in order to represent it (13). Schweblin achieves this through the urgent dialogue between the protagonists, which both suspends time by detaching it from temporal markers and alerts us to its fleeting nature. David begs Amanda not to waste any time, insisting that it's coming to an end and that she must finish telling her story because she will soon die. The rhythm of the dialogue is frenzied, giving urgency to the narration and making the narrated violence sudden and devastating.

Another technique the novel employs to confront the long time frame of agrochemical pollution is to use children as its principal victims. Both Rebekah Sheldon and Natalia Cecire, who have worked on the figure of the child in environmental discourse, identify the child's importance in relation to temporality. Sheldon begins with the child as a symbol of the future, so often used by politicians and environmentalists, adding that the use of this figure expresses an underlying concern for the extinction of humanity. The child, in addition to the future, represents survival, according to Sheldon (vii). For her part, Cecire notes that temporality is a central trope of environmental discourse, given the necessity of getting people to act. Children offer a human scale to slow violence's temporal scale, says Cecire, facilitating the transformation of environmental innocence into responsibility (167).

This is certainly true in *Distancia de rescate* where the children not only personify our concern with the survival of humanity and the planet but are also the agents of this change. David, an innocent child, is the only character concerned with investigating the origins of the poisoning. Neither his parents, nor the other inhabitants, nor even the nurses who look after the poisoned children—some of whom are their own offspring—show David's impatience to find the cause of the poisoning. Even the way the nurses ignore the problem shows a mixture of ineptitude and complicity. In this context, it's the child, the monstruous child, who must reveal and warn about the hidden dangers of biotechnology and the use of agritoxins in the countryside. It's also the child's task to remind us that the time to act is running out, that we are facing an emergency.

In his quest for the source of the tragedy, David tells Amanda: "Buscamos gusanos, algo muy parecido a gusanos, y el punto exacto en el que tocan tu cuerpo por primera vez" (42) ("we're looking for worms, something very much like worms, and the exact moment when they touch your body for the first time" 52). David must rely on the physical sensations that the toxins produce, since agrochemical poisoning is invisible to the human eye. But even the sensations themselves are indescribable, so he has to use the simile of worms, capturing the horror of the intrusion of foreign substances in the body through a disturbing image that denotes death and putrefaction. From this description and throughout the novel, Schweblin adopts the themes, aesthetic and rhetoric of the Gothic genre to give shape to the ineffable and invisible destruction of agrochemical pollution, while calling attention to these attributes of toxic aggression.

In the array of devices that Schweblin uses to this end, one of the main ones is the child monster that, through its connection to the eco-zombie, materializes the alienation that results from tampering with the environment. Taking into account Cohen's affirmation that cultures can be read from the monsters they engender (3), it's worth asking, what is the main fear made manifest in David's monstrosity, and what does it announce for our culture?

Schweblin has created a monster that exposes our anxieties around environmental damage, anxieties that revolve around fears of loss and dehumanization. But the main fear established through the boy's monstrosity is that, faced with the slow aggression of agrochemical pollution, our only options are to die or be condemned to live an existence in which our identity is so transformed that we are no longer recognizable to ourselves. And that may well be the monster that our culture has created: a monster that warns us about the perils of the slow aggression of agrochemical pollution by tensing and eventually breaking the sisal thread that ties us to who we are as humans.

Notes

1. This chapter was delivered as a paper at the 2018 congress of the Latin American Studies Association, Barcelona, May 24. The title of my paper was "*Distancia de rescate* de Samanta Schweblin o la narración eco-gótica del desastre agroquímico," and it was part of the panel "The Dimensions of Disaster: Scale, Circulation, and the Specular Economy in Latin American Disasters. Part II."
2. I would like to thank Dominique Russell for her translation and her generous comments on this essay.
3. As Charles M. Benbrook explains, "No-till farming in South America lowers machinery and labor costs, and reduces soil erosion, but at the expense of heightened reliance on herbicides for weed control, and other pesticides to control insects and fungal pathogens."
4. Argentina is the third producer of soy worldwide, after the United States and Brazil (USDA). According to a report by the Buenos Aires Grain Exchange (Bolsa de Cereales de Buenos Aires) in March 2018, the production of soy for the 2017/18 period is estimated at 42 million tons, showing a decline from the 2016/17 production of 57.5 million tons. Despite this, soy represents about a quarter of its foreign trade (Leguizamón 684). Transgenic soy, or soy that has been genetically modified (GM) to tolerate glyphosate-based pesticides, accounts for 100% of the soybeans planted in Argentina, as reported by the American Soybean Association in 2014. These crops now cover approximately 50% of the arable land of the country, displacing other crops and transforming Argentina's agrarian production (Beilin y Suryanarayanan 2017). As a consequence of the massive expansion of transgenic crops, the use of glyphosate-base pesticides "rose from 821,000 kilograms in 1996, when only 6 per cent of the soybeans planted were GM, to 88,000,000 kilograms in 2014 applied over an area of 20 million hectares of GM soy" (Leguizamón 688).
5. For a list of scientific research undertaken within and outside of Argentina that relates the exposure to pesticides, glyphosate included, to the increase in genetic malformations, cancers, miscarriages, and hormonal disorders, see Medardo Ávila-Vazquez's study, "Agricultura tóxica y pueblos fumigados en Argentina."

6. Critics such as David Punter (1–8) and Fred Botting (57–8), among others, situate the origin of the Gothic in 1764 with the publication of Horace Walpole's *The Castle of Otranto*.
7. In her translation, Megan McDowell chooses the word "stuck" (5) for "anclada," but I prefer "anchored" because it points to the presence of an object or an external force that prevents Amanda from moving forward in her story. Amanda sees it this way when she wonders whether her inability to continue talking is caused by the nurse's injections, not knowing yet that the culprit is the agritoxins.
8. The presence of mothers and children in horror fiction is not, however, a new device. As Rodrigo Ignacio González Dinamarca explains, the use of child figures in horror fiction is so common as to constitute a cliché. Mothers who question the identity of their own children and believe them to be a supernatural other are also part of the female Gothic tradition (Arnold 92).
9. McGregor questions the validity and the effectiveness of maternal rhetoric in ecofeminism and proposes, instead, to see women's environmental activism as an expression of citizenship: "I argue for a project of *feminist ecological citizenship*. I believe that it is a project worth pursuing because citizenship, defined in feminist terms, offers a way to develop ecofeminist positions that are non-essentialist, democratic, and oppositional" (6).
10. I use Nora Domínguez's definition: "Politizar implica entonces demandar justicia y encauzar una acción ética" (285; "Politicize implies demanding justice and channeling ethical action").
11. I want to thank Ilka Kressner for pointing out the uselessness of the maternal instincts in the novel and Elizabeth M. Pettinaroli for the observation about Schewblin's use of the word "*distancia*" as a measure of length that ends up being a failed measure, both in a quantitative and a qualitative sense, to protect the child from harm.
12. Various critics coincide in pointing to these two attributes as the main characteristics of the monster. Mabel Moraña claims that the figure of the monster is "metáfora de la *hibridez* y de la *diferencia*" (31; "a metaphor of hybridity and difference"), Noël Carroll emphasizes the unnatural mix of opposing attributes in the monster as its most important element (43), and David Gilmore concurs with Carroll in that the hybridity of the monster is its most common and revealing characteristic (189). Jeffrey Jerome Cohen focuses on the otherness of the monster when he affirms "The monster is difference made flesh" ("Monster Culture (Seven Theses)" 7).
13. In this I differ with Rodrigo Ignacio González Dinamarca, who emphasizes David's external appearance as proof of his monstrosity (94).
14. Sarah Juliet Lauro also recognizes the plurality of the zombie figure that "is always revising and reworking its own tropes, as well as the themes evident in other genres of living dead. The tropological zombie . . . continually restages its own incarnation of the interzone that makes murky a distinction between the living and the dead, the natural and the unnatural" (55).
15. Oloff explains that the Haitian zombie also has a subversive aspect, given its association with the rebellion of the oppressed, but "the zombie—whether docile or rebellious—is a figure whose roots in the experience of brutal enslavement and exploitation are readily discernible" (33).
16. Sarah Juliet Lauro and Karen Embry trace the evolution of the zombie from a typographical distinction: the Haitian *zombi*, related to slavery in the plantations, becomes *zombie* when imported to American culture, becoming "evil, contagious, and plural" (88) and representing various social preoccupations.
17. Interestingly, as Lauro points out, there is ample precedent in film of presenting pesticides as a "zombie-making agent" (60), citing *Raisins de la mort*

(1978), *The Living Dead at Manchester Morgue* (1975), and *Toxic Zombies* (1980) as examples.

Works Cited

Ackerman, Hans W., and Jeanine Gauthier. "The Ways and Nature of the Zombie." *Journal of American Folklore*, vol. 104, no. 414, 1991, pp. 466–94.

American Soybean Association. "International: Adoption of Biotech Enhanced Seedstock." *SoyStats*, 2018, soystats.com/international-adoption-of-biotech-enhanced-seedstock/.

Arancibia, Florencia. "Challenging the Bioeconomy: The Dynamics of Collective Action in Argentina." *Technology in Society: An International Journal*, vol. 35, 2013, pp. 79–92.

Arnold, Sarah. *Maternal Horror Film: Melodrama and Motherhood*. Springer, 2016.

Ávila-Vázquez, Medardo. "Agricultura tóxica y pueblos fumigados en Argentina." *Revista de Extensión Universitaria +E*, vol. 4, 2014, pp. 28–34, doi:10.14409/extension.v1i4.4586.

Beilin, Katarzyna Olga, and Sainath Suryanarayanan. "The War between Amaranth and Soy: Interspecies Resistance to Transgenic Soy Agriculture in Argentina." *Environmental Humanities*, vol. 9, no. 2, 2017, pp. 204–29.

Benbrook, Charles M. "Trends in Glyphosate Herbicide Use in the United States and Globally." *Environmental Sciences Europe*, vol. 28, no. 3, 2016, doi:10.1186/s12302-016-0070-0.

Bolsa de cereales de Buenos Aires. "Estimaciones agrícolas campaña 2017/18." 13 Mar. 2018, www.bolsadecereales.com/imagenes/informes/2018-03/27-expoagro.2018.pdf.

Botting, Fred. *Gothic*. Routledge, 1996.

Carroll, Noël. *The Philosophy of Horror or Paradoxes of the Heart*. Routledge, 2003.

Cecire, Natalia. "Environmental Innocence and Slow Violence." *Women's Studies Quarterly*, vol. 43, 2015, pp. 164–80.

Cohen, Jeffrey Jerome. "Monster Culture (Seven Theses)." *Monster Theory: Reading Culture*, edited by Jeffrey Jerome Cohen. U of Minnesota P, 1996, pp. 3–25.

———. "Preface: In a Time of Monsters." *Monster Theory: Reading Culture*, edited by Jeffrey Jerome Cohen. U of Minnesota P, 1996, pp. vii–xiii.

Del Principe, David. "Introduction: The EcoGothic in the Long Nineteenth Century." *Gothic Studies*, no. 1, 2014, pp. 1–8.

Domínguez, Nora. *De donde vienen los niños: Maternidad y escritura en la cultura argentina*. Beatriz Viterbo Editora, 2007.

Freud, Sigmund. "The Uncanny." https://web.mit.edu/allanmc/www/freud1.pdf. Accessed 12 Apr. 2019.

Gilmore, David D. *Monsters: Evil Beings, Mythical Beasts, and All Manner of Imaginary Terrors*. U of Pennsylvania P, 2012.

González Dinamarca, Rodrigo Ignacio. "Los niños monstruosos en *El Orfanato* de Juan Antonio Bayona y *Distancia de rescate* de Samanta Schweblin." *Brumal: Revista de Investigación Sobre Lo Fantástico/Research Journal on the Fantastic*, vol. 3, no. 2, 2015, pp. 89–106.

Haber, Baron. "'We Destroyed It Undiscovered': Slow Violence, the Gothic, and Neocolonialism in Mahasweta Devi's Imaginary Maps." *Darkmatter Journal: In the Ruins of Imperial Culture*, vol. 13, 2016, pp. 1–20.

Hurley, Kelly. "British Gothic Fiction, 1885–1930." *The Cambridge Companion to Gothic Fiction*, edited by Jerrold E. Hogle. Cambridge UP, 2002, pp. 189–207.

IARC, International Agency for Research on Cancer. "Evaluation on Five Organophosphate Insecticides and Herbicides." *IARC Monographs*, vol. 112, 2015, www.iarc.fr/wp-content/uploads/2018/07/MonographVolume112-1.pdf.

Lauro, Sarah J., "The Eco-Zombie: Environmental Critique in Zombie Narratives." *Generation Zombie: Essays on the Living Dead in Modern Culture*, edited by Wiley Lenz and Stephanie Boluk. McFarland, 2011, pp. 54–66.

Lauro, Sarah Juliet, and Karen Embry. "A Zombie Manifesto: The Nonhuman Condition in the Era of Advanced Capitalism." *Boundary 2*, vol. 35, no. 1, 2008, pp. 85–108.

Leguizamón, Amalia. "Environmental Injustice in Argentina: Struggles Against Genetically Modified Soy." *Journal of Agrarian Change*, vol. 16, no. 4, 2016, pp. 684–92.

MacGregor, Sherilyn. *Beyond Mothering Earth: Ecological Citizenship and the Politics of Care*. UBC Press, 2011.

Moraña, Mabel. *El monstruo como máquina de guerra*. Iberoamericana, 2017.

Nixon, Rob. *Slow Violence and the Environmentalism of the Poor*. Harvard UP, 2011.

Oloff, Kerstin. "'Greening' the Zombie: Caribbean Gothic, World-Ecology, and Socio-Ecological Degradation." *Green Letters*, vol. 16, no. 1, 2012, pp. 31–45.

Pengue, Walter A. "Cuestiones económico-ambientales de las transformaciones agrícolas." *Problemas del desarrollo, Revista latino Americana de economía*, vol. 40, no. 157, 2009, pp. 137–61.

Punter, David. *The Literature of Terror: A History of Gothic Fictions from 1765 to the Present Day*. Longman, 1996.

Rath, Gudrun. "Zombi Narratives: Transatlantic Circulations." *Reshaping (g) local Dynamics of the Caribbean: Relaciones y Deconexiones—Relations et Déconnexions—Relations and Disconnections*, edited by Anja Bandau, Anne Brüske, and Natascha Ueckmann. U Publishing, 2018, pp. 385–96.

Rudd, Alison. *Postcolonial Gothic Fictions from the Caribbean, Canada, Australia and New Zealand*. U of Wales P, 2010.

Schweblin, Samanta. *Distancia de rescate*. Penguin Random House Grupo Editorial España, 2015.

———. *Fever Dream: A Novel*. Translated by Megan McDowell, First Edition. Riverhead Books, 2017.

Sedgwick, Eve Kosofsky. *The Coherence of Gothic Conventions*. Gothic Studies and Dissertations. Arno Press, 1980.

Sheldon, Rebekah. *The Child to Come: Life After the Human Catastrophe*. U of Minnesota P, 2016.

USDA. United States Department of Agriculture. Foreign Agricultural Service. "Table 07: Soybeans: World Supply and Distribution." https://apps.fas.usda.gov/psdonline/app/index.html#/app/downloads

Yang, Sharon Rose, and Kathleen Healey. "Introduction: Haunted Landscapes and Fearful Spaces: Expanding Views on the Geography of the Gothic." *Gothic Landscapes: Changing Eras, Changing Cultures, Changing Anxieties*, edited by Sharon Rose Yang and Kathleen Healey, Cham. Palgrave Macmillan, 2016, pp. 1–18.

2 Toxic Nature in Contemporary Argentine Narratives

Contaminated Bodies and Ecomutations[1]

Gisela Heffes

I would like to begin this chapter with what can be defined as the "rural turn" taken by recent Argentine narratives. Contrasting with a literary production that throughout the twentieth century has mainly prioritized the urban landscape, in the last few years, Argentine literature (and I venture to say films as well) has been produced such that the primary featured setting is the countryside. At the beginning of the twenty-first century, the *pampa*, a space that harbors so much in the Argentine imaginary—from debates about national constitutions to disputed heritage and traditions—acquires the traits of a reconfigured and resemanticized space. A few examples of this phenomenon are the novels *La inauguración* (2011) by María Inés Krimer; *El viento que arrasa* (2012) by Selva Almada; *La omisión* (2012) and *Desmonte* (2015) by Gabriela Massuh; *Matate, amor* (2012) and *La débil mental* (2015) by Ariana Harwicz; *La vi mutar*, by Natalia Rodríguez (2013); *Distancia de rescate* (2014) by Samanta Schweblin; *Un pequeño mundo enfermo* (2014) by Julián Joven (pseudonym of Cristian Molina); *Las hamacas de Firmat* (2014) by Ivana Romero; *El rey del agua* (2016) by Claudia Aboaf; and *Las estrellas federales* (2016) by Juan Diego Incardona, among many others. Significantly, all of these novels were published in the years between 2011 and 2016. Nonetheless, Pedro Mairal's novel *El año del desierto* (2005) prefigures some of the topics that recently became more visible in the Argentine cultural scene. With this, I refer to a certain dissolution of the classic dichotomies faced antagonistically by spaces and ideologies—and here I am thinking of Argentine literary critic Josefina Ludmer's formulation of "islas urbanas" ("urban islands"). If, at the beginning of the nineteenth century, urban spaces constituted the key locus for rebuffing and cleansing the "barbarity" deeply rooted in rural territories, as emerged in paradigmatic texts like Domingo F. Sarmiento's *Facundo* (1845), this disjunction changed by the end of the twentieth century and the beginning of the twenty-first. As Ludmer accurately suggested in her article "Territorios del presente: En la isla urbana" (2004), contemporary literature is now urban: it absorbs rurality and becomes barbarous.[2] While Ludmer considered a wide number of narratives in which the privileged diegetic site is the Latin American city, the juxtaposition that blurs and redefines the traditional boundaries that confronted urban and countryside spaces may be read, as well, as a twofold mechanism where the rural also absorbs urbanity, becoming a disciplined and tamed territory. Therefore, if the city becomes "barbaric" and erases spatial frontiers, likewise the rural landscape is no longer untamed; rather, it has become domesticated by the unfettered use of monocultures, be they soy or wheat, and by the use of pampean soil as an artificial laboratory where the global economy and an increasingly unregulated state intervene, thereby objectifying it. This reversal, which marks the emergence of a new rurality, one in which the countryside is anthropogenically intervened, trimmed, exploited, and domesticated, questions assumptions that assign both the urban and rural landscape defined and exclusive traits. Not only is it the case that contemporary Argentine literature is no longer urban, but it is also true that aesthetic expressions that define the rural depart from previous representations of the pampean landscape, thus reconfiguring the natural world.

How is the countryside—the country, rurality, the *pampa*—represented in contemporary Argentine literature? What specific traits emerge alongside the implementation of neoliberal policies in a space that has long served as a symbolic reservoir of wishes and projects and as a site of cultural disputes? And, even more importantly, what happens to the bodies traveling through this space of erasure and intersections? What is their physiognomy? What are they made of?

Some recent works of criticism, such as those of Lucía de Leone and Dinorah Cossío,[3] explore the transformation of the natural countryside and rural spaces into a setting traversed by what Lawrence Buell has defined as a toxic discourse. Buell attributes the origin of contemporary toxic discourse to Rachel Carson's book *Silent Spring* (1962). In Buell's words, the first chapter of Carson's book, "A Fable for Tomorrow," introduces one of the key discursive motives for understanding this idea of toxicity. In the book, a town in the heart of the United States wakes up one fine spring day without any birds or insects. This fictionalized town, according to Carson, "might easily have a thousand counterparts in America or elsewhere in the world" (3); affected by a "grim specter" that arrives inadvertently, a tragedy can become a cruel reality about which it is necessary to create awareness (3). It's well-known that Carson's invective was against the use of DDT (dichlorodiphenyltrichloroethane), along with other pesticides that have left an indelible mark on both flora and fauna. The reception of Carson's work was immediately very significant. Not only did her book sell more than 2 million copies, but it also rendered visible a problem that until then had never been presented so simply and concisely: if humanity poisons nature, nature will in due course poison humanity (Griswold). For Carson, both the destructive actions of humans and ordinary day-to-day blunders enter into the vast cycles of the earth, such that they eventually come back, adding threats and danger to our lives. Without a doubt, Carson's legacy is the dissemination of modern ecology.

In the Argentine rural landscape, the discourse of toxicity Buell refers to is articulated through the emergence and increasing predominance of agrochemicals. For example, as Leone suggests, the cultivation of transgenic soy has become both a profitable (albeit monopolistic and short-term) business and a source of environmental and collective health problems. It redesigns the "use of fertile land, interpersonal and affective relationships, practices and social mobility," in cases such as unemployment and migratory issues, and establishes "new thematic repertoires" (65). In a recent article about interspecies war, especially between soy and amaranth, Katarzyna Beilin and Sainath Suryanarayanan point out that Roundup Ready (RR) soy is genetically modified to resist the herbicide called Roundup, which is produced by the Monsanto Company, the main ingredient of which is glyphosate.[4] What is remarkable about this herbicide is that it eliminates all undesired plants except for the primary crop, which mutates into an immunologic one. The "magnificent" crop not only tolerates both insecticides and herbicides alike but also causes the chemical substances to increase their intensity as the crop becomes more resistant. If the soy is genetically modified, so is the milieu where this practice takes place. In addition to the recurrent spraying of the soil with fumigants, the landscape itself is being altered by deforestation. In order to make room for more farming, the physical space is transformed, as its waters, soil, and air are being contaminated (Cossío 10).

The connection between the human and nonhuman world consists of relationships inextricably tied to varied cosmovisions and epistemologies. These visions neither merely correspond to a specific historical moment nor reflect a particular ecological concern that may not always be present. For a naturalist like Alexander von Humboldt, the earth consists of a single great living organism within which everything is connected and whence an audacious vision of the natural world is conceived. This perspective, as Andrea Wulf suggests, still influences to a degree the way in which we understand the natural world (2017). It isn't that Humboldt lacked a fascination for scientific instruments, systems of measurement, or processes of observation but rather that he thought nature, aside from being analyzed and cataloged, should also be protected. If the natural world is seen as a web made of interconnected threads, its vulnerability becomes obvious: everything works together to form a framework, and if one of the threads is torn, the entire structure could collapse (5). For this reason, Wulf highlights that, when Humboldt visited Venezuela in 1799 and saw the devastating environmental effects of the colonial plantations on Lake Valencia (or Lake Tacarigua), he became the first scientist to expose the dangers of climate change caused by human actions: deforestation had left the earth sterile, the water levels in the lake were decreasing, and, with the disappearance of the undergrowth, torrential rains washed and pushed the earth from the slopes around the mountains (5). Thirty years later, in 1829, Simón Bolívar also realized the danger hiding in the exploitation of the forests and emphasized that "throughout the region we are experiencing excessive harvesting of wood, dyes, quinine, and other substances, especially in the forests belonging to the state, with disastrous consequences" (199).

Sarmiento's vision, temporally not too distant from Humboldt's, offers an ideological project that distances itself from Humboldt's position. Obsessed with a national agenda, Sarmiento attempts to preserve the scientific aspect we saw in Humboldt but discards the synergy between the natural and the human worlds. For the *sanjuanino*, the fundamental problem was that natural space, the rural territory of the countryside, lacked *res publica* (71). Instead, it constituted, in the nineteenth-century dichotomy he had founded in *Facundo: Civilización y barbarie* (1845), a barbarous space where "wild nature" resided, making it part of the "uncultured plains" (68). According to Sarmiento, in the natural landscape of the grasslands there were no vestiges of civilization. He conceived of this space as governed by "the predominance of force, the preponderance of strength, the limitless, irresponsible authority of those in command" and as one in which justice was administered "without form nor argument" (63). To sum up, this rural space forges a literary, historical, and cultural genealogy to which the Argentine imagination returns like a charm to ward off civilization, to protect against immigrants, and to defend against the city. It also serves as a repository for a nostalgia and lament for an ever more inexistent past,[5] which will now become, as I'll demonstrate, the space where a different violence resides, one that is invisible and slow and that will destabilize national metaphors, proposing imaginary alternatives without precedent. What's more, at a time when an eschatological discourse has come to be a recurrent aesthetic and mediatic account, I will inquire on the fate of the body, the subjective and objective body that inhabits this chemically contoured territory, and suggest that both human and nonhuman bodies have become an "economic resource," a disturbing metaphor of an enduring exploitation of the also modified natural world.

While Mairal's *El año del desierto* is not centrally relevant to the narratives I analyze in greater detail, it functions as a precursor to the four texts I will analyze in more detail later, insofar as it displays key characteristics and predicts some of their most relevant features. The protagonist of *El año*, María Valdés Neylan, narrates her experience during the year in which Argentina is razed by a strange phenomenon called the *Intemperie*.[6] In the story, the *Intemperie* refers to a peculiar occurrence during which "the narrative chronology advances (covering a year from the novel's onset to its conclusion)," but history "insensibly 'moves backward'" from the "tumultuous beginning of the twenty-first century" to the sixteenth century, as Juan Pablo Dabove and Susan Hallstead observe in the new critical edition of the novel (Dabove and Hallstead X). With the emergence of this new regressive phenomenon, the *Intemperie* arrives "desde el fondo de la pampa" ("from deep in the pampa") and surrounds "el centro de la ciudad" ("the center of the city"), advancing upon it (IX). Because of this sweeping experience, María crosses paths with events, places, and figures of the past, while symbols of civilization and progress, which were clearly present at the start of the novel, begin to vanish. Thanks to the *Intemperie*, "the city begins to disappear and the desert commences to reconquer," as Dabove and Hallstead suggest, "that which always belonged to it" (IX). This allusion refers specifically to the urban disposition of Buenos Aires, with its neoliberal economy and technology at the service of big international corporations. Under these conditions, María struggles to survive in an increasingly hostile environment. As the story unfolds, she finds herself ever more tied to the cruelty of nature and subject to the will of men. *El año* makes evident that, when it comes to narrating rurality, the contours of a space that emerges and reemerges within the Argentine literary tradition, what is at stake is the "invention of a cultural past" both as the construction of "a cultural mythology" as well as a "personal aesthetic project," as Graciela Montaldo has accurately asserted (14).

In the three narratives, *Distancia de rescate*, *Las estrellas federales*, and *La vi mutar*, as well as the poetry collection *Un pequeño mundo enfermo*, which form the center piece of this analysis, the representation of bodies defines a relationship between a subject and the natural world, replacing a discourse about *Buen Vivir* (good living) with a narrative about what I call bad living (*mal vivir*). While the notion of *Buen Vivir* is understood as part of a long search for alternatives of life forged in the heat of humanity's struggles for emancipation and survival, what these narratives make evident is that a rurality anchored in the notion of Nature does not necessarily adhere to *Buen Vivir*'s postulates.[7] Uruguayan sociologist Eduardo Gudynas specifies that this concept is not traditional but rather new and that the diversity of meanings attributed to *Buen Vivir* aims to create an alternative to the dominant idea of development, as well as to support the idea that Nature is a right-bearing subject capable of contesting the founding principles of Western anthropocentrism and capitalism (17). On the contrary, the narratives to be analyzed in this chapter consist of a refutation of this notion, demonstrating that the Argentine rural space has become a *locus* that not only seeks economic growth but also does so to the detriment of all living organisms (people, soil, air, water) that reside nearby.

The aesthetic production analyzed in this chapter effectively appeals to a reflection about environmental justice that considers the damage and imminent deterioration of all bodies (be they human or not, from plants to animals, both organic and inorganic). The first work, by Samanta

Schweblin, tells the story of Amanda and her daughter Nina, who come from the city to spend their summer in the countryside, and that of Carla and her son David, who live in the countryside. David has been poisoned before the beginning of the narrative, and his story sets off the conflict. David was intoxicated by coming into contact with water from a stream where herbicides and pesticides had been disposed of, albeit this is never described directly in the novel. Such poisonings occur frequently, and they also take control of the bodies of Amanda and Nina. A conversation with David on the novel's first page, which sets the story in motion, describes the effect of the poison as "*gusanos*" (worms); that is, an invisible substance all the town's residents experience, describing it as "gusanos, en todas partes" (11; "worms, everywhere"). As a result of being poisoned, Amanda's body, reclined and prostrate, is immobilized and unresponsive. Like that of the horses, ducks, dogs, or any other animals in the surrounding area, Amanda's body is inserted into a space "rodeado de sembrados" ("surrounded by cultivated fields") where there is a town in which children wait to be admitted to a precarious and doctorless hospital since they can no longer write because "no controlan bien sus brazos . . . su propia cabeza, o tienen la piel tan fina que, si aprietan demasiado los lápices, terminan sangrándoles los dedos" (86; "they can't control their arms . . . their head, or their skin is so fine that if they hold their pencils too tightly, their fingers start to bleed"). This is a story of altered chronologies. By this, I refer to the temporal dimension of these spatial effacements, a both temporal and spatial chronology that alters the environment, as well as the divisions and/or similarities between spaces: the city and the countryside, the notion of civilization and barbarism, the body and the landscape. It is a chronology where pesticides like diazinon and malathion and herbicides like glyphosate infiltrate through all the channels necessary for human survival. They pervade the bodies (both human and nonhuman), altering their physiognomy, the landscape, and the constructed environment by denaturalizing them. It is a space and a time in which the "chicos son extraños" ("children are strange"), very few are born healthy, and most are deformed—"no tienen pestañas, ni cejas, la piel es colorada . . . y escamosa también" (108; "they don't have eyelashes nor eyebrows, and their skin is colored . . . and scaly too"). David's body, in spite of being treated by a local healer, becomes—according to the story—a "monstrosity," an aberration caused by the implacable and invisible machine of wild biocapitalism. Note that here is where the "wild" should be, no longer in a Sarmientian notion of "nature"—which pushes both humans and nonhumans toward an abyss of mutations and possessed bodies.[8]

In Juan Diego Incardona's novella, *Las estrellas federales*, the "mutants"—the name is attributed to them from the start—are part of a circus caste that inhabits the province of Buenos Aires. The story, which begins in 1989, takes place in the past, but it also represents a dystopian future. According to Incardona, it is a postapocalyptic world of discards and of ruins, where mutations from an environment that suffered a crisis mainly make do with the nature of the place, especially the loss of work. Besides unemployment, the crisis is also characterized by foreclosures and marginality.[9] The novel alludes to a "phenomenon" which, like the *Intemperie* in Mairal's novel or the reference to the "worms" in Schweblin's, occurs unexpectedly. However, "[n]o se sabe si las semillas ya estaban esparcidas desde antes o si el viento matancero las levantó, desterrándolas del campito" y creando las condiciones favorables para que una plaga de poinsettias o estrellas federales cubran la vasta zona de Villa Celina "con un rojo furioso" o "punzó" (Incardona 21–2; "it is unknown whether the seeds were already sown beforehand or if the wind from Matanzas lifted them up, unearthing them from the fields," creating favorable conditions for a plague of poinsettias to cover the vast area of Villa Celina "with a furious red" or "punzó").[10] The fabulous event consists of an apocalyptic spectacle that announces not only the expected presence of "un hongo nuclear" ("a nuclear mushroom") but also the arrival of the "criaturas más fantásticas del mundo" ("most fantastic creatures in the world"): the "Circo de las Mutaciones" or "Circus of Mutations" (22–3). The year the narration took place, 1989, is significant, as it marks the beginning of the presidency of Carlos Saúl Menem, who inaugurated a neoliberal agenda in Argentina based on economic adjustments, deregulations, and an obsequious relationship to the International Monetary Fund.[11] During the emergency caused by the phenomenon, the narrator-protagonist of the story seeks refuge under the circus's tent, where he finds employment working for the "Hombre Regenerativo" ("Regenerating Man of La Tablada"), a man who cuts off his own limbs, which then grow back as if he had never lost them. Other mutants also inhabit the circus: for instance, the "Mujer Lagartija" ("Lizard Woman") "con su cola larga y puntiaguda" ("with her long and pointy tail"); the "Infracaballos" ("Underhorses"), "dos equinos del tamaño de hormigas" ("two equines the size of ants"); and the "Petiso Orejudo" (Big-Eared Small Man), among many others (23–4).

References to Peronism abound, and the histories of Argentina—social, economic, literary, and cultural—are interwoven into the plot like returning ghosts who never really left. The story features a game of ambivalences where the red punzó of the poinsettias (better known in Argentina as "estrellas federales" or ["federal stars"])[12] articulates a series of emblematic relations. On one hand, red is reminiscent of "blood," "fire," and "living flesh," elements that together allude to the ecological disaster caused by the "phenomenon": "explotaban hongos químicos sobre la Matanza, rompiendo bielas, pistones y cojinetes, desatándose correas y engranajes para finalmente derrumbarse y caer, en bloques de hierro y fundición, sobre nuestras casas, nuestras escuelas, nuestras iglesias y nuestros clubes" (56) ("chemical mushrooms exploded over La Matanza, breaking cranks, pistons, and bearings, untying straps and gears until finally collapsing and falling, in blocks of iron and foundry, over our houses, our schools, our churches, and our clubs"). On the other hand, the red punzó is reminiscent of the nineteenth-century symbol that defined the federal party led by Juan Manuel de Rosas, the mortal enemy of Sarmiento. Thus, this broad metaphor can be read on various levels, although, undoubtedly, irony and sarcasm prevail as discursive critiques.[13]

In *Las estrellas*, the environmental disaster becomes devastating with the arrival of sulfuric acid rain. With explicit references to the "rain of fire" from the eponymous story by Argentine writer Leopoldo Lugones (*La lluvia de fuego*), the "nubes saturadas de dióxido de azufre" ("clouds saturated with sulfur dioxide"), which form an "arcoíris petroquímicos" ("petrochemical rainbow"), ambush the suburbs of Buenos Aires and melt all the houses, parks, people, and animals in their path (55). This catastrophic phenomenon has a clear origin: "gases escapados de los depósitos sin mantenimiento, de los tanques abandonados por las empresas, de las fugas del cementerio de fábricas" ("gases escaped from deposits in disrepair, from tanks abandoned by companies, from the leaks in the factory's cemetery"), the industrial residues that have been partially dismantled and that transform the neighboring populations into "risk societies," as defined by Ulrich Beck. For Beck, in late modernity, the social production of goods is systematically accompanied by the social production of risks. Therefore, problems and conflicts related to distribution in a society of scarcity are juxtaposed with the problems and conflicts that emerge from the production, definition, and distribution of risks created through technological and scientific means (Beck 19). It's no coincidence, then, that in a society of risk, both the unknown and unintended consequences become historically and socially dominant forces (22).

The disparity between the social production of goods and its opposite, the social production of risks, reinforces the idea that neoliberal biocapitalism has consequences, as suggested by Kelly Fritsch, that are connected to the way we think about toxicity and corporeity. This applies even more if bodily representation implies both paralysis and disability. Grandjean and Landrigan engage the correspondence between the distribution of goods and risks, along with the question of which populations do and do not become weakened in relation to it (cit. in Fritsch 360). Even though their investigation is based on the United States, the problem of distribution and the weakening of populations constitutes a similar, even worse problem in Latin America and, more specifically, in Argentina. The lack of standards and laws that regulate the production of chemical substances, along with the distribution and environmental exposure to specific toxic particles, can have noxious long-term effects, such as the debilitation of certain populations in relation to others (Fritsch 360). If in *Distancia* disability appears in Amanda's immobility as she lies prostrate, as well as in the blindness of the contaminated characters ("todo está tan blanco" [109; "everything is so white"]), the deformations ("la nena de la cabeza gigante" [108; "the girl with the giant head"]), or even the "dolor de cabezas" ("headaches"), "náuseas" ("nausea"), "úlceras de la piel" ("skin ulcers"), "vómitos con sangre" ("bloody vomits"), and "abortos espontáneos" (23; "spontaneous abortions"), then in *Las estrellas*, the effect of paralysis is even more violent. Subjectively, it's "la estampa de un hombre quemándose vivo" (56; "the imprint of a man being burned alive"); in spatial terms, it's the consequent disappearance of the Buenos Aires suburbs (known as *el campito*), whose existence ends after the exodus of the subjects dismembered by the effects of the sulfuric acid rain: "La comunidad caminaba al sudeste hasta que se le caía la piel, la carne, los huesos, y ya no quedaba nada para la fuerza de gravedad: en matrimonio con la nada, el conurbano se derretía a la hora del reloj de plastilina" (84; "The community walked southeast until their skin, their flesh, their bones, were falling off, and nothing was left to suffer gravity: married to the void, the suburbs melted to the ticking of a clay clock"). Similarly, the rural space vanishes in *Distancia*, although in a very different modality. It is Amanda's husband who returns to the city, turning his back on the countryside and, without looking back, wanting to perhaps erase it from his memory forever:

> No ve los campos de soja, los riachuelos entretejiendo las tierras secas, los kilómetros de campo abierto sin ganado, las villas y las fábricas, llegando a la ciudad. No repara en que . . . hay demasiados coches, coches, y más coches cubriendo cada nervadura de asfalto. Y que el tránsito está estancado, paralizado desde hace horas, humeando efervescente. (124)
>
> He doesn't see the soy fields, the creeks interweaving the dry earth, the kilometers of fields without cattle, the villages and factories, upon reaching the city. He doesn't notice that . . . there are too many cars, cars and cars covering every asphalt nerve. Or that the transit is stopped, paralyzed for hours, effervescently lingering on.

As in *El año en el desierto*, Schweblin's and Incardona's novels allegorically appeal to a phenomenon that stalks the characters and obliterates the spatial-temporal contours that divide, fragment, and classify origins, classes, genders, and races—a phenomenon whose scale transcends traditional borders or, we could say, modern borders. Similarly, Natalia Rodríguez's novel *La vi mutar* revolves around Vito, a young boy whose mother mutates into a "monstrosity" that places her in the hospital, mysteriously disfigured and all covered with flowers. Witnessing the metamorphosis that culminates with her death, Vito notices that the same phenomenon occurs with a group of women whose husbands work in the same factory as his father. The story takes place in an unknown town, in the neighborhood of Los álamos (The poplars), although sarcastic references to the irony of the name abound in the novel, since what was supposed to be a grove has become a poor village without trees:

> Yo vivo en un lugar que se llama Los Álamos, porque en un tiempo había muchos álamos que se cortaron para hacer casas y muebles y

> papel. Mamá dice que los árboles esos fueron mutilados, que es como cortados pero dejándoles la raíz . . . Yo creo que en algún momento las raíces y los árboles van a volver a crecer y a mutilar todas las casas y a todas las personas, de bronca.
>
> (20)

> I live in a place named Los Álamos because a long time ago there were many poplars which were then cut down to make houses and furniture and paper. Mom says that the trees were mutilated, that they were sliced through but their roots left in the ground . . . I think one day the roots and the trees will grow back and will mutilate all the houses and the people, out of rage.

While the cause of the mutation in the town is unknown, some clues suggest chemical exposure. Vito's best friend, El Guille, has his house raided by the police several times because his father allegedly has a clandestine laboratory. Although it is unclear, a number of scenes imply that the illness produced by exposure to the chemicals is contagious. Vito's mother rests in a confined room, and, like Amanda in *Distancia*, she is unable to move. Entering the room where she is being monitored requires wearing "un traje de astronauta" (26; "an astronaut-like outfit"). Right after Vito's mom's death, several women from the same town begin to mutate. Their husbands have become unemployed due to the authorities' shutting down of the factory where they worked. Laid off and poor, the men decide to open a freak show, similar to the one in *Las estrellas*, where they display their wives in cages. While it is not explicit whether the town is near cultivated land, several cues suggest as much. For instance, when the mutated women attend a meeting at Vito's house organized by his father, Vito notes the "olor raro" ("strange smell"), like "tierra mojada" ("moistened soil"), the same smell he sensed when he sat with his friend Julieta in the courtyard of her house and they watched "la misma nube" ("the same cloud") over and over, experiencing that "olor a caracol, a tierra revuelta" ("smell of snail, of stirred soil") and "aire podrido" (445; "rotten air"). In the same vein as *Las estrellas*, the apocalyptic end leaves the characters in chaos. Nature comes back to avenge its mutilations; now toxic, it "[e]xplotó" ("exploded") and "sangró" (76; "bled") ceaselessly. According to Vito's account, the "desastre" ("disaster") came from the "desechos mutantes . . . los mismos líquidos putrefactos que el papá de Guille daba de tomar todos los días a las señoras" (76; "mutants' residues . . . the same putrid liquids that Guille's father gave to the women every day to drink"). The mutated women were presented in the circus as an outcome of "nature," a dead, contaminated and polluted nature, deeply rooted in the soil of a town, any town, as the one allegorically described in Carson's essay. Nature's toxicity in *La vi mutar* challenges a long-standing engrained imaginary that traverses Argentine cultural tradition from the nineteenth century. Toxicity, not modern subjectivity, alters bodies and landscapes alike. Furthermore, if a rural cultural production has functioned as a "residue," as Graciela Montaldo has shown in her now classic *De nuevo el campo*, not only is that "excess" now gaining presence within the national literary imaginary, but it is also embodying, literally and metaphorically, an objectified and experimentalized space that serves both the local and global economy. Similarly to neoextractivism, a practice that both progressive and neoliberal governments have been exploiting all over Latin America, the soy boom, along with other forms of agriculture and crop growing in Argentina's rural areas, will have an enduring effect on the nature of the landscape, and the landscape of nature, as well as on the organic bodies that inhabit it.[14]

Julián Joven's poetry collection also makes reference to another "phenomenon": this time, "el Mal," or "Evil." The collection of poems engages in opportune dialogue with previously analyzed novels. From its onset, the poetic voice establishes a connection between the rural space, the body, and toxicity:

> Abrieron el cajón
> y salieron moscas
> de la nariz del cadáver.
>
> No había nada que dijera qué
> o quién
> solo una plaquita metálica con él
> empotrado en un sombrero de paja
> y con una pala en la mano.
>
> El campo atrás.
>   Soja
> mucha mucha Soja.
>   Y Trigo.
>
> (12)
>
> They opened the coffin
> and flies came out
> from the nose of the body.
>
> There was nothing that said what
> or who
> only a small metal plaque with him
> built-in with a straw hat
> and a shovel in hand.
>
> The field behind.
>   Soy
> lots and lots of Soy.
>   And Wheat.

The "small sick world" alluded to in the title is a world where cancer, asthma, and other illnesses related to agricultural production and the use of agrochemicals slowly take over the characters, weaving another framework, one of toxic discourse, which renders visible that which for many lies forgotten or lost in darkness. In an interview soon after the release of Joven's book, the poet referred to a "metaphorical and literal" cancer that is more and more present, especially "en los pueblos, o en el campo, espacios que ya no están tan alejados de la ciudad en ningún sentido" ("in towns, or in the countryside, in spaces that aren't that far from the city anymore, in any sense"), and points out that these are the same spaces that, also until recently, seemed to have disappeared from the agendas of contemporary writings, absorbed by the eminent presence and interest in urban spaces (Molina). According to Joven, there are still those who live in little towns or in the countryside in the twenty-first century, and there too are "malestares que los atraviesan . . . que no son ya los del corral decimonónico o de inicios del siglo veinte: el crecimiento inaudito de las tasas de cáncer es uno de ellos y es también uno de los malestares de nuestra cultura—aunque no el único" (Molina; "malaises abound, and these aren't the discomforts of the nineteenth century corral or the start of the twentieth century: the silent growth of cancer rates in one of them is also one of the ailments of our culture—although not the only one").

In his poetry collection, Joven registers these high "rates" in the concrete bodies of the characters (the pa, the ma, the aunt, but also the horse, and many dead, and flies, and worms) to whom his poems give voice.[15] As in the multiple episodes that reemerge again and again in *Distancia*, the "caballo apareció reventado en medio de las vía" ("horse appeared as if it had burst on the rails"), and inside it, the veterinarians found "una pasta verde/idéntica a la de las Chinches" (37; "a green paste/identical to that of the *Chinches*"). But the small world in the poetry collection is an expanding world, with a sickness that attacks and encompasses entire populations. Those who daily suffer its effects must "run" to avoid being contaminated:

> Ella destendía la ropa
> y de golpe
> un aluvión de olores empezó a sofocarla
> las chinches caían y explotaban en el césped
> como la plaga de Egipto
> y con sus ácidos verdes y pegajosos
> en los pelos y en la ropa.
>
> (73)

> She unfolded the clothing
> and suddenly

an alluvium of smells began to suffocate her
the beetles were falling and exploding on the grass
like an Egyptian plague
and with their green and sticky acids
in her hairs and on her clothing.

The women ran "desesperadas dentro de la casa" ("desperately inside the house") from fear because "las partículas vaporosas" ("vaporous particles") and some "semilla pelada" ("peeled seed") "se les iba a meter adentro" ("were going to get inside them") (75). The poison penetrates the body through the air, the same air that intoxicates the pa, "atacado sin respiración en pleno asma" ("attacked without breath in mid asthma"), who ironically attempted to "aspirar el aire fresco" ("breathe fresh air") while in the background "se oían los motores . . . de las cerealeras en la madrugada" (29; "one could hear the motors . . . of the grain silos in the early morning").

In *Las estrellas*, the air's toxicity also acquires a certain ubiquity: "un sol contaminado despidiendo luces hacia el espacio, cargadas de venenos químicos" ("a contaminated sun radiating light toward space, charged with chemical venoms") such that "en las orillas de otras cuencas y riachuelos . . . las plantas hicieran fotosíntesis de nuestros desechos" (701; "on the shores of other basins and creeks . . . the plants conduct photosynthesis with our waste"). The ecological principle that characterizes the ecosystem, meaning the chain of relations that interconnects different elements that was notably described by Humboldt, reveals in these tales–narrative and poetic–that the nineteenth-century disjunction that civilization once faced with barbarity now reemerges in a new form, no longer inverse or confronted like a binomial but instead juxtaposed. If the ecological crisis is a trans-spatial, transnational, and transcontinental crisis, then contemporary Argentine narratives allow us to glimpse a toxic discourse that calls into question traditional stigmatizations, offering instead new corporeal ones: be they the stigma of illness, of deformity, or of monstrosity.

The descriptions examined here signal a displacement within the aesthetic representations of the ecological crisis. It is remarkable that the global imaginary portraying the environmental catastrophes we are now facing has adopted an eschatological tone anchored in a discourse about the end: the end of species, the end of forests, the end of glaciers, the end of the mountains, and the end of clean oceans. It is not the end of nature as the end of a world "in which the natural environment disappears," as British sociologist Anthony Giddens has defined it, but the end of nature itself, as there are now "few if any aspects of the physical world untouched by human intervention" (206). Stemming from "the intensification of technological change" (206), the natural world has been thus replaced by a "postnatural" one (McKibben 126).

The devastation this now postnatural world has suffered has been documented by photographers such as Edward Burtynsky and Chris Jordan. Both Burtynsky and Jordan have portrayed the impact that the production of oil, extractivism, and other modern manufactures (Burtynsky) and the ongoing production of waste (Jordan) have had on the global environment and on human and nonhuman lives. The images captured by Argentine photographer Pablo Ernesto Piovano also reveal the materiality of the catastrophe. The same year that Schweblin and Molina published their texts, Piovano embarked on three journeys that covered 15,000 km in order to show the effects of direct contact with pesticides (though not indirect contact, as can occur, for example, through food). In a photo series titled *El costo humano de los agrotóxicos* (2014–17), Piovano attempted to break the vow of silence the media has established, which omits the effects of spraying pesticides on the bodies and lives of neighboring populations, keeping in mind that, according to the photographer, 370 million agrochemical products are used per year, a fact that goes unnoticed by the general public.[16] Images such as the "crystal boy"—the caption says his name is Lucas Techeira and he is three years old—reveal the slow destructive effect of the agrochemicals, rendering visible what remains by and large invisible. These toxins resist being captured by the eye, but their continuous employment—while impalpable or presumably immaterial—pierces the materiality of the body, all bodies alike, to the extent that they resurface as a physical distortion. Lucas Techeira was born with ichthyosis, a disorder that causes dryness of the skin. His mother was in contact with glyphosate during her pregnancy. While not intended to be morbid, the images are disturbing, fluctuating between horror and surrealism. Toxicity, one may argue, entails a materially eschatological framed reality. The images are living examples of the effects of agrochemicals on the human body. They visualize bodies that have been exposed, both metaphorically and literally, to a "large material world" that, as Stacy Alaimo states, is "penetrated by all sort of substances and material agencies that may or may not be captured" (4).[17] Paradoxically, while invisible, these substances emerge in the body as a protuberance, turning the invisibility of the spraying into a material, concrete, and physical reality. Furthermore, the exposure of these bodies conveys a "state of total unprotectedness" similar to those exposed to nuclear explosions and analyzed by Adriana Petryna in *Life Exposed: Biological Citizens After Chernobyl* (Alaimo 4). However, unlike Petryna's examination of a fast and devastating violence, the exposure to agrochemicals permeates slowly and sharply the contours of the material body as well as the physical landscape to the point of an outright tangible, measurable both deformation and mutation.

The immateriality of the slow destructive effect of the agrochemicals reminds us of what Ursula Heise has described as "riskspaces." Heise refers to chemical toxins as the most crucial of these risks, "as agents that effectively blur the boundaries between body and environment, domestic and public spheres, and between beneficial and harmful technologies" (177). By turning the impalpable or immaterial into a physical visible reality through the representation of a mutated human and nonhuman organism, the images captured by Piovano do not only connect to Rob Nixon's conceptualization of slow violence but also to the notion of hyperobjects, theorized by Tim Morton.

Rob Nixon defines slow violence as "a violence that occurs gradually and out of sight, a violence of delayed destruction that is dispersed across time and space" (2). While violence is usually conceived as an event "that is immediate in time, explosive and spectacular in space," slow violence, on the contrary, is not frequently "viewed as violence at all" (2). It is through this gradual mechanism of violence that the immateriality and invisibility of toxins grow into a tangible, palpable phenomenon whose main imprint is a disfigured organism. In the same vein, Morton refers to hyperobjects as objects that, due to their immensity, transcend the specificity of space and time; these encompass emissions in the atmosphere and oceans, as well as the global use of pesticides by the agricultural exportation industry. They are hyper because they don't allow for a clear division between the global, the local, and the body. The hyperobjects Morton refers to emerge in a moment of ecological crisis and erase material boundaries. To an objectified space are now added objectified bodies, bodies rendered exposed and vulnerable by these hyperobjects through Nixon's slow violence. In a time when the very notion of the future is open to speculation, when a discourse of the ends has become a frequent aesthetic and mediatic narrative, what happens with the body, the subjective and objective body that inhabits this chemically manicured space? What happens when environmental decay permeates both bodies and landscapes until it transforms corporeality itself into a mutation? Illnesses, deformities, monstrosities, aberrations, othernesses: they all emerge discursively and cross the space of the natural world, dividing healthy and sick bodies. Facing this ecological crisis that modernization has unleashed and continues to exacerbate, can we talk about a corporeal crisis, understanding that our body constitutes the first environment, as has been argued by Sarah Jaquette Ray and Jay Sibara? Taking into account the relationship between the subject and its habitat, I would like to propose, perhaps tentatively, possibly in a search for a new definition, that bodies that have come to be a mere "economic resource" consist of horrifying metaphors of the continued exploitation—agricultural, mining, oil, floricultural, nuclear, among many other kinds—of a natural world that has also been reconfigured. In this sense, it's important to preserve alongside these notions the concept developed by Rachel Carson of the ecology of the human body, which can serve as a starting point for reflecting on the relationship between humans and the environment. For

Carson, bodies do not constitute limits; on the contrary, they are vulnerable and affected by the global use of chemicals (agricultural or not), in the same way the human and nonhuman worlds are. In this sense, all forms of life are more similar than dissimilar.

The novels of Samanta Schweblin, Juan Diego Incardona, and Natalia Rodríguez, along with the poetry collection of Julián Joven, all predict, through the representation of transformed and mutated bodies, varied bodies (young and old, feminine and masculine, rich and poor), the deterioration of the latter as part of the last boundary of environmental pollution, be it through the increasing amount of urban pollution or the use of agrochemicals in agricultural production within contemporary rural spaces. Not only is the correlation between the subject and his natural environment redefined, but landscape—like the body itself—bears witness to this metamorphosis. Bodies ravaged by diseases (respiratory ailments, migraines, cancer) are also incapable of reproducing, hence aborting the idea of a future. These are bodies that exacerbate the apocalyptic speculations the Anthropocene warns us about and that therefore incarnate the same aberration of human intervention within the largest framework in history.

Notes

1. This chapter is an augmented version of "Narrativas del 'mal vivir' en América Latina: cuerpos inóculos y ecomutaciones," in *La invención de la naturaleza latinoamericana. Genealogía discursiva y funcionalidad sociocultural*, eds. Wolfgang Matzat and Dr. Sebastian Thies (Iberoamericana/Vervuert Verlag), forthcoming. I am extremely thankful to my dear friend Ryan Long, who has reviewed the essay more than once and sent me invaluable feedback to improve the chapter both aesthetically and argumentatively. I am also thankful to Rice University undergraduate student Mariana Nájera for helping me with the translation of this chapter. All the translated quotations are mine unless indicated.
2. According to Ludmer, contemporary literature has become urban, has absorbed the rural, and has become "barbarian," contesting the old traditional opposition civilization versus barbarism.
3. The work of Dinorah Cossío is currently unpublished. Her work is the qualifying paper presented during the process of obtaining her doctorate degree from the University of Texas at Austin, where I am a member of the evaluation committee.
4. The World Health Organization (WHO) said glyphosate was "probably carcinogenic" in 2015, which was merely a confirmation of what the scientific community already knew. Now that glyphosate has to be relicensed in the European Union, the WHO has decided to take a step backward. How come the International Agency for Research on Cancer (IARC)—which is part of the World Health Organization—retracts something it earlier claimed? How reliable does the WHO demonstrate itself to be? See interview with photographer Pablo Ernesto Piovano: www.lifegate.com/people/news/pablo-ernesto-piovano-interview

5. A good example of this imagination would be *Don Segundo Sombra*, by Ricardo Güiraldes (1926).
6. This term is not directly translatable but is similar to "wilderness" or "outdoors." The term conveys the notion of "exposed to the elements" with a strong connotation of being "battered by the weather."
7. I capitalize "Nature" to echo Alberto Acosta's call that "Nature" with a capital letter addresses a major issue, much larger and transcendent than just naming it and what we understand in the Western world: namely, to assign nature a mere economic resource role. "Nature" with a capital letter refers thus to Nature as an entity with political rights (11–19).
8. I use the concept of "biocapitalism" following Kelly Fritsch's definition in reference to the economization of life, which leads to a manner of speaking about different lives as having more or less value in more economic than exclusively biological terms (368).
9. See "Hoy estamos nuevamente en una época de mutaciones," an interview with Juan Diego Incardona in *Página 12* (October 17, 2016): www.pagina12.com.ar/diario/suplementos/espectaculos/4-40309-2016-10-17.html. Accessed March 22, 2019.
10. Please note that "punzó" is a Gallicism for bright red.
11. In the story, references to the Peronist movement crisscross the narrative from start to finish, constituting an indelible mark of its literary production. The presence of Peronism and, more specifically, "la patota menemista" (referring to Carlos Saúl Menem) as "emblematic of the violence of 90's liberalism, which attacks authentic Peronism" was analyzed in depth by Sandra Contreras (5).
12. In the United States, the "poinsettias" (*Euphorbia pulcherrima*) are known as Christmas Star.
13. However, because this concerns a saga whose narrative locus is Villa Celina (*The attack on Villa Celina* [2007], *Villa Celina* [2008], *El campito* [2009], and *Barrial Rock* [2010]), these metaphoric mutations can be understood using the explanation and hypothesis presented by the author in the introduction: "Many tales about Villa Celina occur in the 90's, pre-2001, and I think there lies the explanation about why the stories break with realism and become fantastic narratives where ghosts, monsters, and mutants appear. *The entire era is a great metamorphosis*" (Incardona 13; the emphasis is mine).
14. In *Derechos de la naturaleza*, Eduardo Gudynas argues that these neoextractivist policies have inspired new conflicts, in which progressive governments are confronted with their former allies. Furthermore, in Latin America, these ecoterritorial conflicts (sometimes called social-environmental conflicts) have become the main reason for conflict in the continent, reflecting how the dispute over common goods and territories are defining the future of the region (9).
15. The "pa" is a colloquial reference to the "padre" or "papá," the father of the poetic voice, in the same way that the "ma" refers colloquially to the "madre" or "mamá," the mother of the poetic voice.
16. See note 5. In addition, Argentine filmmaker Pino Solanas just released a documentary on the same topic, *Viaje a los pueblos fumigados*.
17. It should be noted that these pesticides are often illegal in Europe and the United States. However, the latter is not exempt from the risks and catastrophes we have just described: just the classic Steven Soderbergh film, *Erin Brockovich* (2000), which is based on a real story and tackles similar problems, is a very good example; the more recent case of Flint, Michigan, which suffers from a water contamination crisis, constitutes another paradigm of negligence and environmental injustice.

Works Cited

Aboaf, Claudia. *El rey del agua*. Alfaguara, 2016.

Acosta, Alberto. "Prólogo. Los Derechos de la Naturaleza o el derecho a la existencia." Gudynas, Eduardo. *Derechos de la Naturaleza. Ética biocéntrica y políticas ambientales*. Programa Democracia y Transformación Global; Red Peruana por una Globalización con Equidad; CooperAcción; Centro Latino Americano de Ecología Social, 2014, pp. 11–19.

Almada, Selva. *El viento que arrasa*. Mardulce, 2012.

Alaimo, Stacy. *Environmental Politics and Pleasures in Posthuman Times*. University of Minnesota Press, 2016.

Beck, Ulrich. *Risk Society: Towards a New Modernity*. Sage Publications, 1992.

Beilin, Katarzyna Olga, and Sainath Suryanarayanan. "The War Between Amaranth and Soy: Interspecies Resistance to Transgenic Soy Agriculture in Argentina." *Environmental Humanities*, vol. 9, no. 2, Nov. 2017, pp. 204–29.

Buell, Lawrence. "Toxic Discourse." *Critical Inquiry*, vol. 24, no. 3, Spring 1998, pp. 639–65.

Carson, Rachel. *Silent Spring*. Mariner Books, 2002.

Contreras, Sandra. "Economías literarias en algunas ficciones argentinas del 2000 (Casas, Incardona, Cucurto y Mariano Llinás)." *Orbis Tertius*, vol. 16, no. 17, pp. 1–14.

Cossío, Dinorah. *La maternidad tóxica y el campo industrializado en la narrativa argentina del siglo XXI*. Doctoral Qualifying Paper, U of Texas, Austin, 2017.

Dabove, Juan Pablo, and Susan Hallstead. "Introduction." *El año del desierto*, edited by Pedro Mairal. Stockcero, 2012, pp. vii–xxxiv.

De Leone, Lucía. "Campos que matan. Espacios, tiempos y narración en *Distancia de rescate* de Samanta Schweblin." *452°F. Revista de teoría de la literatura y literatura comparada*, no. 16, 2017, pp. 62–76.

Fritsch, Kelly. "Toxic Pregnancies: Speculative Futures, Disabling Environments, and Neoliberal Biocapital." *Disability Studies and the Environmental Humanities: Toward an Eco-Crip Theory*, edited by Sarah Jaquette Ray and Jay Sibara. U of Nebraska P, 2017.

Giddens, Anthony, and Christopher Pierson. *Conversations with Anthony Giddens: Making Sense of Modernity*. Polity Press, 1988.

Griswold, Eliza. "How 'Silent Spring' Ignited the Environmental Movement." *New York Times Magazine*, 21 Sep. 2012, www.nytimes.com/2012/09/23/magazine/how-silent-spring-ignited-the-environmental-movement.html.

Gudynas, Eduardo. *Derechos de la Naturaleza. Ética biocéntrica y políticas ambientales*. Programa Democracia y Transformación Global; Red Peruana por una Globalización con Equidad; CooperAcción; Centro Latino Americano de Ecología Social, 2014.

Harwicz, Ariana. *Matate, amor*. Lengua de trapo, 2012.

———. *La débil mental*. Mardulce, 2014.

Heise, Ursula K. *Sense of Place and Sense of Planet: The Environmental Imagination of the Global*. Oxford UP, 2008.

Incardona, Juan Diego. *Las estrellas federales*. Interzona, 2016.

Joven, Julián [Cristian Molina]. *Un pequeño mundo enfermo*. La bola editora, 2014.

Krimer, María Inés. *La inauguración*. El Ateneo, 2011.

Ludmer, Josefina. "Territorios del presente—En la isla urbana." *Pensamiento de los confines*, no. 15, Dec. 2004, pp. 103–10.

Lugones, Leopoldo. "La lluvia de fuego." *Las fuerzas extrañas*. Arnoldo Moen y Hermano, Editores, 1906.
McKibben, Bill. *The End of Nature*. Bloomsbury Publishing, 2003.
Mairal, Pedro. *El año del desierto*. Interzona, 2005.
Massuh, Gabriela. *Desmonte*. Adriana Hidalgo, 2015.
———. *La omisión*. Adriana Hidalgo, 2012.
Molina, Cristian. "Esa idea de Deleuze del escritor como médico." Interview by Pablo Chacón, Télam, 9 Sep. 2014, www.telam.com.ar/notas/201409/77534-esa-idea-de-deleuze-del-escritor-como-medico.php.
Montaldo, Graciela. *De pronto el campo*. Beatriz Viterbo Editora, 1993.
Morton, Timothy. *Ecology Without Nature: Rethinking Environmental Aesthetics*. Harvard UP, 2007.
———. *Hyperobjects: Philosophy and Ecology After the End of the World*. U of Minnesota P, 2013.
Nixon, Rob. *Slow Violence and the Environmentalism of the Poor*. Harvard UP, 2011.
Perrone, Tommaso. "The Human Cost of Agrotoxins: How Glyphosate Is Killing Argentina." *Lifegate*, 11 Nov. 2015, www.lifegate.com/people/news/the-human-cost-of-agrotoxins-glyphosate.
Ray, Sarah Jaquette, and Jay Sibara, eds. *Disability Studies and the Environmental Humanities: Toward an Eco-Crip Theory*. U of Nebraska P, 2017.
Rodríguez, Natalia. *La vi mutar*. Editorial Wu Wei, 2013.
Romero, Ivana. *Las hamacas de Firmat*. Editorial Municipal de Rosario, 2014.
Sarmiento, Domingo F. *Facundo: Civilización y barbarie*. Alianza Editorial, 1988.
Schweblin, Samanta. *Distancia de rescate*. Random House, 2014.
Wulf, Andrea. *The Invention of Nature: Alexander Von Humboldt's New World*. Alfred A. Knopf, 2017.

3 The Ruins of Modernity[1]

Synecdoche of Neoliberal Mexico in Roberto Bolaño's *2666*

Diana Aldrete

Roberto Bolaño's monumental novel *2666* (2004), situated in the fictional border town of Santa Teresa, centers its attention on the unsolved murders of women and girls. Without offering a solution, its depiction expounds a critique on the oppressive effects of neoliberalism and globalization. Bolaño problematizes the many "violences" that are seen and unseen, or ignored, ultimately positioning "our own complicit participation in the systems that perpetuate injustice" (Velasco and Schmidt 107). This essay explores the effects of neoliberalism that have led to the deterioration of humanity and the environment in Bolaño's novel. While the representation of "spectacular" violence in *2666* has been studied extensively by scholars, I propose an ecocritical analysis using Rob Nixon's concept of "slow violence," in order to explore problems of environmental justice[2] and to contribute to the already existing body of critical work on the novel. More specifically, through the exploration of Nixon's denunciation of "capitalism's innate tendency to abstract in order to extract" (41), fiction allows for writers like Bolaño to play an interjecting role in helping to counter this tendency and oppose the layered invisibility of environmental violence and its stake in the lives of the poor. This chapter presents the multiple effects that capitalism and its violence have on Mexican communities, often dismissed with arguments such as "collateral damage" in the name of "progress," as mirrored in Bolaño's *2666*. The novel depicts a system that dehumanizes and materializes bodies into tools of production and in turn disavows the value of people and the environment in their exploitation.

In *2666*, the city of Santa Teresa, a literary replica of Ciudad Juárez, is located on the edge of the state of Sonora and directly south of the city of Tucson, Arizona. The city, described as apocalyptic in nature, connects all the books/chapters[3] in the more than 1,000 pages of this novel;[4] it is a place where the Global North and South are in constant reference with each other. Unlike the many tourists and visitors whose privilege allows them to cross borders (Bauman; Augé), its poor inhabitants live within the confines of the city, unable to escape what has been considered a necropolis (Franco). By borrowing the symbol of the ruin, as an "impending breakdown of meaning" and "as a uniquely flexible and productive trope for modernity's self-awareness" (Hell and Schönle 6), Bolaño's novel clings to the "apocalyptic tradition" to explore what happens at the border between Mexico and the United States, where cities become dystopian futures. Indeed, Santa Teresa, like Ciudad Juárez, serves as a synecdoche of the failures of modernity and as a gravesite for those affected by capitalism.

Roberto Bolaño, a "synecdochal figure" (Hoyos 7) himself,[5] used the book *Huesos en el desierto* (2002) as reference for *2666*, as the essential mirroring of our neoliberal present and future. Sergio González Rodríguez, a friend of Bolaño's who was harassed and who endured several murder attempts because of his denunciation of feminicides and of organized crime, was made into a character in *2666*. For Bolaño, *Huesos en el desierto* represented evil and corruption, the metaphor of Mexico's past and the uncertainty of Latin America's future. It is a book "no en la tradición aventurera sino en la tradición apocalíptica, que son las dos únicas tradiciones que permanecen vivas en nuestro continente, tal vez porque son las únicas que nos acercan al abismo que nos rodea" (*Entre paréntesis* 215; "not in the adventurous but in the apocalyptic tradition, which are the only two traditions that remain alive in our continent, perhaps because they are the only ones that bring us closer to the abyss that surrounds us"). As in Ciudad Juárez, Santa Teresa serves as "un retrato del mundo industrial en el Tercer Mundo . . . un *aide-mémoire* de la situación actual de México, una panorámica de la frontera" (373) ("a sketch of the industrial landscape in the third world . . . a piece of *reportage* about the current situation in Mexico, a panorama of the border" 294–5). In both places, the real and the fictitious, the lives of many are enmeshed in a rhizomatic network where corruption, impunity, murder, exploitation, and degradation are identifiers for this volatile space. These two cities see the effects of capitalism where other forms of power networks see avenues of exploitation within legal and illegal markets: maquiladoras, drug trafficking, snuff movies, and sex/human/organ trafficking.

The experience of the "apocalyptic" Latin American city is what Esperanza López Parada explains as the monstrosity of the American city. One that grows like a tumor and whose postmodern collapse "ha ocurrido ya, se está viviendo permanentemente. Sería una tierra de nadie, un territorio 'postapocalíptico o postcatastrófico' . . ., puesto que habría soportado de antemano la decadencia occidental de los viejos contenidos y la defunción de los grandes relatos modernos" (*La ciudad imaginaria* 224; "has already happened, it is being lived permanently. It would be a no-man's land, a 'post-apocalyptic or post-catastrophic' territory . . ., since it would have endured in advance the western decadence of the old contents and the demise of the great modern stories"). This is what Ciudad Juárez/Santa Teresa[6] points to, a multifaceted city that occupies the backdrop of the economic world, yet through its capitalist enterprises it fragments the lives of the people who live in it.

As Edward Soja asserts, "[I]nternational trade and flows of capital, information, and people tends, without significant intervention, to lead to the continuing redistribution of wealth from the poor countries to the rich, from the periphery to the core" (57). In rewriting the neoliberal city, *à la* Ciudad Juárez, Santa Teresa becomes the epitome of a globalized problem, where the boundaries that separate the North from South are cognizant to the power structures that make it apparent that, although human migration is limited to the North, the influx of capital is the exception.[7]

In *2666*, the inhabitants of Santa Teresa are in a constant confrontation with the abject. This is presented in two ways: by the spectacular display of abjected female bodies and by the abjected Mexican land. This abjection is part of the criticism of capitalism's predisposition to disintegrate lives in order to dehumanize and consider them as abstract parts for a larger structure. In *Powers of Horror*, Julia Kristeva argues that literature permits the representation of "the ultimate coding of our crises, of our most intimate and most serious apocalypses. . . . [L]iterature may also involve not an ultimate resistance to but an unveiling of the abject" (208). This notion goes along with Rob Nixon's challenge of the representation of slow violence, in that "narrative imagining of writer-activists may thus offer us a different kind of witnessing: of sights unseen" (15). Therefore, Bolaño's novel is not a mere complacency of literary endeavor but rather, in its representation of the dismal effects of modernity, a call to pay attention to the horrors of capitalism and possibly our own complicity.

Similarly to Rob Nixon's analysis of the "environmental picaresque" in Indra Sinha's novel *Animal's People*, where neoliberal globalization is put on display, *2666* shows the three defining characteristics of the neoliberal order at play in portraying (1) "the widening chasm . . . that separates the megarich from the destitute," (2) the "ecological degradation that impacts the health and livelihood of the poor most directly," and (3) "the way powerful transnational corporations exploit under cover of a free market ideology the lopsided universe of deregulation" (46). This denunciation allows the reader to readdress that which continues to be invisible in our society, a slow violence that affects the poor who experience and are affected by it over time and space. Yet this reality is obscured by the media's attention on the spectacular and visible violence at the border.[8] By focusing on literature with an ecocritical approach, the novel engages in a process that materializes that which capitalism intends to convert to abstraction in order to extricate (Nixon 41). It further expands the scope in looking at environmental justice as it tests the "boundaries of realism and temporality, not [as] a route of escapism" but rather "to better understand why and how the exploitation of people of color, women, and the environment are linked, historically and systemically" (Sze 173).

Therefore, fragmentation as a method by which capitalism threatens social unity[9] is contested in Bolaño's *2666*, in the examples that follow in this essay, as ways of looking at the effects of capitalism on the Global South through a vision of slow violence. As Edward Soja maintains in his chapter "On the Production of Unjust Geographies" in his seminal text *Seeking Spatial Justice*, "such terms as North–South, First–Second–Third Worlds, the international division of labor, core and periphery, developed-industrialized versus developing-industrializing countries express the unfairness, inequality, and injustice of global geographies" (56). A notable method of representation in *2666* is the use of cartography by Bolaño to describe the city of Santa Teresa. This geographical mapping allows for the distinction of the overall power relations that exist within the city. This urban plan is guided by the first impressions of the visitors at the beginning of the first chapter "La parte de los críticos" ("The Part About the Critics"). The visitors' omniscient gaze directs readers to visualize the observations of the three critics on the city of Santa Teresa that pay attention to the precarity of its inhabitants,[10] its chaos,[11] its enormity,[12] and its foreignness.[13]

From the beginning, readers are presented with an abstract, cartographic perspective that conveys the abstract experiences, while the main characters provide a direct connection to the different violences. However, through the gaze of the characters, Bolaño fills in the gaps so as to orient and connect the reader back to the overall critique of the different violences experienced in Santa Teresa. In the north, aside from the border between the United States and Mexico, there are shopping centers, hotels, assembly plants, and the desert. In the east, there are more shopping centers, where the middle and upper classes live, and the university. However, if one kept going east, "llegaba un momento en que los barrios de clase media se acababan y aparecían, como un reflejo de lo que sucedía en el oeste, los barrios miserables, que aquí se confundían con una orografía más accidentada: cerros, hondonadas, restos de antiguos ranchos, cauces de ríos secos que contribuían a evitar el agolpamiento" (*2666* 171) ("there came a moment when the middle-class neighborhoods ended and the slums began, like a reflection of what happened in the west but jumbled up, with a rougher orography: hills, valleys, the remains of old ranches, dry riverbeds, all of which went some way toward preventing overcrowding" 129). In the south, there were two highways leading out of the city, the "maquiladoras" and "un barranco que se había transformado en un basurero" (171) ("a gully that had become a garbage dump" 129). The western part of the city is where the very poor inhabitants live; the roads are unpaved, and there are waste materials and shanty towns. In the middle is the old abandoned city, yet crowded and a constant point of crossing.

Beyond the city is the desert, which serves as a parameter that entraps those within. The desert is a constant reminder of death to the community, as some of the murdered women are found in it. The desert is also mentioned in the epigraph of the novel, "an oasis of horror in a desert of boredom," a line from the poem "The Voyage" by Charles Baudelaire from his book of poems *The Flowers of Evil*. In this poem, the entire stanza describes a tedious latency of our time: "Bitter wisdom one gleans from travel! the world, monotonous and/small, today, yesterday, forever, gives us back our image: an oasis/of horror in a desert of ennui!" (181). Our image reflected back is that of the reality of modernity. Bolaño explains that, ultimately, horror is evil: "hoy, todo parece indicar que sólo existen oasis de horror o que la deriva de todo oasis es hacia el horror" ("Literatura + enfermedad = enfermedad"; "today, everything seems to indicate that only the oases of horror exist or that the drift of every oasis is toward horror"). Similarly to the indication by Baudelaire's poem, Bolaño suggests that our evil reality, rooted in capitalism, is ever so present and without an end in sight. And that within this system, the only alternative in our modernity is to become complicit, "o vivimos como zombis, como esclavos alimentados con soma, o nos convertimos en esclavizadores, en seres malignos" ("literatura + enfermedad = enfermedad," or we live like zombies, like slaves fed with soma, or we become enslavers, evil beings). What freedom and modernity have brought, in that "oasis of horror in a desert of boredom," is "self-destruction through the quest for pleasure that leads to boredom or worse" (Franco 235), perhaps murder or the destruction of the environment.

Santa Teresa is the epicenter that connects all characters and chapters. However, some of the very protagonists in the chapters are foreigners, and all see themselves moving throughout the city with ease. For example, the European critics who are searching for the elusive writer Benno von Archimboldi in "La parte de los críticos" ("The Part About the Critics"); Óscar Amalfitano, the Chilean professor who arrives to the University of Santa Teresa in "La parte de Amalfitano" ("The Part About Amalfitano"); Quincy Williams, known at work as Oscar Fate, the American journalist who arrives to Santa Teresa on assignment to cover a boxing match in "La parte de Fate" ("The Part About Fate"); and again the mysterious writer Benno von Archimboldi, who is really Hans Reiter an ex-German soldier of the Eastern Front in "La parte de Archimboldi" ("The Part About Archimboldi") and who is the uncle of Klaus Haas, believed to be one of the serial killers in "La parte de los crímenes" ("The Part About the Crimes"). This freedom defines them as consumers since "the consumer is a person on the move and bound to remain so" (Bauman 85), which gives them the ability to cross borders, something that the inhabitants are unable to do. In addition, globalization, as described by Zigmunt Bauman, "is geared to the tourist's dreams and desires" (93), again reinforcing consumption at the expense of those who remain invisible.

According to Marc Augé's argument in *Non-Places: Introduction to an Anthropology of Supermodernity*, supermodernity produces *non-places* that relieve people from their identity and confine them to their activity, such as travelers or consumers. Although Ciudad Juárez/Santa Teresa would be considered a "modernized city," it also exemplifies what I believe to be a *non-place*. In other words, it is a place where "the realities of transit" are fixed and the crossroads are outlined by "the passenger (defined by his destination) with the traveler (who strolls along his route—. . . [and] where people do not live together and which is never situated in the center of anything" (Augé 107–8). As *non-places* reveal much of today's reality, in that "the concrete reality of today's world, places and spaces, places and non-places intertwine and tangle together" (Augé 107), the mapped description of the center of Santa Teresa exemplifies the *non-place*:

> En el centro la ciudad era antigua, con viejos edificios de tres o cuatro plantas y plazas porticadas que se hundían en el abandono y calles empedradas que recorrían a toda prisa jóvenes oficinistas en mangas de camisa e indias con bultos a la espalda, y vieron putas y jóvenes macarras holgazaneando en las esquinas, estampas mexicanas extraídas de una película en blanco y negro.
>
> (171)

> The city center was old, with three- or four-story buildings and arcaded plazas in a state of neglect and young office workers in shirtsleeves and Indian women with bundles on their backs hurrying down cobblestoned streets, and they saw streetwalkers and young thugs loitering on the corners, Mexican types straight out of a black-and-white movie.
>
> (128–9)

The many residents in *2666* are described as trapped in impoverished communities or shanty towns that resemble camps for gypsies or refugees (*2666* 149, 111) or in "un mar de casas construidas con rapidez y materiales de desecho" (170–1) ("a sea of houses assembled out of scrap" 128). Both the visitors and the inhabitants mix in this environment, turning the city into the center of intersection and the core of interaction. The contrast that the narrator observes is the limit of the Mexican space opposed to the movement of others in their freedom. Like the notion of *non-place*, the city is in constant intersection, where many people cross paths even if they do not interact, only of power and economy. Yet the effects of these exchanges are felt in the environment that slowly deteriorates.

Poverty is the reality for many of the citizens who live in the city. *2666*'s mapping reveals the tremendous discrepancy among the inhabitants of the city, as it shows which neighborhoods have allocated resources. This is clearly exposed in the dichotomies of east and west; the poor live on unpaved roads and waste materials in the west, juxtaposed by the shopping centers, the university, and the houses of the middle and upper class to the east. However, north and south seem to mirror each other as defined by the limits of the "maquiladoras." Beyond the perimeter of the north is the border between Mexico and the United States, a boundary that is foreseen in the south in the description of the depth of emptiness from the ravine converted into an industrial dump from the maquiladoras. This representation of the "dump," which I will describe in depth later in the chapter, plays into the symbolism of the demarcation of the Global South as representative of the exploitative ventures of capitalism.

In the last three decades, Ciudad Juárez has come to symbolize a city rooted in violence and a site that has exercised control of its border area between Mexico and the United States. Its geographical location has been of key importance in Mexican history, from the French Intervention and the Mexican Revolution, all the way to the enactment of the North American Free Trade Agreement (NAFTA) more recently in 1994. The Border Industrialization Program of 1965 initiated "maquiladoras,"[14] factories established by international/multinational companies where raw materials are assembled into finished or exportable products. This program served as the model for NAFTA, in which greater incentives such as low tariffs, lax environmental regulations, and exceptions to minimum wage policies are granted to multinational companies. What emerged from this is an increase in work for people who were willing to travel to this border area in order to get work immediately. However, at the core, "maquiladoras" became systems for the exploitation of poor communities, disguised as "opportunity ventures" where both the workers and the environmental resources intersected in an equally exploitative nexus. And yet what this neoliberal model reinforces is the history of Latin America's colonial past in the displacement of people and land as a new world order. As Julia Sze explains, "[C]ontemporary critics argue that the current crisis of corporate globalization is the newest manifestation of 'old' problems: the first wave, colonialism, and the second, development. The extraction of natural and labor resources links these disparate political, geographic and economic contexts" (171).

As shown in *2666*, the inhabitants of Santa Teresa remain in limbo, where they are either exploited or killed or must suffer the slow violence of their spatial presence. Rob Nixon describes what he calls "displacement without moving," in order to rethink the temporal and physical displacements involved in slow violence against the poor. That is, "instead of referring solely to the movement of people from their places of belonging, refers rather to the loss of the land and resources beneath them, a loss that leaves communities stranded in a place stripped of the very characteristics that made it inhabitable" (Nixon 19). While the residents have seen maquiladoras move in, they too have seen the resources of their environment diminish—with scarcity of water and the low air quality of the dusty surroundings.

The "maquiladoras" factories, where most of the murdered women worked, undoubtedly have had a harmful effect on the environment in border towns (Grineski et al.; Grineski and Collins). The gradual decline of the border's ecosystem and its towns is not a recent phenomenon, but it is "space and not time that hides consequences from us" (Nixon 45). The maquiladoras are clearly an integral part of Santa Teresa, as journalist Chucho Flores notes: "tenemos de todo. Fábricas, maquiladoras, un índice de desempleo muy bajo, uno de los más bajos de México, un cártel de cocaína, un flujo constante de trabajadores que vienen de otros pueblos, emigrantes centroamericanos, un proyecto urbanístico incapaz de soportar la tasa de crecimiento demográfico" (*2666* 362) ("We have everything. Factories, maquiladoras, one of the lowest unemployment rates in Mexico, a cocaine cartel, a constant flow of workers from other cities, Central American immigrants, an urban infrastructure that can't support the level of demographic growth" 286). For some of its inhabitants, those who arrive every year in search of work, the promises of job opportunities feed into the need to advance in their lives. For the majority, however, progress has not reached their spaces. Those who live in misery, in houses made of cardboard—the same cardboard that the factory discards as waste—find themselves at the whims of powerful corporations.

These factories rely on the necessity of employment to assert their dominance, they also "have the power to quell community resistance to their operations at the local level due to the steady supply of labor available along the border, close relationships with municipal authorities, and the vast economic resources of their transnational parent companies, which far outweigh local resources and can be marshaled toward halting mobilizations" (Grineski et al. 3). The character of Yolanda Palacios in *2666* points to the city's high levels of production, with the lowest unemployment rate in Mexico. This is because "aquí casi todas las mujeres tienen trabajo. Un trabajo mal pagado y explotado, con horarios de miedo y sin garantías sindicales, pero trabajo al fin y al cabo, lo que para muchas mujeres llegadas de Oaxaca o de Zacatecas es una bendición" (710) ("All the women have work. Badly paid and exploitative work, with ridiculous hours and no union protections, but work, after all, which is a blessing for so many women from Oaxaca or Zacatecas" 568). And in the industrial park General Sepulveda, "sólo una de las maquiladoras tenía cantina para los trabajadores. En las otras los obreros comían junto a sus máquinas o formando corrillos en cualquier rincón. Allí hablaban y se reían hasta que sonaba la sirena que marcaba el fin de la comida. La mayoría eran mujeres" (449) ("only one of the maquiladoras had a cafeteria. At the others the workers ate next to their machines or in small groups in a corner, talking and laughing until the siren sounded that signaled the end of lunch. Most were women" 358). The exploitation of the worker into what Foucault calls the "docile body" in order to control and become automatons of production,[15] is used to again erase the individuality of the person in the name of capitalism's predisposition of abstraction for extraction purposes (Nixon 41).

In addition to the exploitation found inside the maquiladoras, other biological hazards in border cities put the health of the inhabitants at risk, such as landfills (Grineski and Collins 253). In May of 1993, the year when *2666* begins to record the deaths of women, a dead woman was found in a dump in the industrial park of General Sepúlveda, "en el basusrero donde se encontró a la muerta no sólo se acumulaban los restos de los habitantes de las casuchas sino también los desperdicios de cada maquiladora" (449) ("in the dump where the dead woman was found, the trash of the slum dwellers piled up along with the waste of the maquiladoras" 358). The area surrounding this industrial park, which housed four maquiladoras, describes the contrast between the surrounding environment and the maquiladoras:

> Entre unas lomas bajas, sobresalían los techos de las casuchas que se habían instalado allí poco antes de la llegada de las maquiladoras y que se extendían hasta atravesar la vía del tren, . . . En la plaza había seis árboles, uno en cada extremo y dos en el centro, tan cubiertos de polvo que parecían amarillos. En una punta de la plaza estaba la parada de los autobuses que traían a los trabajadores desde distintos barrios en Santa Teresa. Luego portones en donde los vigilantes comprobaban los pases de los trabajadores, tras lo cual uno podía acceder a su respectivo trabajo.
>
> (449)

> Amid some low hills, were the roofs of shacks that had been built a little before the arrival of the maquiladoras, stretching all the way to the train tracks and across. . . . In the plaza there were six trees, one at each corner and two in the middle, so dusty they looked yellow. At one end of the plaza was the stop for the buses that brought workers from different neighborhoods of Santa Teresa. Then it was a long walk along dirt roads to the gates where the guards checked the workers' passes, after which they were allowed into their various workplaces.
>
> (358)

These factories are driven not only by the greed of its corporations but also by the subjugation of workers through exploitative tactics. Furthermore, besides being exploited for work purposes, the poor suffer the degradation of their physical environment and the risks to their health. When describing the southern area of the city, where the "maquiladoras" are located, the ravine serves as the dumpster for these factories: "un barranco que se había transformado en un basurero, y barrios que crecían cojos o mancos o ciegos y de vez en cuando, a lo lejos, las estructuras de un depósito industrial, el horizonte de las maquiladoras" (171) ("a gully that has become a garbage dump, and neighborhoods that had grown up lame or mutilated or blind, and, sometimes, in the distance, the silhouettes of industrial warehouses, the horizon of the maquiladoras" 129).

Too often is Mexico described as the "backyard" of American production, where everything is produced, to be later consumed by those in the North. According to Sergio González, this backyard becomes "a metaphor for private territoriality and subsidiary domain" (22).[16] This obscured reality is what Rob Nixon calls the "superpower parochialism," that is, "a combination of American insularity and America's power as the preeminent empire of the neoliberal age to rupture the lives and ecosystems of non-Americans, especially the poor, who may live at a geographical remove but who remain intimately vulnerable to the force fields of U.S. foreign policy" (34). This "superpower parochialism" is fueled by the idea of the "backyard" concept where everything that is happening "there" does not affect those that are "here." This distancing "is shaped by the myth of American exceptionalism and by a long-standing indifference" (Nixon 35) that allows for those in the Global North to disavow any responsibility for the livelihood of those in the South.

Dust, described as encapsulating many of the areas surrounding the city especially around the maquiladoras, symbolizes the way in which reality can be obscured, while still offering the metaphor of toxicity. In 2002, Mexican writer Elena Poniatowska made a special report for the newspaper *La Jornada* about Ciudad Juárez. Entitled "Ciudad Juárez: matadero de mujeres" ("Ciudad Juárez: The Slaughterhouse of Women"), Poniatowska's account describes a city drowned by dust:

> Juárez es una ciudad tomada por la chatarra, un inmenso cementerio de automóviles. Allí, entre la herrumbre de las salpicaderas, las cajuelas y las portezuelas, tratan de respirar los habitantes. . . . Ahogados por hierros retorcidos y llantas ponchadas, los extraterrestres (o casi) que viven en esta franja de tierra cumplen con el precepto: "polvo eres y en polvo te convertirás." Un polvo gris, mortuorio, todo lo ensucia, los escasos árboles se cubren de polvo, los cadáveres de 300 muchachas se desintegran enterrados en el polvo, el espíritu de 500 desaparecidas se va perdiendo como ánima en pena convertido en polvo. (Poniatowska; Juárez is a city taken over by scrap metal, an immense car cemetery. The inhabitants try to breathe in between the rust of the fenders, the trunk and the doors. . . . Drowned by twisted irons and flat tires, the aliens (or almost) living in this strip of land comply with the precept: "you are dust and unto dust you shall return." A gray mortuary dust, everything is dirt, few trees are covered with dust, the corpses of 300 girls disintegrate buried in the dust, the spirit of 500 missing is lost as souls in grief turned into dust).

Like Ciudad Juárez, Santa Teresa works as a well-oiled machine, according to the character Chucho Flores; his assessment of Santa Teresa reiterates the warning on the failures of capitalism as he compares the city with Detroit—once believed to be the mecca of industrialization with its automobile industry: "sólo nos falta una cosa . . . Tiempo para que esta mierda, a mitad de camino entre un cementerio olvidado y un basurero, se convierta en una especie de Detroit" (362) ("there's just one thing we haven't got, . . . Time for this shithole, equal parts lost cemetery and garbage dump, to turn into a kind of Detroit" 286). Oscar Fate, from the chapter "La parte de Fate" ("The Part About Fate") knows Detroit as earlier in *2666* it was described how he had covered the story of Barry Seaman after his mother died. One of his first observations when he arrived to the city was that of a "barrio [que] parecía un barrio de jubilados de la Ford y de la General Motors" (307) ("neighborhood [that] looked like a neighborhood of Ford and General Motors retirees" 241). Yet he quickly noticed the collapse of the city as he observed "un lote baldío lleno de malezas y de flores silvestres que ocultaban los cascotes del edificio que antes se levantaba allí" (307) ("a vacant lot full of weeds and wildflowers growing over the ruins of the building that had once stood there" 241). According to George Steinmetz, "unlike many cities in the eastern United States, Detroit was a low-rise metropolis of working class houses" (314).[17] Thus, it is quite reasonable that *2666* links the city of Detroit, depicted after its collapse of the automobile industry, with that of a border city such as Santa Teresa, where the failure and devastation, after NAFTA, is already observed from the beginning of the novel.

In addition to the symbolic gesturing of the border as a backyard, so is the allegory of the dumpster represented in various sections of the novel to reinforce the level of toxicity that inhabitants have to endure. While maquiladoras take advantage of the different and often ineffective regulations thanks to NAFTA, environmental law violations from foreign-owned companies are facilitated by poor enforcement, lack of adequate environmental legislations, and weak institutional frameworks (Grineski and Collins 252–3), making Mexico complicit in its own slow violence. In the novel, "El Chile" is described by its monumental overtake of the city as:

> El mayor basurero clandestino de Santa Teresa, más grande que el basurero municipal, en donde iban a depositar las basuras no sólo los camiones de las maquiladoras sino también los camiones de la basura contratados por la alcaldía y los camiones y camionetas de la basura de algunas empresas privadas que trabajaban con subcontratos o en zonas licitadas que no cubrían los servicios públicos. (752)

> The biggest illegal dump in Santa Teresa, bigger than the city dump, where waste was disposed of not only by the maquiladora trucks

> but also by garbage trucks contracted by the city and some private garbage trucks and pick-ups, subcontracted or working in areas that public services didn't cover.
>
> (602)

Similar to other dumps described as associated to several maquiladoras, "El Chile's" surrounding area is isolated, "donde hasta los matorrales estaban cubiertos por una gruesa capa de polvo, como si por aquellos lugares hubiera caído una bomba atómica y nadie se hubiera dado cuenta, salvo los afectados" (752) ("where even the brush was covered with a thick layer of dust, as if an atomic bomb has dropped nearby and no one had noticed, except the victims" 602–3) and whose residents are being swallowed by it (466; 372). Thus, the space contaminates the environment and affects the residents: "los habitantes nocturnos de El Chile son escasos. Su esperanza de vida, breve. Mueren a lo sumo a los siete meses de transitar por el basurero. . . . Todos, sin excepción, están enfermos. . . . La población permanece estable: nunca son menos de tres, nunca son más de veinte" (*2666* 466–7) ("the night residents of El Chile were few. Their life expectancy was short. They died after seven months, at most, of picking their way through the dump. . . . All, without exception, were sick. . . . The population was stable: never fewer than three, never more than twenty" 372–3). This contradicts the claim that "casualties are postponed, often for generations" (Nixon 3), as the contamination is rapidly affecting the inhabitants of the city. What this shows is that "the casualties of slow violence—human and environmental—are the casualties most likely to be seen, not to be counted" (Nixon 13) in neoliberal modernity.

At "El Chile," the metonymy of Santa Teresa, criminal violence and industrial neglect overlap; bodies are only coincidentally found, while fires are constantly reported in the dump, and it is unknown if they are set on purpose. They flare up by chance, or they are the scene of a crime (*2666* 466; 372). It's in this dump where the materiality of bodies share the same space with industrial materials, reinforcing the idea that women's bodies, as well as the spaces of Santa Teresa, are waste. In Roberto Bolaño's novel, the repetition of unsolved murders has a rhetorical function. The horror of the constant confrontation of abjected bodies in relation to the citizens, speaks to the contamination where the corpses appear throughout the city, in the streets, the garbage dumps, the desert, the vacant lots, behind the schools, near the maquiladoras, and even the drainage of the city.[18] The effect of and reaction to finding the corpse are frightening, and in the scenes of greatest impact, the corpses subsist with the citizens of the city. This repetitive horror is further exacerbated in the scenes of "El Chile," where the neighbors visit on several occasions and dead women are found. The biological risks due to the contamination from the maquiladoras assist in exploring the slow violence at the border. Like the corpses left in different places, the temporal reality shows that the environment too has been forgotten, and thus body and space are united in an allegory of the social depravity in Santa Teresa.

Roberto Bolaño highlights the networks that connect us with the invisible. The use of fragmentation as a literary technique resembles the modus operandi that has been implemented in Mexican novels over the years. As Carol Clark D'Lugo notes, "[T]he nation's fragmented social and political reality is consistently exposed in novels that dramatize a lack of cohesion, urban atomization, or disparities of class, race and gender" (D'Lugo 1). A key example of fragmentation in *2666* is the presentation of vignettes in the different cases and lives of murdered women. The enumeration of deaths not only determines a violent effect that exists in the city but also, by detailing the files of each victim, humanizes them.

Border cities are volatile spaces devastated by violence, a cemetery for hundreds of people every year, who are entrenched either in the violence of the city or in their attempt to cross the border.[19] What is particular to this border region is that it presents "a telling microcosm of North–South relations, revealing the forms, consequences, and tensions of global economic and cultural integration" while "it offers especially fertile terrain to assess the international dimensions of environmental justice in Latin America" (Carruthers, Where Local Meets Global . . ., 137). Santa Teresa is a "ciudad que dibuja semi-derruida, aislada, desértica, repleta de galpones industriales, pequeñas villas habitacionales y basurales. . . . Ciudad frontera, donde las fuerzas del orden no reaccionan, una urbe envejecida y desvencijada que va quedando en el olvido salvo por algún que otro hecho de sangre" (Rivera de la Cuadra 179–80; "city that draws semidemolished, isolated, deserted, full of industrial sheds, small villas and garbage dumps. . . . Border city, where the forces of order do not respond, an aging and dilapidated city that is being forgotten, except for the reality of blood"). And for Bolaño, the image of Santa Teresa is that of hell on earth: "Como Ciudad Juárez, que es nuestra maldición y nuestro espejo, el espejo desasosegado de nuestras frustraciones y de nuestra infame interpretación de la libertad y de nuestros deseos" (*Entre paréntesis* 339; "Like Ciudad Juárez, which is our curse and our mirror, the restless mirror of our frustrations and our infamous interpretation of freedom and our desires"). In this interview, the use of "we" by Bolaño is descriptive of the complicit reality that also becomes invisible within globalization. The global market makes it possible that we all benefit from the labor and exploitation of women's bodies and foreign natural resources through commodities.

The novel *2666* helps to formulate the intention of representing our reality. Bolaño explains that "all literature, in a certain sense, is political. I mean, first, it's a reflection on politics, and second, it's also a political program" (*Roberto Bolaño, The Last Interview: and Other Conversations* 588). This becomes our abjection, our discomfort and our rejection. But as noted by Marcela Valdés, it is also a statement that Bolaño leaves to his readers:

> In his 1998 acceptance speech for the Rómulo Gallegos Prize, Bolaño revealed that in some way everything he wrote was "a letter of love or of goodbye" to the young people who died in the dirty wars of Latin America. His previous novels memorialized the dead of the 1960s and '70s. His ambitions for *2666* were greater: to write a postmortem for the dead of the past, the present and the future.
> (*Roberto Bolaño: The Last Interview* 109).

The enumeration of the crimes listed in *2666* formulates a requiem for those victims who are forgotten, whether of spectacular violence or slow violence.

For Roberto Bolaño, Ciudad Juárez comes to represent that chaotic image of hell, which as a result becomes a reflection of our contemporary reality (*Entre paréntesis* 339). This image is transferred faithfully to Santa Teresa as a border city that is liquidated from all its resources for the benefit of others. The city becomes a fragment within the whole of the globalized world. It is no surprise, as later revealed in Ignacio Echeverría's "Nota a la primera edición" ("Note to the First Edition"), that the number in the title, *2666*, is "un cementerio olvidado debajo de un párpado muerto o nonato, las acuosidades desapasionadas de un ojo que por querer olvidar algo ha terminado por olvidarlo todo" (1124) ("a forgotten cemetery under the eyelid of a corpse or an unborn child, bathed in the dispassionate fluids of an eye that tried so hard to forget one particular thing that it ended up forgetting everything else" 897). Although this may suggest "not only an end to memory but a radical amnesia, the suspension of consciousness" (Franco 235), this number also represents a grandiose cemetery that houses the holocaust[20] of the neoliberal world: the ruins of modernity.

Notes

1. As if by happenstance, when I first wrote the title for this essay, I was not aware it was also the name of the anthology edited by Julia Hell and Andreas Schönle in their publication with Duke University Press (2010). Upon reading, I now borrow some of its pages in this chapter as the trope of ruins can stand in as a metaphor for "the reflexivity of a culture that interrogates its own becoming" (Hell and Schönle 7).
2. "Environmental justice movements call attention to the ways disparate distribution of wealth and power often leads to correlative social upheaval and the unequal distribution of environmental degradation and/or toxicity" (Adamson et al. 5).
3. In the first note that opens the novel, "Nota de los herederos del autor" ("A Note from the Author's Heirs"), the disclaimer reveals that, prior to Roberto

Bolaño's death, he left specific instructions for his novel *2666* to be published in five separate books in order to offer financial support to his children. However, upon his death, Ignacio Echeverría, editor and friend of Bolaño, decided to publish *2666* "en toda su extensión en un solo volumen, tal como él habría hecho de no haberse cumplido la peor de las posibilidades que el proceso de su enfermedad ofrecía" ("in a single volume, as he would have done had his illness not taken the gravest course").

4. This number is in reference to the original version published in Spanish through Anagrama (2004). However, the translated quotes referenced in this essay come from the English translation by Natasha Wimmer (2004).
5. "In the twenty-first century, the synecdochal figure has been Roberto Bolaño, who in many circles has come to represent the entirety of contemporary Latin American literature" (Hoyos 7).
6. I use the interchangeable image of Ciudad Juárez with that of Santa Teresa here due to the already noted disclaimer by Bolaño in that Santa Teresa is a literary representation of Ciudad Juárez.
7. As Julie Sze's analysis of Karen Tei Yamashita's novel *Tropic of Orange* (1997) equally observes, "[T]he reality of free trade: the 'right' to the free movement of goods, and for corporations to move factories to low-wage nations, is accompanied by the restrictive movement of people, and xenophobia" (169).
8. As described by Rob Nixon, "the insidious workings of slow violence derive largely from the unequal attention given to spectacular and unspectacular time" (6).
9. In referring to native communities, David Carruthers explains that such forces "threaten to fragment them, displace them, and drive them toward cultural disintegration" (10).
10. "Entraron por el sur de Santa Teresa y la ciudad les pareció un enorme campamento de gitanos o de refugiados dispuestos a ponerse en marcha a la más mínima señal" (*2666* 149) ("They drove into Santa Teresa from the south and the city looked to them like an enormous camp for gypsies or refugees to pick up and move at the slightest prompting") (111).
11. "Antes de volver del hotel dieron una vuelta por la ciudad. Les pareció tan caótica que se pusieron a reír" (*Ibid.* 150) ("Before they went back to the hotel they took a drive around the city. It made them laugh it seemed so chaotic" 112).
12. "La ciudad, como toda ciudad, era inagotable. . . . Tuvieron la certeza de que la ciudad crecía a cada segundo" (*Ibid.* 171) ("The city, like all cities, was endless. . . . They were convinced the city was growing by the second" 129).
13. "Sus movimientos fueron medidos y discretos, como los de tres astronautas recién llegados a un planeta donde todo era incierto" (*Ibid.* 172) ("Their movements were measured and cautious, as if they were three astronauts recently arrived on a planet about which nothing was known for sure" 130).
14. The "maquiladora" industry is not a recent invention. As detailed by Sara E. Grineski et al., the "phenomenon traces its roots to the Bracero Program, started by the US government in 1942. . . . When the program ended in 1964, several hundred thousand Mexican workers were returned to Mexican border cities. In an attempt to alleviate overcrowding and unemployment in these cities, the Mexican government created the Border Industrialization Program to promote industrial development and employment (Liverman and Vilas 2006). As a result, the *maquiladora* industry grew tremendously during the 1970s. In 1970, Mexico had 72 factories, and by 1979, it had 620. Today, there are approximately 3,000 *maquiladoras* in Mexico's northern border region" (2; emphasis in the original). And although the extent of the damage

on the environment caused by "maquiladoras" is unclear, there is a consensus among environmental activists and researchers that the overall growth of factories along the Mexican side of the border has caused environmental degradation and amplified health risks (Grineski et al. 2; Grineski and Collins 253).

15. On the question of docility, Michel Foucault has theorized on the manipulation of the body: "a body is docile that may be subjected, used, transformed and improved" (136) and that, through this docility, discipline forces the bodies' utility and obedience (138). It is no wonder Leslie Salzinger used this to explore the methods of gender production inside the factories, where femininity guarantees "docile bodies"; "docile labor cannot be bought, it is produced, or not, in the meaningful practices and rhetorics [sic] of shop-floor life. It is in the daily routines of the shop floor that gender shapes possibilities, profits, and transnational production" (16).
16. González adheres to the image of Juárez as the backyard with a postapocalyptic tone, condemned to misery: "the dump-desert city, the metropolis in ruins where human-machine-beasts, vacant lots, and junk survive as a generalized condemnation: the kingdom of rust that moves along a slithering plane, to pure materiality no longer thinkable that the norms and procedures of the city's past tend to be nothing more than post-human information. In Juárez, the fluidity of years gone by is now halted by army checkpoints, police, gunfire, gated communities, and the anti-violence protests of its citizens" (*Feminicide Machine* 22–3). In his book *The Femicide Machine*, Sergio González Rodríguez further establishes that Ciudad Juárez stands as a city with multiple purposes and identities throughout its history: (1) as a border town as sin city where U.S. citizens could participate in decadent tourism and as a backyard; (2) as the assembly/global city where women's labor aided in the global production of neoliberalism; and (3) as the war city, where the global market has exchanged its producers from the factories to illicit activities of cartels.
17. Detroit had a rapid growth, "peaking at a population of around two million in the mid-1950s. And just as the rise of Fordism created twentieth-century Detroit, the demise of Fordism has been responsible for Detroit's extreme impoverishment and for peculiarities of its ruination, such as the large number of abandoned high-rise office buildings in the downtown. Detroit is thus in many ways the ultimate museum and ruin of Fordism" (Steinmetz 314).
18. The body of the disappeared Penélope Méndez Becerra, "lo encontraron unos funcionarios de Obras Públicas de Santa Teresa en un tubo de desagüe que recorría bajo tierra la ciudad desde la colonia San Damián hasta la barranca El Ojito, cerca de la carretera a Casas Negras, pasado el vertedero clandestino del Chile" (*2666* 506) ("was found by some city maintenance workers in a drainage pipe that ran beneath the city from Colonia San Damián to the El Ojito ravine, near the Casas Negras highway, past the clandestine dump El Chile" 404).
19. As described by Osvaldo Zavala, author of the recent exposé *Los cárteles no existen: narcotráfico y cultura en México*, "*2666* se estructura entonces como un gradual acercamiento a la compleja materialidad de la vida en la frontera entre México y Estados Unidos, zona de conflicto donde convergen los vectores históricos de la violencia sistémica occidental moderna" (150; *2666* is then structured as a gradual approach to the complex materiality of life on the border between Mexico and the United States, a zone of conflict where the historical vectors of modern Western systemic violence converge). I see this materiality of life, including that of the nonhuman and ecological system, as part of what Argentinian anthropologist Rita Laura Segato calls the

"pedagogies of cruelty," that is, "all the acts and practices that teach, accustom, and program subjects to turn forms of life into things" (209). Although Segato specifically explores the exploitation of women's bodies within feminicidal violence, the same can be said of the space that is equally disadvantaged in the "extractive enterprise set up in the fields and small towns of Latin America to produce commodities for the global market" (209).

20. As Jean Franco demonstrates in her book *Cruel Modernity*, Bolaño's novel "is a monumental act of mourning not only for the generation born in the 1950s but for the Holocaust dead, for the Russian dead, and for the young women born in the 1970s and 1980s" (236).

Works Cited

Adamson, Joni, Mei Mei Evans, and Rachel Stein. "Introduction: Environmental Justice Politics, Poetics, and Pedagogy." *Environmental Justice Reader: Politics, Poetics, and Pedagogy*, edited by Joni Adamson, Mei Mei Evans, and Rachel Stein. U of Arizona P, 2002, pp. 3–14.

Augé, Marc. *Non-Places: Introduction to an Anthropology of Supermodernity*. Verso, 1995.

Baudelaire, Charles. *The Flowers of Evil*. Translated by Keith Waldrop. Wesleyan UP, 2006.

Bauman, Zygmunt. "Tourists and Vagabonds." *Globalization: The Human Consequences*. Columbia UP, 1998, pp. 77–102.

Bolaño, Roberto. *2666*. Editorial Anagrama, 2004.

———. *2666*. Translated by Natasha Wimmer. Farrar, Straus and Giroux, 2004.

———. *Entre paréntesis: Ensayos, artículos y discursos (1998–2003)*. Editorial Anagrama, 2004.

———. "Literatura + enfermedad = enfermedad." *Página/12*, Sep. 2003, www.pagina12.com.ar/diario/suplementos/libros/10-750-2003-09-28.html. Accessed 16 May 2013.

———. *Roberto Bolaño: The Last Interview: And Other Conversations*. Translated by Sybil Pérez. Introduction by Marcela Valdés. Melville House Publishing, 2009.

Carruthers, David V. "Introduction: Popular Environmentalism and Social Justice in Latin America." *Environmental Justice in Latin America: Problems, Promise, and Practice*, edited by David V. Carruthers, MIT Press, 2008, pp. 1–23.

———. "Where Local Meets Global: Environmental Justice on the US-Mexico Border." *Environmental Justice in Latin America: Problems, Promise, and Practice*, edited by David V. Carruthers, MIT Press, 2008, pp. 137–60.

D'Lugo, Carol Clark. *The Fragmented Novel in Mexico: The Politics of Form*. U of Texas P, 1997.

Foucault, Michel. *Discipline and Punish: The Birth of the Prison*. Translated by Alan Sheridan. Vintage Books, 1977.

Franco, Jean. *Cruel Modernity*. Duke UP, 2013.

González Rodríguez, Sergio. *The Femicide Machine*. Semiotext(e); Distributed by MIT Press, 2012.

Grineski, Sara E., Timothy W. Collins, and Maria de Lourdes Romo Aguilar. "Environmental Injustice Along the US–Mexico Border: Residential Proximity to Industrial Parks in Tijuana, Mexico." *Environmental Research Letters*, vol. 10, 2015, pp. 1–10.

Grineski, Sara E., and Timothy W. Collins. "Exploring Patterns of Environmental Injustice in the Global South: 'Maquiladoras' in Ciudad Juárez, Mexico." *Population and Environment*, vol. 29, no. 6, 2008, pp. 247–70.

Hell, Julia, and Andreas Schönle. "Introduction." *Ruins of Modernity*, edited by Julia Hell and Andreas Schönle. Duke UP, 2010, pp. 1–14.

Hoyos, Héctor. *Beyond Bolaño: The Global Latin American Novel*. Columbia UP, 2015.

Kristeva, Julia. *Powers of Horror: An Essay on Abjection*. Translated by Leon S. Roudiez. Columbia UP, 1982.

López Parada, Esperanza. "El mapa del caos: ciudad y ensayo en Hispanoamérica." *La ciudad imaginaria*, edited by Javier de Navascués. Iberoamericana Vervuert, 2007.

Nixon, Rob. *Slow Violence and the Environmentalism of the Poor*. Harvard UP, 2011.

Poniatowska, Elena. "Ciudad Juárez: matadero de mujeres I." Special Report on the Killing of Women in Ciudad Juárez, *La Jornada*, 26 Nov. 2002, www.jornada.com.mx/2003/06/02/esp_juarez/0060.htm. Accessed 24 Apr. 2019.

Rivera de la Cuadra, Patricia. "Santa Teresa: Ciudad-moridero en *2666*." *Revista de Filología Romántica*, 2008, pp. 179–86.

Salzinger, Leslie. *Genders in Production: Making Workers in Mexico's Global Factories*. U of California P, 2003.

Segato, Rita Laura. "A Manifesto in Four Themes." Translated by Ramsey McGlazer. *Critical Times*, vol. 1, no. 1, 2018, pp. 198–211.

Soja, Edward. *Seeking Spatial Justice*. U of Minnesota P, 2010.

Steinmetz, George. "Colonial Melancholy and Fordist Nostalgia: The Ruinscapes of Namibia and Detroit." *Ruins of Modernity*, edited by Julia Hell and Andreas Schönle. Duke UP, 2010, pp. 294–320.

Sze, Julia. "From Environmental Justice Literature to the Literature of Environmental Justice." *Environmental Justice Reader: Politics, Poetics, and Pedagogy*, edited by Joni Adamson, Mei Mei Evans, and Rachel Stein, U of Arizona P, 2002, pp. 163–80.

Velasco, Juan, and Tanya Schmidt. "Mapping a Geography of Hell: Evil, Neoliberalism, and the Feminicides in Roberto Bolaño's *2666*." *Latin American Literary Review*, vol. 42, no. 83, 2014, pp. 97–116.

Zavala, Osvaldo. "'El secreto del mundo' en Ciudad Juárez: *2666*, ética y política en la modernidad." *La modernidad insufrible: Roberto Bolaño en los límites de la literatura latinoamericana contemporánea*, edited by Oswaldo Zavala. U of North Carolina P, Chapel Hill, 2015, pp. 145–208.

Part II

Econarratives and Ecopoetics of Slow Violence

4 The Representation of Slow Violence and the Spatiality of Injustice in *Y tu mamá también* and *Temporada de patos*

Laura Barbas-Rhoden

There is a key sequence in *Y tu mamá también* (2001): the trio of young protagonists returns to their campsite from a boating trip with a local fisherman and his family, with whom they have spent the day exploring an idyllic coast. Tenoch, Julio, and Luisa met the fisherman, Chuy, on their trip to Boca del Cielo beach, a destination the youths from the capital invented (and then happened upon as a real place) in order to convince Luisa, the Spanish wife of Tenoch's cousin, to join them on a road trip. Upon their return from the outing with Chuy to the place where they have pitched their tents, the trio discovers that escaped pigs from a nearby hog farm have invaded their campsite and are rooting through their tents and wreaking havoc. Twice the track of diegetic sounds—the ocean, the pig squeals and grunts, human voices shouting—is cut, and a voice-over states plainly what happens next (and which will not be represented by the film): Chuy, the fisherman, will lose his livelihood when a hotel project buys up coastal property to commercialize it, and, additionally, various people in the nearby community will get sick after eating contaminated pork because the swine at the hog farm are sick.

The sequence is one of many in recent feature films suggesting that, in the globalized world of the early twenty-first century, nothing remains outside the complex processes of control. In the vision presented by the sequence from *Y tu mamá también*, there is little or no possibility of being "outside the system" and removed from its associated risks: diseases, contagions, and malaises, from porcine ailments to human cancers, depression, and alienation. The world is one of ecological and psychological vulnerability, as well as environmental violence, much of which is the result of invisibilized processes that unfold over time (Nixon 8). In the campsite sequence and throughout the film, the voice-over is one of the techniques by which the director expands interconnections across both time and space and opens the possibility for viewers to apprehend slow violence in the processes of globalization.

Y tu mamá también by Alfonso Cuarón (2002) and the less widely circulated *Temporada de patos* (2001) by Fernando Eimbcke are two works of Mexican cinema that actively engage with the representation of slow violence and its consequences at different levels of scale, from the scale of the human body and psyche (sites in which violence is experienced differently according to the positionality of diverse subjects) to the macro level of nation-scapes, where intersectional forces shape lived experiences (Crenshaw). *Y tu mamá también* is a mainstream art film that enjoyed immediate global circulation. *Temporada de patos* was Fernando Eimbcke's small-budget *opera prima*; it screened first in Mexico and then was picked up for wider international distribution (Poblete 53). Importantly, both films are situated in an aesthetic and market context in which, John Waldron argues, "the globalizing imaginary" has replaced the "pre-globalizing episteme" (11) in Mexican cinema and in which new critical approaches are required "to see beyond the popular aesthetic" and to look for "emergent fragments and new articulations within a text that is already coded as a commodity" (Waldron 13). My ecocritical approach, grounded in spatial justice theory (understood in decolonial terms), engages the films in precisely such a way as to see beyond the popular aesthetic.

Though they are not explicitly environmental in focus, *Y tu mamá también* and *Temporada de patos* visibilize the alienation, environmental destruction, and death that are wrought by overlapping regimes of control.[1] They also convey a sense of distress, malaise, and a nascent desire for liberation vis-à-vis the status quo of globalized modernity. The young protagonists of *Y tu mamá también* traverse both urban and rural landscapes in Mexico; those of *Temporada de patos* are limited to the confines of an urban apartment complex. Notwithstanding the difference in setting, the world that both films depict is one in which the co-occurrence of biocide and ecocide is a result of the intertwining of social and natural processes that are coproduced spatially, materially, and temporally. The directors' emplacement of scenes in particular sociohistorically and temporally marked landscapes emphasizes this interconnection. Furthermore, given that both films persistently associate psychological unease with the sociospatial realities of a globalized world, their "witnessing authority" (Nixon 16) insists that systems and processes contrary to life (both human and nonhuman) operate simultaneously at multiple levels of scale, from the troubled psyche (or diseased body) to the vast fragmented, commodified, privatized, and secured cityscapes and landscapes of the age of neoliberalism.

How exactly do the films effectively engage with what Rob Nixon (following Raymond Williams) calls the challenge of "rendering visible occluded, sprawling webs of interconnectedness" (45)? They do so by means of the presentation and foregrounding of an ecopsychosocial nexus. That nexus, in fact, is at the heart of the representation of globalization in the two films. In order to understand the ecopsychosocial nexus in these Latin American films, it is first helpful to explain and weave together two theoretical streams, spatial justice and decolonial theory, and then apply those theoretical filters to the interpretation of the films.

Theorizing Slow Violence in Latin American Contexts: Spatiality, Coloniality, and the Ecopsychosocial Nexus

There is a sociospatial dialectic, Edward W. Soja argues, and it is capable of producing injustices at multiple levels of scale, from the individual human (body and psyche) to the level of the planet (31). Ideologies shape both social relations and the spatial-material world, and there is interplay between social relations and spatial-material ones. Though the sociospatial dialectic is dynamic, "socially constructed geographies" and built environments persist and endure, sometimes for centuries, and so the sociospatial dialectics of the past have lasting implications over time (Soja 89). Even if (or when) ideologies change, physical and political structures (land/property tenure systems, regional and city plans, edifices and structures, for example) persist and replicate patterns and relationships from previous eras.

The sociospatial dialectic of previous eras has, in most parts of the world, been profoundly shaped for the last five centuries by what Aníbal Quijano calls "the coloniality of power" (533). In Latin America, as in many other parts of the world reorganized materially, spatially, and socially by settler cultures from Western Europe, the violence of subjugation and exploitation predicates and permeates many relationships; violence always already exists in the substrate upon which contemporary society takes shape. What is more, because Western ontologies and epistemologies generally present the human and social as separate and distinct from the material world, interconnections and interdependencies are invisibilized, and so therefore, too, are the violences by which interconnections are ruptured in a repeated and iterative process. Arturo Escobar (following Claudia Von Werlhof) explains that ruptures are key to exploitative control, which is exerted by means of (1) the fragmentation of relational ties among humans, as well as between humans and the material and spatial foundations of life, (2) the reintegration of entities (human, material, spatial) in units that are built and optimized for the production of financial gain, and (3) the destruction of what is not useful for immediate gain (Escobar 10).

Spatial justice theory, in combination with decolonial theory, can provide a useful theoretical framework for an ecocriticism capable of rendering visible not just interconnections but the mechanisms and enactments of injustice born of processes of coloniality and modernity. The interweaving of theories is important for ecocriticism; as a critical practice, ecocriticism must be able to engage with the realities of parts of the world outside of those in which ecocriticism as an academic field initially came to be known by that name. Importantly, for example, ecotheory and ecocriticism by Latin Americanists has co-occurred with a decolonial turn in theory in Latin America. Ecocriticism in Latin American contexts is informed by works in cultural studies, philosophy, anthropology, and literary studies that theorize, explore, and illuminate works of human imagination in a region of rich heterogeneity. This heterogeneity is comprised of ontologies and epistemologies, ideologies and imaginations, that are expressed in social, political, and cultural forms, in built environments, and in constructed orders for governing relationships (among humans, between humans and the rest of the material world, among all these and the more-than-human world) and that take shape by means of vast and multiple connections occurring over time and space. In its interrogation of philosophy and the social sciences, Latin American theory has questioned logics that appear to be "natural" from what Santiago Castro-Gómez calls the "hubris of the zero point" (79) anchored in a positionality of Western European patriarchal modernity. Enrique Leff argues, in a similar vein, that the environmental crisis is fundamentally a crisis of knowledge, of ordering the world according to the implications of a particular epistemology (27). As decolonial cultural studies theorists have repeatedly underscored, the human, social, material, and spatial realities of Latin America have fundamentally taken shape as the expression of the dynamics of modernity, and that modernity was cocreated with a colonial world order, including in the fields of knowledge production.

Because the processes of colonization entailed the destabilization, invisibilization, destruction, and/or suppression of alternative ways of knowing, being, and experiencing the world, ecocriticism in Latin American contexts must not only take into consideration just material, social, and spatial dimensions of change over time but also explore the psychological dimensions of the slow violence that shapes such change. If one imagines levels of scale from the global to the individual, the psyche, like the human body, is important because it is the place upon and in which multiple forces interact; the sociospatial dialectic shapes it, too, in profound ways.

In the West, the notion that there is an ecopsychosocial dimension to human experience has its roots in theoretical and empirical work in psychology and the environment and in the interventions that ecopsychology has made with regard to the biopsychosocial model in Western medicine. The biopsychosocial model is a focus on health and disease postulating that biological, psychological, and social factors play an important role in the context of illness or infirmity (*Oxford Reference* 2015). Expanding upon this understanding, ecopsychologists have asserted not only that biological, psychological, and social factors are in play in health but also that physical and material surroundings—environmental and spatial realities exterior to the biology of the human subject—are important factors in the well-being or malaise, whether physical or psychological, of human beings in any given moment (Fisher; Kahn and Hasbach; Kuo). Attention restoration theory, for example, holds that being in natural spaces is restorative and relieves mental fatigue (Kaplan). Of course, imaginations of human-spatial-material relations are not all Western ones, and the decolonization of theory is promising for the area of psychology and environment. In the meantime, decolonial spatial justice theory presents a framework by which to begin to explore phenomena that in the West have been conditioned to be seen as separated into social, psychological, and material spheres; it is also a tool here for reading film critically in order to reveal the processes by which the directors aim to slow violence and its consequences, from levels of scale of the psyche and body to those of the nation.

Reading Film Through Spatial Justice Theory

Film is a genre with visual and auditory modes, and it represents, imagines, and constructs worlds by means of the replication, disruption, reorientation, and/or subversion of existing cultural codes, both cinematic and sociohistorical. The application of spatial justice theory to films like *Y tu mamá también* (And Your Mama, Too) and *Temporada de patos* (Duck Season) invites an interpretive approach that moves along three axes: a narrative one, from a specific scene or sequence to narrative arc; a sociohistorical one that considers an image, word, sound, or trope and its cultural history; and a cinematic one that considers a particular technique in light of its history or as an innovation or disruption of convention. Injustices (and when present, resistances or liberatory efforts) may be signaled along any one of these axes or, in the case of the two films considered here, along all of them in combination.

The metaphors and narrative arcs of each film, analyzed with attention to the spatial dialectic, including the eco-psychosocial dimension, point to an ethical and ideological stance vis-à-vis processes and places that are "normalized" in globalization. They underscore human needs for ecopsychosocial well-being which simply cannot be realized in spaces structured by a modernity that is nearly totalizing in its capacity to organize and structure spaces, relationships, and time. In both films, human psychosocial malaise co-occurs with the physical and social reorganization of the world according to dominant economic agendas, those that persist from the past, for example, in development projects, including urbanization projects like those of the post–World War II era featured in *Temporada de patos* or more recent ones, like the global financial flows that bring changes to Chuy's coastline in *Y tu mamá también*. The films depict distinct worlds that exist within a national territory and that interact or conflict or coexist as a result of various processes: instrumentalist rationalities, political systems, the internal contradictions of nation-building, the processes of neoliberal globalization and resistances to it. Latin America, this representation suggests, is a palimpsest of ideologies, political structures, and built landscapes that, with the passage of time, have overlapped and interweaved themselves and, in doing so, have slowly eliminated spaces and places of well-being and increased the numbers of those for whom the possibility of self-realization and thriving is increasingly remote. And importantly, the films depict the co-occurrence of displacement, deterritorialization, and commodification with alienation and psychic pain; they underscore an association between the ordering of life in the pursuit of self-serving gain with unease and disquiet, sometimes latent, sometimes overtly expressed. Likewise, sequences in which tensions are relieved appear to exist outside the regulated, ordered, surveilled spaces of the megalopolis where the protagonists live and in which, in these spaces of escape from the processes of control and death, an ethics of care is affirmed, class differences become minimized (temporarily), and market values appear to cede to relational ones.

The poetic sensibility of the films thus highlights diverse spaces (cityscapes, rural areas, shorelines) and insists that their inhabitants be seen and be noticed. At the same time, they also suggest that ethics of care and solidarity persist (even if they lie dormant) in the face of relentless commodification, dehumanization, and depredation. Finally, each film concludes with a commentary on agency, space, and place—and the possibility of addressing the malaise of modernity. My reading begins here with *Y tu mamá también*, in which there is no resolution of tension in the conclusion, and then proceeds to *Temporada de patos*, the conclusion of which suggests that the odyssey of human beings is not yet concluded and that art and solidarity have some role to play in the search for some other way of being than the status quo.

Y tu mamá también

Y tu mamá también received an Oscar nomination for best original screenplay and enjoyed international commercial success. Ernesto Acevedo-Muñoz asserts that the film both "passes" as international cinema in the new globalized marketplace and continues the "historical trajectory of Mexican cinema," particularly in its treatment of topics of politics, class, and the economy" (39). Baer and Long argue that the film, while marked by similarities with other productions in an era of globalization, insists upon "the specificity of Mexican national-historical memory in its temporal and spatial dimensions" (151). Multiple social signifiers locate the film in time and space: the place is Mexico (specifically, the territory between the capital and the southern Pacific coast), and the narrative time of the film is the summer of 1999 (Baer and Long 161). Importantly, the year is immediately prior to the electoral defeat of the Partido Revolucionario Institucional (PRI) in Mexico, five years into both the era of the North American Free Trade Agreement and the Zapatista uprising, and squarely in the middle of a period in which "the Washington Consensus" set the terms of economic policy in international development. So the plot unfolds in the temporal center of processes of globalization and at a moment of national political transition in Mexico.

In *Y tu mamá también*, innovative techniques in montage impede the consumption of the film as the story of youthful sexual escapades or a "road trip" film and visibilize the death, displacement, and destruction wrought by multiple, slow, intersectional forces that converge upon one site (a migrant crossing a road, a stretch of coastline). In particular, the film is marked by the use of the "distracted" camera (in the hands of award-winning filmographer Emmanuel Lubezki), in combination with the extradiegetic voice on the soundtrack that interrupts what is understood as the primary narrative, the road trip. From an ecocritical perspective, what is most striking about the combination of the "distracted" camera and extradiegetic narration is that this combination effectively prevents viewers from consuming the film without critical reflection upon what they have been conditioned to see as background. The technique brings the sociospatial dialectic and, in particular, the ecopsychosocial nexus into the foreground. The "interstitial scenes," as Maria Saldaña-Portillo calls them (752), force the spectator to "think slowly" (Kahneman 20–21). The interspersing of shots by the distracted camera, in co-occurrence with an interruption of diegetic sound, draws attention to stories and dramas unperceived or minimally perceived by the protagonists as they pursue their own aims, in dramas in which they are the protagonists. As the egocentric narrative of Julio and Tenoch unfolds, they are depicted as seeing very little or nothing of their surroundings, just as they give no indication of perceiving the sadness and emotional distance of Luisa (who is, from time to time, drawn to details of her surroundings that Julio and Tenoch are not). Thus the film constructs a world in which there are (1) three main characters, two young men who want a sexual adventure and an older woman who has just received (and not disclosed) a diagnosis of terminal cancer, for whose favor and body they compete, and (2) multiple other people and landscapes whose histories and stories the distracted camera notices and that the extradiegetic male voice explains in a voice that is measured and clinical in intonation.

The audience sees multiple worlds within the world of the film, and must respond in some way to the dissonance their juxtaposition produces. Each interstitial scene interrupts the apparently "main" action of the road trip in the same way: the camera becomes distracted from the banalities of the boys and appears to take on agency of its own by looking out of a window, turning a corner, or following what is happening in another room. At the same time that the camera shifts to direct attention to a new object of its gaze, the diegetic sound is cut, and there are approximately two seconds of silence before a male voice calmly relays the stories of the place or person being passed by. The stories are frequently those of tragic death, migration, or displacement, which have occurred or will soon occur in the place toward which the camera gaze is directed. The stories of death and displacement render perceivable the slow violence and the "temporal and spatial occlusions" (Nixon 41) of the sociospatial ordering of infrastructure, labor, and landscapes in complex and vast networks for political and financial gain. In this way, the film depicts the landscape as more than a backdrop for the stories of the human beings who inhabit and traverse it; it is also an agent in and an archive of their stories, which are perceived by the camera and restored to memory by the voice of the extradiegetic narrator.

For its part, the setting and landscape of the film has been the topic of critical commentary, though much of that commentary has been brief.[2] There is substantial critical agreement on the function of the setting as a contrast to the diegetic narrative centered upon the youthful trio, especially as the setting moves to the forefront in interstitial scenes.[3] An ecocritical perspective reframes readings of the setting (without discounting their insights): the foregrounding of social, historical, and spatial contexts in the interstitial moments, for which there is no unifying narrative, allows for a more encompassing (but not totalizing) depiction of spatial injustices, as the attention of the viewer is directed from the pleasure-seeking, pain-avoiding body and psyche (of each of the main characters) to the body politic in the era of globalization. Injustices and inequities with bio-/ecocidal consequences that are generally hidden or occluded are thereby rendered visible; they interrupt the plot line that cinematographic convention would suggest is the primary narrative.

According to Gisela Heffes, certain aesthetic practices produce and insist upon a new critical episteme that combines a concern for the environment and life more generally (32). Alfonso Cuarón's praxis in *Y tu mamá también* is just such an aesthetic practice. Agency and spatiality, according to his filmic imagination, are always already intertwined, and it is impossible to tug on the thread that is psychosocial well-being, or nation-formation, or neoliberalization without also pulling the threads of environmental justice, social justice, economic justice, or the politics and performance of gender, class, and ethnicity. The image presented by the film is of a city and then rural landscape in which the sociospatial structure and processes generated by the coloniality of power, from the earliest days of the Spanish empire (acknowledged with visual references like the art objects in Tenoch's house, as well as his name) through eras of independence, nation-building, revolution, industrial development, and then neoliberalism, have fractured human relationships, eliminated the possibility of humans living lives of dignity and self-realization, and coproduced social and physical death.

The opening sequences of the film—which include one depicting a traffic jam generated by the death of a migrant from Michoacán when he is struck by a vehicle and another portraying the idyllic and ample property and home of Tenoch—serve as a filmic register of the problems and inequities of urbanization. The early sequence involving the traffic jam and the deceased migrant is representative of many Mexican cultural references in the film that draw attention to the way bodies and dreams are emplaced in time, space, and social relationships, and they themselves form part of a layering of stories that shape a culture. Other tropes and symbols function as a communicative code in the film: Tenoch's beloved nanny; the appropriation of indigenous names by the political class; the mural on the roadside with a Benito Juárez quote; the leftist sister of Julio; the esoteric pastimes of Tenoch's mother, pursued on the cultivated grounds of the family estate. Each figure, symbol, or reference serves as shorthand and signals to those familiar with Mexican history a legacy of classism, migration, and nation-building projects, each with spatial, material, and psychosocial dimensions: the impossibility of eking out a livelihood in rural areas, and their subsequent depopulation; the (sometimes fatal) attraction of cities for those who hope for a better life; the isolation and insulation of members of the dominant class from the conditions of life of the majority they govern.

Despite the friction generated by their different positionalities, all the tensions appear to recede into the background when the trio arrives at Boca del Cielo, the invented paradise upon which they stumble completely by chance. On the ocean shore, the sexual competition between Julio and Tenoch subsides, as do the tensions of class between them. They spend an idyllic day with a local fisherman and his family, and after the porcine incursion into their campsite ruins it, Luisa, Tenoch, and Julio relocate to rooms that Chuy's family rents to them. The night ends with what appears will be a reconciliation in a ménage à trois, but Luisa fades from their dance and the boys continue without her. Tensions resolve, but the film does not end with resolution. The image of the boys in bed together cuts abruptly to the city and a crowded intersection. In a conclusion reminiscent of that of José Emilio Pacheco's classic *Batallas en el desierto*, Julio and Tenoch bump into each other by chance in the crowded urban landscape where each is pursuing a destiny that appears to be scripted for their social position, one an upper-class young man, the other middle class. They share an uncomfortable coffee together. Back in the city, there is nothing more between them: no sexual affinity, no common ground of friendship, and no Luisa, who has succumbed to the cancer that was killing her. The intensely lived experiences of their trip and of their competition for Luisa has given way to inexorable, intersectional forces and a path to a predetermined adulthood. Biron notes the plot denies viewers "conventionally satisfying, totalizing conclusions" (59). Baer and Long similarly note that the "emancipatory possibilities put forward by the narrative are revoked by the film's forceful narrative closure" (159). By representing the terminus of the road trip in the anonymizing cityscape and the exodus of the male characters from their space of freedom and rebellion only into socially prescribed paths and roles, *Y tu mamá también* conveys a message about the impossibilities of liberation in a world in which globalization is written over previous forms of organization and control.

The conclusion of the film, just like its opening sequence, underscores the fragmentation of space, the occlusions such fragmentation produces, and the alienation experienced by human beings when technologies and systems become both materially destructive and "fatally disabling of personal and collective autonomy" (Escobar 8). The film depicts a palimpsest of overlapping norms, regulations, and mechanisms for the control of space, from the body to the globe, against which there is (occasionally) a struggle for emancipation and self-realization. In this vision of late twentieth-century Mexico, the space for emancipation—for life and thriving—appears to be narrow and occurs only briefly in interstices, if at all. As Forns-Broggi has noted, the boys go where life takes them, without they themselves deciding where to go (169). Though the narrative arc of the film has involved intertwining narratives that have run along a trajectory, they have all ended in a place of alienation, deterritorialization, or death. The patriarchal pleasure journey of the boys ends not at the beach but back in a reality shaped by norms of gender and class. Luisa's journey—a trip impulsively taken to live a few moments away from a loveless marriage and in the face of a terminal diagnosis—ends in literal, cellular death. And finally the micro narratives, delivered by the extradiegetic voice and visibilized by the inquisitive gaze of the distracted camera, show people and communities that have been "unimagined," as Nixon calls the phenomenon in narratives of nation-formation (150–1), and that are seeking to live even as they are being swept up by the ever evolving, autonomy-annulling complexity of global systems. The brilliance of the film lies in the creative innovation that insists all three narratives be seen and heard.

Temporada de patos

In an interconnected, globalized world, though, is it possible for resistance and resilience not just to survive in the interstices but also to give rise to generative or restorative activity, in spaces regulated by norms, practices, laws, and built environments? Can the possibilities of the interstices become visible and be recuperated by those unaware of them? Are there alternatives to what appear to be foreordained destinies shaped by complex forces that give rise to inequities at every level of scale, from the body and psyche to the megacity?

The narrative arc of *Temporada de patos*, in combination with an engagingly different technique involving shooting from fixed cameras, suggests that the ordinary and quotidian can give rise to inspiration for those who are discontent, uneasy, angry, frustrated, bored, or questioning, like the protagonists of the film. The film insists that viewers take notice of built landscapes and constructed social relations with deep historical roots. It also suggests that resistance, resilience, and legacies of hope can be cocreated in those same spaces and that viewers themselves may be coparticipants in shaping new understandings of ordinary stories (Poblete).

The setting and plot of *Temporada de patos* are deceptively simple. A middle-class apartment in a plain, multistory complex near a highway in a megacity, serves as a microcosm of urban life at the turn of the twenty-first century. In that apartment, two friends—adolescent boys Moko and Flama—look forward to an afternoon of pizza and video games with no parents at home. A neighbor girl, Rita, soon knocks to ask for ingredients to make herself a birthday cake; she is home alone, too, her birthday forgotten by her family. The pizza, delivered by Ulises, arrives just as the promised window of time for delivery is expiring. When the boys argue that the pizza is, in fact, late and that the tardiness means they are entitled to a free pizza, Ulises disputes the claim and refuses to leave the apartment without payment. Thus the cast is assembled and the plot set in motion.

Gustavo García asserts that "mucha bibliografía" is necessary to understand *Temporada de patos* (100; "much bibliography"), and indeed, a broader bibliography expands the interpretive possibilities for the 90-minute film. Intertextuality abounds: in visual elements, in the nod to the Beatles' balcony album cover and the name of the building in which the action occurs; in Ulises's name and journey. The filming privileges the scene, and in doing so, it draws from a rich history, both film history and the historical memory associated with the location.

The film contains visual markers in the opening sequences that fix its spatial location in the cityscape of the Mexican capital, and the "plain, multi-story apartment complex" is, in fact, a place of historical significance: a particular building in the Nonoalco Tlatelolco area of Mexico City. The location is associated with at least two moments of extraordinary public significance: the massacre of protesters, upon orders from the Mexican government, on the eve of the 1968 Olympics, and the collapse of a building and the subsequent rescue efforts by ordinary citizens in the 1985 Mexico City earthquake (Agencia Reforma; Maguire and Randall 134). In his astute reading of the film, Poblete asserts that the action of the film takes place between the public square (the site of the protests and the violence exerted by the state upon the protesters) and the private space of the flat (the site of a slow violence of control wrought by global forces upon the young people inside) (54). Significantly, the trio of young protagonists lives in the Niños Héroes building, named for the young soldiers who, according to lore, defended the Chapultepec castle in the capital during the Mexican-American War.

The opening sequences of *Temporada de patos* thus both establish the generic setting—decaying urban infrastructure, multifamily apartment complexes stained by years of pollution, old basketball courts and swing sets meters away from a highway overpass painted with graffiti—and anchor the story in a particular, historically significant place, especially for "insider" viewers for whom the social signifiers (the building name or neighborhood, for example) would be meaningful. The opening sequence is also significant for its technique. It is reminiscent of a black-and-white slide show: though there is motion within the frame, each frame ends in a dissolve to black before the next image appears. Poblete notes that in *Temporada*, Eimbcke often uses a frame that "shows an empty space, or, to be more precise, a physical place where there is no action because there are no agents" (55). When a human figure enters the scene and performs some minimal action, which is followed by silence, the effect is to slow down temporality and draw attention to details of the setting (Poblete 55). Though the human interactions are the apparent driver of the diegesis, the use of such a frame (empty of humans, then inclusive of them), along with that of the fixed camera that appears to gaze out from walls and appliances in scenes shot within the apartment, disrupts the centrality of the human. The spaces are there, and the objects/camera gazing, even when the human beings are not in the space. This concern with foregrounding spatiality and interconnectedness co-occurs with the decentering and recontextualizing of human agency throughout the film and is especially important in the denouement and conclusion.

After the opening sequence establishes the setting, the film shifts to establish social relationships within it: Flama's mother (his parents are separating) says goodbye, leaving instructions and money for Flama and his friend Moko to order pizza; Rita, the neighbor girl, comes over; and Ulises, the young pizza delivery man, arrives. Subsequently, the drama plays out in a way that foregrounds the sociospatial dialectic in the quotidian: an intermittent series of power outages punctuates the afternoon. These outages both signal an infrastructure inadequate to sustain consumption and also draw attention to the existence of forces beyond the apartment that are shaping a mundane Sunday afternoon for three adolescents and the young Ulises.

The adolescents appear to live a typical, cloistered, screen-mediated existence for middle-class, urban youth in the twenty-first century, marked by suffocating routine and an imagination-numbing freedom from routine that comes in the form of video games, pizza, and candy. They appear to be the products of exacting, complex, and overlapping forces of homogenization: colonialism (which rearranged the place they inhabit and established enduring hierarchies and privileges), economic imperialism, the built legacies of urban development practices, class conventions, and gender norms. In the analysis of Mark Hathaway and Leonardo Boff, such forces produce both objective oppression and psychological consequences, the result of which is the internalization by subjects of the loss of power to effect change (9). In *Temporada de patos*, the possibility of liberation comes from the fortuitous co-occurrence of the boys' psychological uneasiness with the present (which is a product of their anticipated separation; their nascent sexual identities) and a convergence of factors that serve as catalyst: Ulises's presence (and in particular his cross-species empathy), the power outage, Rita's brownies, and the painting of the ducks. The painting, in particular, is crucial. Both banal in appearance (it is a genre of wide distribution) and exceptional in its meaning, it depicts a duck taking flight from among the reeds of a lakeshore, as three ducklings swim behind another duck; the far shore, on the horizon, is marked by mountains.

By means of the simplicity of the plot and setting and the use of fixed cameras, director Fernando Eimbcke insists viewers take a close look at urban reality by reducing it to a basic unit—four walls, a small cast, a limited set of desires—depicted in a way that forces viewers to be interested in them. The filmic representation insists by means of narrative arc, cultural referents, and technique that this microcosm is one in which hegemonic norms and expressions co-occur with misgivings about them, including those misgivings anchored in some sense of alternative ways of being. In this way, the film makes the same observations as leading theorists from different fields. Philosopher Enrique Dussel refers to the coexistence of the Western world with its alternatives as "transmodernity" (221). Anthropologist Arturo Escobar understands it as a pluriverse and argues for a reframing of notions of knowing and being in the West such that the pluriverse becomes visible (Escobar xvi, 15–16). Liberation theologian Leonardo Boff and his coauthor Mark Hathaway argue that there is an important generative capacity in human communities and that the generative capacity persists in the face of the encroachment of ecocidal and biocidal ways of being, which all see in the processes of Western modernity. In *Temporada de patos*, cross-species identification and the co-creation of a new way of seeing the mundane (in the film, represented by the painting of the ducks), opens up emancipatory possibilities.

The use of fixed cameras, which appear to be looking out from paintings, cabinets, and appliances, conveys a sense of entrapment: nothing moves but the quartet of people, within a fixed space and time. The sense of limits, imposed by others, is underscored when we learn that Flama is about to move and it is unlikely he and Moko will be able to play again. The power outages also emphasize limits and an anonymous, routine-shaping agent located outside the apartment, in some distant, massive, complex, and malfunctioning system. Each power outage sets a new rhythm in the apartment, and the narrative of the film unfolds as a series of interconnected (and interrupted) episodes. For example, a video game FIFA match that was going to decide if the boys would pay for the pizza or not ends abruptly when the screen blacks out; tensions and disputes give way to hilarity after everyone eats the brownies Rita has baked and to which she has added her mother's stash of marijuana. In this way, the episodes of boredom that come from the disrupted routine of entertainment appear like minicrises from which ultimately spring the possibilities of reconciliation, resilience, resistance, or rebellion. Ulises tells about how he moved from his rural hometown to the city and how much he loves animals; how because of the economic necessity in his family, he was forced to take a job in a pound for stray dogs (which the film depicts in a flashback in one of the only scenes that takes viewers beyond the apartment and then only to a place of death and psychological trauma); how much his boss and his customers disrespect him in his job as a pizza delivery man; how he feels unease and unhappiness in the city. In another sequence, Flama speaks angrily about the absurdity of the disputes of his parents in their divorce (including their dispute about the painting) and, with his BB gun, shoots the many decorative figures, plates, and ornaments that convey the acquisitive consumerism of the middle class.

Most importantly, first Moko and then everyone is fascinated by the painting of the ducks, which appears to come to life. This "escape" sequence (actually a series of sequences), in which tensions are resolved, suggests that liberation is still a possibility for humans in search of another way to understand and be in the world. After eating the brownies, the four young people see the duck painting come to life; this change in perception is depicted in the film in auditory and visual ways. A camera appears to look out from the place of the painting into the gazing, wondering eyes of the youths, as the sound track changes to lapping water, the call of birds, sounds of the countryside. They all look upon nature-art come to life, and it looks back at them through the return gaze of the painting/camera. In a narration depicted in close-up shots, Ulises then explains the migration of ducks: that first one takes off, then others; that solidarity in the journey allows them to fly far; that they take turns being the lead duck and thus they fly far. Importantly, the work of the camera allows the name plate ("Ulises") on his uniform to move briefly into focus and thus the narrative of animal migration joins the story of the myth of the search for home and gives the key to the interpretation of the conclusion.

The sequence occurs in the last moments of the film and serves as a metaphor for a contestatory stance vis-à-vis ecocidal and biocidal forces, especially those that make human beings, like Ulises, complicit in a machinery of death (that of the dogs he was to euthanize for pay; the psychological death he feels in that job and in the pizza delivery job). Art, nature, and solidarity, the film suggests, can liberate the spirit, restore or awaken agency, and give strength for the search for or creation of other ways of being in the world. The painting in fact produces a visible change in Ulises: he is depicted "inside" the reality of the painting come to life, apparently nude, and when Flama extends to him the phone, he quits his job. Just a few moments later, the film cuts to its final sequence, which takes place outside the apartment, using more standard cinematographic technique rather than the fixed cameras.

After listening to one another and cocreating meaning, itself grounded in interspecies identification, the youths act on that understanding. Ulises exercises a right to mobility, and he leaves with (what is apparently) the gift of the painting. The competition of the video games and the boys' standoff with Ulises give way to an act of solidarity (Poblete 63). And the painting itself is liberated; it is no longer (just) a kitschy art piece or contested possession in the divorce proceedings of Flama's parents but rather a symbol and inspiration for liberation from the status quo. Adaptive creation, first in the form of the marijuana-enhanced brownies and second in the interpretation of the painting, offers emancipatory possibilities (Poblete 62).

The last sequence centers upon Ulises, who continues his odyssey in search of his place. It is night, and the viewer sees, from behind, a figure on a motorcycle on a highway. It becomes clear that the figure is Ulises (the helmet reads "Telepizza"), and on his back he carries the painting of the ducks. The motorcycle moves away from a camera that cannot keep up, and the music of the soundtrack changes, increasing in tempo and syncopation. Art is what Ulises takes with him; it is bound to him, literally, as he embarks on what is next. As the credits scroll, the music continues, eventually fading, then fades, and the sounds of birds chirping and water lapping are heard, as the credits make visible this line: "Como Juan Díaz Bordenave y Eduardo Galeano, la producción sigue creyendo, contra toda evidencia, que los patos unidos jamás serán vencidos" ("Like Juan Díaz Bordenave and Eduardo Galeano, the production continues to believe, against all evidence, that ducks united will never be defeated").

And so it is that the rather ordinary genre of the coming-of-age film, in Eimbcke's hands, poses an open question about being and whether it might be possible to have a world in which many worlds fit (to borrow the Zapatista phrase), in which objects see, and in which what might have been assumed to be background (like a painting) or diversion (like a film) might actually have and convey meaning for those who are able to see it. Samuel Steinberg, in a brief footnote, compares *Temporada de patos* with *Rojo amanecer* (Red Dawn), which was also filmed on location in Tlatelolco, and observes that *Temporada* "more subtly intimates the futures of emancipatory politics" (25). Poblete asserts that *Temporada de patos* is a cinematic intervention that explores "forms of liberatory film spectatorship and subordinated memories in neoliberal times" (51) and that the film is, in fact, about critical or participatory spectatorship (52). What is clear in *Temporada de* patos, for those who are able to see it, is that slow violence is not just made visible, it is also challenged by means of the reimagination of spaces and agencies and historical (perhaps even biological) memory in ways that are restorative.

Coda

Without misgivings about the present, people rarely seek alternative ways of being in the world. Art can be a means of turning attention to the malaise felt in the present, as is the case with *Y tu mamá también*, which closes off emancipatory possibilities in the diegesis, in the same way it disrupts a viewing of the film as a (or just a) road trip movie. Art can also be a means of drawing attention to the fissures in the façade of the routine and ordinary, to explore histories, imaginaries, and processes in which humans are a part and that shape us, and a new understanding of which can open up emancipatory possibilities, as is the case in *Temporada de patos*.

Because it can compress time and represent processes in multispatial ways, film (including popular film) is adept at making visible forms of slow violence that have been normalized and troubling viewers to perceive reality differently. What both Cuarón and Eimbcke have done with these two works is to make visible the spatiality of injustice and the co-occurrence of biocide, ecocide, and psychological malaise in the ordinary and mundane: built environments like apartments and highways, relationships like those of friends and couples, "natural" environments like coastlines and rural areas. The films illuminate the human, psychic, and material consequences of the regulation over time of matter and space (territory), as well as social relationships, by values and forces that have powerfully ordered the modern world. And the directors leave viewers with questions about what persists in the interstices of routine, in reorganized landscapes, in the forging and nurturing of relationships, such that biocide, ecocide, and psychological malaise give way to a new way of being in the world.

Notes

1. "Regimes of control" refers in this case to a wide variety of mechanisms for regulating and/or incentivizing behavior: laws codified at all levels (from municipalities, provinces, and states to national and supranational levels); protocols, practices, and procedures (within organizations or institutions, for example); social norms and expectations (within particular groups or subgroups).
2. Baer and Long assert that the film relies upon "stereotyped images of Mexico" (156). Shaw states that the film conveys "a vision of teenage travels in a folkloric, rural Mexico" (118) and "presents an image of Mexico that will be familiar to tourists/travellers, an image that includes a depiction of rural Mexico for Mexican city dwellers" (119). Saldaña-Portillo comments on the "bucolic Mexican landscape" (751) presented in panoramic shots, while Biron draws attention to the "fragmented and disordered" cityscape the film portrays (59). Acevedo-Muñoz comments that the depiction of the youths' "trip of discovery" through a portion of national territory is "deconstructive of conventional Mexican cinema topics and ideology" (41). Ecocritic and environmental writer Roberto Forns-Broggi, for his part, asserts in a brief film review that the landscape is not just "un fondo ajeno y pintoresco, también gana peso y voz" (170;

"a foreign and picturesque background; it takes on weight and voice"), an assertion with which I am in agreement.

3. In the road trip portion of the film (in contrast to the beginning and end of the film set in the capital), the interstitial scenes often depict what is happening outside the car "to random people or to the actual landscape" (Saldaña-Portillo 752). Forns-Broggi notes that the geography and drama outside the windows of the moving car contrast with that of the boys as they clumsily vie for Luisa (168). Emily Hind asserts that *Y tu mamá también*, like other turn-of-the-century Mexican films, portrays "province as a permissive space that facilitates social freedom" (26) and that "*provincia* functions as a picturesque national space that the capital dwellers often ignore, occasionally visit, and generally control" (41). Deborah Shaw holds that "the digressive journey structure allows the Cuarón brothers to make an allegorically-based film that presents a specific vision of Mexico" that both relies on and subverts genres (120). Jeff Menne reads the film as an allegory in which social signifiers, from names to details of the setting, have meaning in a particular narrative of the nation, and that they function to subvert the narrative (80). Lahre-Vivaz agrees the film is allegorical and argues that it queers the national romance (80). José Ballesteros holds that the interstitial moments move the narrative from a personal realm into a political one (110).

Works Cited

Acevedo-Muñoz, Ernesto. "Sex, Class, and Mexico in Alfonso Cuarón's '*Y tu mamá también*'." *Film & History*, vol. 34, no. 1, Jan. 2004, pp. 39–48.

Agencia Reforma. "Huella imborrable la del terremoto de 1985 en Tlatelolco." *Hoy*, www.chicagotribune.com/hoy/ct-hoy-8467015-huella-imborrable-la-del-terremoto-de-1985-en-tlatelolco-story.html. Accessed 25 Sep. 2018.

Baer, Hester, and Ryan Long. "Transnational Cinema and the Mexican State in Alfonso Cuarón's *Y tu mamá también*." *South Central Review*, vol. 21, no. 3, Oct. 2004, pp. 150–68.

Ballesteros, José R. "*Y tu mamá también*." *Film and History: An Interdisciplinary Journal of Film and Television Studies*, vol. 32, no. 2, 2002, pp. 109–10.

"Biopsychosocial." *Oxford Reference*, www.oxfordreference.com.library.wofford.edu/view/10.1093/oi/authority.20110803095507427. Accessed 11 Dec. 2015.

Biron, Rebecca E. "Death in El D.F.: Urban Fantasy in Aguilar Mora, Ramírez, Fuentes, and Blanco." *Discourse: Journal for Theoretical Studies in Media and Culture*, vol. 26, no. 1–2, 2004, p. 58.

Castro-Gomez, Santiago. "Decolonializar la Universidad: la hybris del punto cero y el diálogo de saberes." *El giro decolonial*, eds. Santiago Castro-Gómez y Ramón Grosfoguel. Siglo del Hombre Editores, 2007, pp. 79–91.

Crenshaw, Kimberlé. "Mapping the Margins: Intersectionality, Identity Politics, and Violence Against Women of Color." *Stanford Law Review*, vol. 43, no. 6, July 1991, pp. 1241–99.

Cuarón, Alfonso. *Y tu mamá también*, Producción Anhelo, Bésame Mucho Pictures, 2002.

Dussel, Enrique D. "World-System and 'Trans'-Modernity." *Nepantla: Views from South*, vol. 3, no. 2, 2002, pp. 221–44.

Eimbcke, Fernando. *Temporada de patos*, Cine Pantera, Lulú Producciones, Instituto Mexicano de Cinematografía (IMCINE), 2005.

Escobar, Arturo. *Designs for the Pluriverse*. Duke UP, 2017.

Fisher, Andy. "What Is Ecopsychology? A Radical View." *Ecopsychology: Science, Totems, and the Technological Species*, edited by Peter H. Kahn, Jr. and Patricia H. Hasbach. MIT Press, 2012, pp. 79–114.

Forns-Broggi, Roberto. "*Y tu mamá también*." *Chasqui: Revista de literatura latinoamericana*, vol. 31, no. 1, 2002, pp. 167–70.

García, Gustavo. "Los Nuevos Centuriones." *Nexos*, vol. 35, no. 422, 2013, p. 99.

Hathaway, Mark, and Leonardo Boff. *The Tao of Liberation: Exploring the Ecology of Transformation*. Orbis Books, 2009.

Heffes, Gisela. "Para una ecocrítica latinoamericana: entre la postulación de un ecocentrismo crítico y la crítica a un antropocentrismo hegemónico." *Revista de crítica literaria latinoamericana*, vol. 79, 2014, pp. 11–34.

Hind, Emily. "'Provincia' in Recent Mexican Cinema, 1989–2004." *Discourse*, vol. 26, no. 1/2, Jan. 2004, pp. 26–45.

Kahn, Peter H., Jr., and Patricia H. Hasbach. *Ecopsychology: Science, Totems, and the Technological Species*. MIT Press, 2012.

Kahneman, Daniel. *Thinking, Fast and Slow*. Reprint edition, Farrar, Straus and Giroux, 2013.

Kaplan, Stephen. "The Restorative Benefits of Nature: Toward an Integrative Framework." *Journal of Environmental Psychology*, vol. 15, no. 3, 1995, pp. 169–82.

Kuo, Frances E. "Coping with Poverty: Impacts of Environment and Attention in the Inner City." *Environment and Behavior*, vol. 33, no. 1, 2001, pp. 5–34.

Lahr-Vivaz, Elena. "Unconsummated Fictions and Virile Voiceovers: Desire and the Nation in *Y tu mamá también*." *Revista de Estudios Hispánicos*, vol. 40, no. 1, 2006, pp. 79–101.

Leff, Enrique. *Aventuras de la epistemología ambiental: De la articulación de ciencias al diálogo de saberes*. Siglo XXI, 2007.

Maguire, Geoffrey, and Rachel Randall. *New Visions of Adolescence in Contemporary Latin American Cinema*. Springer, 2018.

Menne, Jeff. "A Mexican Nouvelle Vague: The Logic of New Waves Under Globalization." *Cinema Journal*, vol. 47, 2007, pp. 70–92, doi:10.1353/cj.2007.0054.

Nixon, Rob. *Slow Violence and the Environmentalism of the Poor*. Harvard UP, 2011.

Pacheco, José Emilio. *Batallas en el desierto*. Ediciones Era, 1981.

Poblete, Juan. "The Labor of Attention and Spectatorship in Latin American Cinema: Fernando Eimbcke's *Temporada de Patos* (Duck Season, 2004)." *Global South*, vol. 8, no. 1, 2014, pp. 51–68.

Pratt, Mary Louise. "Language and the Afterlives of Empire." *PMLA*, vol. 130, no. 2, 2015, pp. 348–57.

Quijano, Aníbal. "Coloniality of Power, Eurocentrism, and Latin America." *Nepantla*, vol. 1, no. 3, 2000, pp. 533–80.

Saldana-Portillo, Maria Josefina. "In the Shadow of NAFTA: *Y tu mamá también* Revisits the National Allegory of Mexican Sovereignty." *American Quarterly*, vol. 57, no. 3, 2005, pp. 751–77, doi:10.1353/aq.2005.0051.

Shaw, Deborah. "(Trans)National Images and Cinematic Spaces: The Cases of Alfonso Cuarón's *Y tu mamá también* (2001) and Carlos Reygadas' *Japón* (2002)." *Iberoamericana: América Latina-España-Portugal*, vol. 11, no. 44, 2011, p. 117.

Soja, Edward W. *Seeking Spatial Justice*. U of Minnesota P, 2010.

Steinberg, Samuel. "Re-Cinema: Hauntology of 1968." *Discourse: Journal for Theoretical Studies in Media and Culture*, vol. 33, no. 1, 2011, p. 3.

Waldron, John V. "Introduction: Culture Monopolies and Mexican Cinema: A Way Out?" *Discourse*, vol. 26, no. 1/2, 2004, p. 5.

Werlhof, Claudia von. "Using, Producing, and Replacing Life? Alchemy as Theory and Practice in Capitalism." *Modern World-System in the Longue Durée*, 17 Nov. 2015, doi:10.4324/9781315633428-10.

5 The Voice of Water

Spiritual Ecology, Memory, and Violence in *Daughter of the Lake* and *The Pearl Button*

Ida Day

Daughter of the Lake (2015) by Ernesto Cabellos Damián and *The Pearl Button* (2015) by Patricio Guzmán communicate a deep spiritual connection that exists between indigenous populations and water in Peru, Bolivia, and Chile. These Latin American documentaries revive a traditional awareness of the sacredness of water as a life-giving source. The sea and the lakes are presented here as living beings, who have a spirit, memory, and voice. Such an attitude toward nature, advocated by spiritual ecology, serves as a model of resistance against unlimited economic growth and modern forms of economic colonialism.

Spiritual ecology is a growing, interdisciplinary field focused on the revival of human connection to the environment in the face of ecological crisis. The essential work in this area of study, *Spiritual Ecology: A Quiet Revolution* by Leslie E. Sponsel (2012), considers that the crisis has not only a physical dimension but also a spiritual one, manifested by "human alienation from nature combined with disenchantment, objectification, and commodification of nature" (xv). The author refers to indigenous nations as the "original spiritual ecologists," which, in contrast to Western ethics of dominance over nature, emphasize a mutual respect of local ecosystems.[1]

These indigenous values and attitudes have been gaining strength globally and have played a central role in the rise of the theory of degrowth. Proponents of degrowth call for a transformation from the extractivist paradigm to a value system based on care and respect for the Earth—a living planet, which would ultimately improve quality of life and promote environmental regeneration. These values, incorporated in the indigenous philosophy of *Buen Vivir* in Latin America, offer a deeper understanding of ecological crisis and inspiring guidance for transforming our relationship with the Earth.[2]

In recent years, several films have discussed ecological concerns related to the global water crisis, for example, *Even the Rain* (2010), *White Water Black Gold* (2011), *Watermark* (2013), *Water and Power: A California Heist* (2017), *The Power of Clean Water* (2018). They focus on violence against local people and the environment, which in the Global South is intimately linked to extractive practices, such as mining and industrialized agriculture. This kind of violence, defined by Rob Nixon as "slow violence," occurs gradually and out of sight, producing "slowly unfolding environmental catastrophes" (2). In *Daughter of the Lake* and *The Pearl Button*, slow violence is reflected in the increasing droughts, deforestation, loss of biodiversity, loss of traditions, and the displacement of indigenous communities. All these environmental and social problems are the result of the model of development based on the continuous exploitation of natural resources and their sale on global markets. These extractive activities provoke many tensions and controversies because of their devastating effects on ecosystems, indigenous people, and traditional economies. By addressing the colonial historical context and globalization, *Daughter of the Lake* and *The Pearl Button* criticize the exploitation of gold in Peru and Bolivia and the destruction of the marine culture of western Patagonia, respectively.

Damián and Guzmán have chosen, as their storytelling strategy, the poetic documentary, which, as defined by Bill Nichols in *Introduction to Documentary*, "stresses mood, tone, and affect much more than displays of knowledge or acts of persuasion" (103). Poetic documentary has its own prominent voice, which represents reality and conveys the message through "a series of fragments, subjective impressions, and loose associations" (103). The poetic voice-overs in *The Pearl Button* and *Daughter of the Lake* play a vital role in transmitting an environmental message and expressing the basic unity of all elements in the universe. Through a captivating rhythm, metaphors, and shifts in time, they create images of the natural world and reflect on the philosophical and spiritual aspects of water, conveying an ecological sensibility.

Poetry can play a vital role in environmental activism, since in order to find solutions, a new ecological awareness and imagination are needed. This essay examines how the poetic mode adequately communicates the objectives of spiritual ecology and expresses Nixon's idea of slow violence. First, poetry acts on our sense of intuition and inner truth; hence it communicates effectively the connection between the human being and the natural world to Western audiences. As John Felstiner stated in *Can Poetry Save the Earth?* (2009), "poems make us stop, look, and listen long enough for imagination to act, connecting, committing ourselves to the only world we've got" (13). Poetry shapes our consciousness, inspires, enlightens, and awakens our awareness of being an integral part of the universe. *Pearl Button* and *Daughter of the Lake*, which connect human experience to the natural world, promote the freedom of imagination necessary for this awareness. The poetic style of narration blurs the rigid division between notions of reality and imagination commonly sponsored by Western worldviews. The directors integrate the voices of nature, ancestors, and spirits in the films to demonstrate the material and spiritual bonds between indigenous people, their traditions, and nature.

Secondly, the poetic documentary is subtler than other cinematographic forms, such as overtly violent feature films with spectacular effects or expository documentaries that favor a more authoritative and direct discourse and a strong point of view. In contrast to expository documentaries, which rely on well supported arguments, data, and statistics to convey knowledge, *Daughter of the Lake* and *The Pearl Button* reach the viewer through non-Western logics and rhetorical modes. According to Nichols's definition of poetic documentary, they "open up the possibility of alternative forms of knowledge" (103). Characterized by nonlinear structure and slow-paced rhythm, the films provoke reflection on slow violence by focusing on experiences and impressions. All these elements are brought together by poetic narration, which gives the viewer the experience of being addressed personally. The slower, lingering, and methodical representation of images and voices in *Daughter of the Lake* and *The Pearl Button*, as opposed to traditional news programs and Internet sound bites, reflects the sense of gradualness and slow unfolding of catastrophic events. Therefore, the viewers gain an absorbing, experiential, insightful, and uniquely empathetic sense of how the events in the films occur and of their impact on the human beings and environments depicted in them.

In his book, *Slow Violence and the Environmentalism of the Poor* (2011), Rob Nixon addresses our lack of attention to disasters that occur slowly and gradually (thawing cryosphere, toxic buildup, deforestation, biodiversity loss, acidifying oceans, etc.). He points out that in an age that venerates sensational news and instant messaging, "slow violence is deficient in the recognizable special effects that fill movie theaters and boost ratings on TV" (6). Because it is so easily discounted, slow violence is especially catastrophic in effect, which is reflected in increasingly dire environmental problems. Nixon's book questions the media bias toward spectacular and explosive disasters (earthquakes, avalanches, volcanoes, tsunamis, etc.) and calls for a reexamination of slow violence: "How can we turn the long emergencies of slow violence into stories dramatic enough to rouse public sentiment and warrant political intervention, these emergencies whose repercussions have given rise to some of the most critical challenges of our time?" (3).

The powerful individual stories in *Daughter of the Lake* and *The Pearl Button* produce similar emotional responses to mobilize the sentiments and intervention alluded to by Nixon and bring to light the oppression and injustice suffered by local communities for centuries. The films include interviews with indigenous people who talk about their territorial dispossessions and loss of traditions. Both directors address the role of slow violence in Latin America in the context of climate change, demonstrating how the processes of ecocide and ethnocide have been unfolding over the centuries. As Eduardo Galeano pointed out in *Open Veins of Latin America*, "The human murder by poverty in Latin America is secret: every year without making a sound, three Hiroshima bombs explode over communities that have been accustomed to suffering with clenched teeth. This systematic violence is not apparent but is real and constantly increasing" (5). This kind of suffering, reflected in *Daughter of the Lake* and *The Pearl Button*, can be traced back to colonial times, when the local traditions and technologies in the Global South were destroyed by the North.

Daughter of the Lake portrays the ecological degradation of an Andean community of Cajamarca through the eyes of a young activist, Nélida Ayay, fighting for the indigenous rights to water, which is being used for the extraction of gold and copper by a mining company—Minera Yanacocha. This company, owned by the U.S.-based Newmont Mining Corporation and supported by Buenaventura of Peru and the World Bank, started the Conga Project in 2010, which was expected to exploit 3,069 hectares of land—"the water source for more than 30,000 people in 200 communities spanning three provinces" ("Conga" n. page). The project involved draining two main lakes and pumping the water waste from the mine into the five major rivers, which affected Cajamarca's ecosystem, a high-altitude biologically diverse wetland.[3] Although local farmers were promised jobs and economic prosperity, they were not consulted in the development of mining on their land and continued to live in poverty. Conga was crucial for Newmont's growth since it was expected to produce annually "580,000 to 680,000 ounces of gold and 155 million to 235 million pounds of copper during its first five years" (Trefis Team n. page). The project became a source of debate about whether economic interests (including job creation and tax revenues) should be a priority over local opposition and environmental concerns. In the following months, many protests and rallies were held across the country, which led to the suspension of the project in November of 2011.[4] At the forefront of the opposition was Máxima Acuña de Chaupe, a Peruvian farmer, environmentalist, and one of the protagonists in *Daughter of the Lake*. The Chaupe family has cultivated land of Cajamarca for over 20 years, and when Newmont's agents attempted to evict them from their farm to expand mining projects, Máxima stood up against this multinational corporation. In 2016, she won the Goldman Environmental Prize for defending environmental and human rights.[5] (See Figure 5.1.)

It is important to note that the extraction of precious metals in Latin America has been an issue since the European conquest; therefore, the economic inequality and the marginalization of indigenous people that we see in the film are the manifestations of a violence that is "incremental and accretive, its calamitous repercussions playing out across a range of temporal scales" (Nixon 2). Galeano traces this kind of violence to the colonial silver mining in Potosí, Bolivia, which in three centuries consumed 8 million lives. He refers to this genocide as "the bleeding of the New World," pointing out that the exploitation still continues, as the

Figure 5.1 Nélida Ayay in *Daughter of the Lake* (2015).
Source: With permission from film director Ernesto Cabellos Damián.

ideological justifications are never in short supply—first, it was in the name of the Christian faith, then for economic progress (*Open Veins* 41). In his later work, *Úselo y tírelo* (Use It and Throw It Away), Galeano continued to explore the links between resource exploitation and the degradation of local ecosystems, bringing into dialogue the divide between the Global North and the Global South: "La divinización del mercado permite atiborrar de mágicas chucherías a las grandes ciudades del sur del mundo, drogadas por la religión del consumo, mientras los campos se agotan, se pudren las aguas que los alimentan y una costra seca cubre los desiertos que antes fueron bosques" (13; "The divinization of the market fills with magical trinkets the big cities of the Global South, drugged by the religion of consumerism, while the fields get depleted, the waters that sustain them become putrid, and a dry crust covers the deserts that used to be forests"). The extractive capitalist policies, criticized by Galeano, are profit driven, indifferent to the environmental consequences of growth, and oppressive to the communities in the Global South. They are pursued by the governments claiming to alleviate poverty; however, the ways of life of indigenous people outside the cities are being endangered and their ecosystems destroyed. Naomi Klein, in *This Changes Everything: Capitalism vs. The Climate*, describes these areas outside the cities as "sacrifice zones"—places that can be "poisoned, drained, and otherwise destroyed, for the supposed greater good of economic progress" (169). *Daughter of the Lake* portrays these "sacrifice zones" in Peru and Bolivia, where the impact of extractivism is manifested by land erosion, biodiversity loss, pollution, and drought.

Cabellos Damián's film focuses on the personal experiences of the protagonist, Nélida. The injustices against her community inspire her to move to the city to study law to defend her community within the modern legal system. Overwhelmed by the impersonal interactions of the urban setting, she reflects nostalgically on her life in the country, which is no longer the same after the mining project destroyed the soil and caused the shortage of water: "Extraño la tierra, los cantos de los pájaros. Extraño mis animales. En la ciudad, no es vivir bonito" ("I miss the land. I miss the songs of the birds. I miss my animals. Life isn't good in the city"). She talks about the disappearance of certain bird species, such as *quilina* that leave the area in search for water, and recalls certain indigenous practices, such as digging potatoes and sowing crops. Although her tone is lyrical and meditative, violence is present and felt throughout the film. The striking images of dried-out wells, scorched landscapes, as well as muddy and polluted industrial areas evoke the feelings of loss, instability, and "a violence of delayed destruction that is dispersed across time and space" (Nixon 2). The last image of the film, following the footage of protests against the mining company, commemorates the people who died defending the water—Paulino Eleuterio García Rojas, Joselito Vásquez Jambo, César Medina Aquilar, Antonio Joselito Sánchez Huaman, and José Faustino Silva Chávez. As a tribute to their lives, Nélida puts their photographs and flowers on the surface of the lake. The personal stories and the visual language of the film leave the audience feeling not only empathy but also the emergency of the situation.

The director also includes a story of a Dutch jewelry designer, Bibi van der Velden, whose ethical awareness is awakened after visiting the gold mine and seeing the apocalyptic conditions in which people work there—kilometers of devastated land, water contaminated with cancer-causing chemicals, and endemic poverty. The designer learns that in order to produce 15 grams of gold, five to six people need to work continuously for 24 hours. During her visit, Bibi reflects on the emotional value of gold: on one hand, she believes that jewelry given for births and weddings immortalize these moments; on the other, she realizes that the way gold is extracted exploits people and destroys the environment: "we all have blood hanging on our fingers, wrists, and necks." This metaphor affects the viewer not only intellectually but also emotionally and viscerally. Bibi's story serves as an example of a slow violence, in which we all participate, often without realizing it.

The producer of *Daughter of the Lake*, Nuria Frigola, stated that the experience described by the jewelry designer can inspire us to question the origin of all the products we consume: "Ya no se trata de los metales, sino de preguntarse de dónde es que provienen todos los productos que consumimos y qué hay detrás de ello" ("It is not just about the metals. We should ask ourselves where all the products we consume come from, and what lies behind this") (*Red Muqui* n. page). This statement emphasizes the importance of environmental, ethical, and spiritual awareness of the human impact on the planet and offers a reflection on material consumption in the context of global ecological limits. The message communicated by the film is the key concept of degrowth and *Buen Vivir*, according to which infinite economic growth is incompatible with the finite resources of our planet. Even though *Daughter of the Lake* has a local focus, its message is global.[6] The film expands our perception of violence, allowing us to apprehend various ecological threats discounted by the media and challenging us to "act ethically toward human and biotic communities that lie beyond our sensory ken" (Nixon 15).

Cabellos Damián commented in an interview that the purpose of the film was not to denounce the mining company but to raise ecological awareness and to promote "el amor al agua, el respeto a Yacumama, la importancia que tiene el agua en la vida" ("the love for water, the respect for Yacumama, the importance that water has in life") (*Red Muqui* n. page). The ecological sensibility conveyed here reflects the poetic documentary mode, which stresses "associative qualities over transferring information or winning us over to a particular point of view" (Nichols 105). The appreciation and respect for water reflected in the film comes from indigenous worldviews which, in contrast to Western/Christian paradigms of human separation from nature, emphasize a relationship of care, respect, and reciprocity with the environment.

The indigenous vision of reality, essentially different form the Western view, was studied by Marisol de la Cadena, in *Earth Beings: Ecologies of Practice Across Andean Worlds* (2015). The author argues that in order to fully understand the relationship between indigenous people and nature, a fundamental rethinking of reality is required. According to this perception, the indigenous world is composed of "socio-natural collectives that do not abide by the divisions between God, nature, and humanity" (de la Cadena 206). She further elaborates that the aforementioned division is a consequence and perpetuation of colonialism, since the indigenous worldviews have been suppressed and dominated by the Western powers. Nélida in *Daughter of the Lake* revitalizes the indigenous perception of the natural world as a living being in the following words: "El agua es la sangre de la tierra, y sin su sangre la tierra no tendría vida" ("Water is the blood of the earth, and without its blood the earth could not live"). This statement reflects an intimate interdependence between the human and nonhuman worlds. The lake in the film is perceived by Nélida as a living being, who receives rituals, respect, and offerings (such as fruit, honey, sugar) in exchange for water: "Siempre hay que llevarte una fruta, una chancona, o un poquito de azúcar. Porque eso te gusta, Mama Yacu. Porque si no, después coges nuestro ánimo hasta pagarte" ("We should always bring you some fruit, honey, or a bit of sugar. Because you like that, Mother Water. Because if not, you take our souls, until we pay you"). Nélida's view on the reciprocity and interconnection with nature challenges extractivism, a form of "a nonreciprocal, dominance-based relationship with the earth, one of purely taking" (Klein 169). This attitude represents indigenous people's moral responsibility to the environment they inhabit. Their reverence toward nature and a sense of belonging to a place have served as influential sources in the field of spiritual ecology.

As Gregory Cajete expressed in "Reflection on Indigenous Ecology," the relationship between indigenous people and their environment "embodies a theology of place, reflecting the very essence of what may be called spiritual ecology" (3). According to this perspective, the natural world is not only a physical place but a spiritual one—an essential part of being, feeling, and understanding life. Cajete's perspective is an analogy that attempts to translate indigenous worldviews to Western sensibilities and that aligns it to the closest possible notion in the Western world, which is spiritual ecology. This worldview is clearly reflected in Nélida's attitude toward the lake; she conceives of this body of water as an animated being. Embodying this holistic perspective, she addresses the spirit of the water in her invocation: "Madre Agua—Mama Yacu, ¿por qué tanta injusticia contra ti? Acaso no entienden que tú eres un ser viviente" ("Mother Water—Mama Yacu, why do they mistreat you? Don't they understand that you're a living being?"). This speech communicates that water is a "mother"—a source of life, a nurturer, and an animated force. Nélida expresses the concept of indigenous kinship with nature, which has a great deal to teach about how to take care of the planet so that future life continues. As Klein observed, these indigenous worldviews on reciprocity and interconnection with the natural world, which are the "antithesis of extractivism," have influenced new generations of activists, who are playing a vital role in antiextraction struggles in Latin America and other regions of the world (182).[7]

Similarly, *The Pearl Button* turns to spiritual ecology and indigenous traditions as a remedy to ecological crisis. The film "captures scenes of nature that frame water as an innocent and uncorrupted character of the Chilean landscape" (Mullen n. pag.). The personalization of water implies an interdependent relationship between the natural and human worlds. Water is an animated force and a protagonist here (it has a voice and memory), and its spiritual link with human beings is expressed through metaphors, images, and symbols.[8] The film offers magnificent images of ocean, fjords, rivers, and glaciers, contemplating water as a sacred life source, a living entity, as well as a protagonist of all life:

> Lo que nosotros somos adentro de nuestro cuerpo somos agua, y las plantas son agua, y todo es agua, con unos pedacitos de un cuerpo más sólido: tierra, piedra, o huesos. Pero todo es agua. Y lo que respiramos también es agua. Entonces no tiene nada raro que los astrofísicos digan que el universo completo está lleno de agua. Todo es agua (Our bodies are made of water, plants are water, so everything's water with bits of something more solid: earth, stones, or bones. But everything is water. What we breathe is also water. So, there is nothing strange in the fact that astrophysicists say that the whole universe is full of water. Everything is water).

The Pearl Button communicates not only the physical connection between water and the universe but also the spiritual one (present in the indigenous traditions of Patagonia) and exposes how it is gradually being lost as the region is threatened with globalization and economic expansion.

Thomas Berry, one of the leading figures in spiritual ecology, refers to the human disconnection from the environment as a "spiritual autism," stating that "we are only talking to ourselves. We are not talking to the rivers, we are not listening to the wind and stars. We have broken that great conversation" (cited in Vaughan-Lee, 59). Similarly, Raúl Zurita in *The Pearl Button* reflects about this "great dialogue" and a "mutual regard" between people and the universe, suggesting that nature is a spiritual being. According to the poet, the sea has a historical memory and a voice that we can hear. This metaphorical statement refers to the spirits of the past and the ravages of colonization, the memory of which is embedded in the natural environment (habitat fragmentation, loss of biodiversity, endangered ecosystems). Apart from the indigenous past, the film recalls the violent events from the 1970s, when the victims of a military dictatorship were secretly murdered and thrown from helicopters into the sea. The immense, awe-inspiring, and impenetrable ocean filmed by Guzmán is presented as a protagonist—a witness of this violence and a guardian of memory. This shift in time and the inclusion of a complex range of historical events are characteristic of the poetic documentary, which "sacrifices the conventions of continuity editing and the sense of a very specific location in time and place" (102). The shots depicting the boundless vistas of water provide an emotional background for the viewer and evoke the feeling of fear. At the same time, the beauty and vastness of the ocean creates a sense of the sublime, associated with infinity, power, and transcendence.

The concept of the sublime, coined by Edmund Burke in the eighteenth century, became a quality central to aesthetics in poetry and visual art. It greatly influenced the romantic poets, and each of them had a slightly different approach to the concept. The sublime was often found in the natural world, which would evoke the feelings of awe and often terror. The human being felt attracted by the magnificent and powerful nature but at the same time alienated from it. Burke perceived the ocean as the most sublime object, evoking the strongest emotions in its beholders: "The ocean is an object of no small terror. Indeed, terror is in all cases whatsoever, either more openly or latently, the ruling principle of the sublime" (cited in Langford and Boulton, 230). This exact emotion is expressed by the narrator in *The Pearl Button*: "Yo no convivo con el mar. El océano me causa admiración, y al mismo tiempo miedo" ("I don't feel close to the sea. For the ocean I feel admiration, and at the same time fear"). The narrator continues his story, describing how his childhood friend was swept away by the sea, and his body was never found. The images of the immense and dark Pacific waves accompanying the story produce the same feeling of terror in the viewer and stand in opposition to the approaches to the natural world present in the ocean-centric cultures. This vision of an intimidating nature is juxtaposed with the indigenous attitudes toward their environment, based on reciprocity, coexistence, and reverence. To contrast the narrator's feelings of the sublime (and the alienation from the ocean), the film presents an interview with an indigenous woman, for whom "el agua forma parte de su familia. Ella acepta tanto los peligros como la comida que el mar le ofrece" ("water is part of her family. She accepts both the dangers and the food that the sea offers her"). By incorporating personal stories of the native inhabitants of the Patagonian coast, the film communicates their holistic view of the world—the links among beliefs, historical memory, traditions, and nature.

The opening quote in *The Pearl Button*, by the poet Raúl Zurita, "Todos somos arroyos de una sola agua" ("We are all streams from one single water"), reflects the holistic perspective of the universe—the inseparable connection of all the things with water. There are multiple layers of meaning in Zurita's words, as the ocean is an important element of Chile's geography and history. An example of this intimate link between the region's tradition and water are the seafaring nomads (the Kawésqar, the Selk'nam, the Aoniken, the Hausch, and the Yamana), who inhabited the Patagonian archipelago already 10,000 years ago, and travelled through the fjords by canoes. They led a hunter-gatherer lifestyle and never farmed the land. *The Pearl Button* explores the oppression faced by these people, when the European settlers arrived in the nineteenth century, destroyed their indigenous ocean-centric tradition, and expanded agriculture inland. Since then, the Kawésqar's territories have been colonized, bringing conflict and infectious diseases: "Después de convivir siglos con el agua y las estrellas, los indígenas sufrieron el eclipse de su mundo" ("After centuries of living alongside the water and the stars, the indigenous people saw their world collapse"). The population declined dramatically, and by 1925, only 150 native Kawésqar were left. The Chilean government relocated the entire population to Puerto Edén in Wellington Island and approved the Kawésqar Protection Law. In 2002, there were "2622 Kawésqar people, representing 0.38% of the country's indigenous population" (Chile Precolombino n. page).

In the film, one of the surviving members of the indigenous tribe tells a story from his childhood about crossing Cape Horn in a small boat with his father during stormy weather, concluding in the following words: "Me gustaría poder volver a navegar de la manera que lo hacíamos. Pero en estos momentos estamos completamente restringidos y no podemos casi ocupar el mar" ("I'd like to be able to travel by boat as we used to. But now, there are too many restrictions, we are barely allowed in the sea"). This story conveys how the spiritual link between the indigenous people of Patagonia and water is gradually being lost as their access to the ocean is restricted. Similarly, another interviewee in *The Pearl Button*—Gabriela, from the Kawésqar nation, talks about the loss of traditions and skills. She recalls her childhood, when she used to paddle around the islands to fetch water from the river and how she learned to dive to find shellfish. In contrast, contemporary Chileans no longer depend on those traditional skills, and communal indigenous practices are threatened by the homogenizing impact of globalization.[9]

Another example of the erasure of the local history, knowledge, and traditions is the disappearance of native languages. This is also a manifestation of slow violence that has been perpetuated since colonial times. Jared Diamond, in *The World Until Yesterday: What Can We Learn from Traditional Societies?* (2014), enlists various ways of eradicating a language, the most direct one consisting of "killing almost all of its speakers" (398). The author gives examples of such events from the nineteenth century, when the European settlers eliminated many languages in the United States and Tasmania. Similarly, *The Pearl Button* links the process of eradicating languages with the collective annihilation of the indigenous Patagonians. The narrator relates the history of colonization, when the settlers arrived in the region in 1883, and started the process of destruction of indigenous languages and cultures: "Les quitaron sus creencias, su lengua y sus canoas" ("They took away their beliefs, their language and their canoes"). The native people were given clothes contaminated with germs, which resulted in epidemics, massive deaths, and cultural and linguistic extinction.

Currently, the Kawésqar language, also known as Alacaluf, is in danger of becoming extinct, since only several speakers of this language are left. The Kawésqar has an extended vocabulary uniquely related to the weather conditions, local environment, fauna, and flora of Patagonia.

The Pearl Button presents an interview with one of the surviving members of the tribe—Gabriela—during which she is asked to translate terms, such as "beach," "water," "storm," and the like. Gabriela is able to recall all these words since they contain environmental knowledge of her native place, to which she feels intimately connected. However, when the interviewer asks her to translate the terms "police" and "God," she responds that there are no equivalents of these in her language. This scene portrays the differences between the two cultures and the role of a language in preserving knowledge and worldviews. As David Grubin explained in his documentary, *Language Matters* (2014), language is the instrument through which we see the world: "Each language is a vision of the world, each language says something different about what it means to be human compared with any other language, and every language that is lost is a loss of a fragment of that vision." Both Diamond and Grubin emphasize how the accelerated disappearance of local languages results in the disappearance of other aspects of local communities, such as oral tradition and ecological knowledge. The authors attract our attention to this form of slow violence—the loss of languages, which are now "vanishing more rapidly than at any previous time in human history (Diamond 370). The loss of languages is parallel to the loss of biodiversity, since both are affected by the expansion of globalized neoliberal capitalism and economic growth.

Ultimately, it is both the poetic documentary form of *Daughter of the Lake* and *The Pearl Button* and each film's dialogue with spiritual ecology that contribute so powerfully to their ecocritical discourse. These works challenge and inspire the audience to resist extractivism, the exploitation and destruction of indigenous peoples, and the capitalist, neoliberal global order, and yet they also manage to affect the viewer experientially in unique and potentially transformative ways. Crafted as visual poems or collective dreams rather than arguments driven by Western logics, *The Pearl Button* and *Daughter of the Lake* seep into the viewer's unconscious and become deeply personal. If indeed, as the narrator of *The Pearl Button* suggests, "everything is water," the voice of the water in these films becomes the viewer's own voice, and as such, its visceral, emotional authority demands a response.

Notes

1. For more details on spiritual ecology and traditional worldviews, see also *Spiritual Ecology: The Cry of the Earth*, ed. Llewellyn Vaughan-Lee.
2. The criticism of extractivism in the Global South from the perspective of degrowth and *Buen Vivir* is addressed by Alberto Acosta, in "Rethinking the World from the Perspective of *Buen Vivir*": "On a final planet there is no room for permanent economic growth. If we continue down this path, we will reach a situation that is no longer environmentally sustainable and is increasingly socially explosive. Overcoming this creed of economic growth, particularly in

the Global North, must be accompanied by abandoning extractivism in the Global South" (n. page).

3. The impact of gold mining in the Yanacocha mine is explored by Gianni Converso in *Open Pit* (2011), a feature documentary on Latin America's most profitable gold operation.
4. The conflict has not yet been resolved, since Newmont and Buenaventura are currently working on a project to extend Yanacocha's copper sulfides production to 2025. Victor Gobitz, chief executive officer, stated in an interview: "That's our priority for now, and it will have a positive impact for the company" (Cespedes, n. page).
5. Since 2011, the Chaupes have been violently attacked, harassed, and threatened; their property invaded and destroyed. In September of 2017, the family filed a lawsuit against Newmont, in the U.S. Federal Court. Máxima insisted on the trial to be held in the United States because of the corruption of Peruvian legal system: "We came here because the courts in Peru are corrupt, and Newmont has corrupted them. The people in Newmont in the United States are green lighting these abuses. That is why we are here" ("Peruvian Subsistence Farmer" n. page).
6. Degrowth and *Buen Vivir* converse with each other; however, it is important to note the difference between them. The first is Latouche's argument to rethink the notion of continuous growth as a foundational notion of modern society. The second is a congeries of indigenous worldviews that conceive the world outside of nature/human structures and is considered as an alternative of "development" and support for the notion of pluriverses. For more information, see *Farewell to Growth* by Serge Latouche and "Rethinking the World from the Perspective of *Buen Vivir*" by Alberto Acosta.
7. Naomi Klein examines a fundamental shift in the approach to economic growth, influenced by the indigenous worldviews: "Space is opening up for a growing influence of Indigenous thought on new generations of activists, beginning, most significantly, with Mexico's Zapatista uprising in 1994, and continuing, as we will see, with the important leadership role that Indigenous land-rights movements are playing in pivotal anti-extraction struggles in North America, Latin America, Australia, and New Zealand" (182).
8. The idea of water as a protagonist has been a theme in various environmental and ecocritical discourses. Shawn William Miller, in *Environmental History of Latin America*, proposed the inclusion of natural resources in environmental history: "Humans will remain at the center stage in our drama. . .; however, the stories of nonhuman life and of the inanimate resources on which life depends will be given place in our plots" (5). As a further development of this attitude toward nature, in 2017, rivers in India, New Zealand, and Colombia have been granted the same legal rights as human beings. For a more detailed discussion of this new law recognizing rivers as legal personalities, see Erin O'Donnell's *Legal Rights for Rivers: Competition, Collaboration, and Water Governance.*
9. The impact of globalization and the free market economy on the Chilean southern coast is portrayed by Andrés Wood, in his film *La fiebre del loco* from 2001. The director exposes the loss of biodiversity (a mollusk called *loco*) and the abuse of nature for economic profit.

Works Cited

Acosta, Alberto. "Rethinking the World from the Perspective of *Buen Vivir.*" *Degrowth in Bewegung(en)*, 30 Aug. 2016.

Cajete, Gregory. "Reflections on Indigenous Ecology." *A People's Ecology: Exploration in Sustainable Living*. Clear Light Publishers, 1999.

Cespedes, Teresa. "Peru's Buenaventura CEO Says No Near-term Conga Gold Mine Revival." *Reuters*, 9 Aug. 2017.

Chileprecolombino. "Native Peoples-Kawáshkar." *Museo chileno de arte precolombino*. http://www.precolombino.cl/en/culturas-americanas/pueblos-originarios-de-chile/kawashkar/

"Conga." *Conga Conflict*. Word Press.

Daughter of the Lake. Dir. Ernesto Cabellos. 2015.

de la Cadena, Marisol. *Earth Beings: Ecologies of Practice Across Andean Worlds*. Duke UP, 2015.

Diamond, Jared. *The World Until Yesterday: What Can We Learn from Traditional Societies?* Viking Penguin, 2012.

Even the Rain. Dir. Icíar Bollaín. 2010.

Felstiner, John. *Can Poetry Save the Earth?* Yale UP, 2010.

Galeano, Eduardo. *Open Veins of Latin America*. Monthly Review Press, 1997.

———. *Úselo y tírelo: el mundo visto desde una ecología latinoamericana*. Planeta, 1994.

"*Hija de la laguna* no es un documental antiminero sino de amor y respeto a la Yacumama." *Red Muqui*. Servicios de Comunicación Intercultural, 3 Sep. 2015.

Klein, Naomi. *This Changes Everything: Capitalism vs. the Climate*. Simon & Schuster, 2014.

La fiebre del loco. Dir. Andrés Wood. 2001.

Langford, Paul, and James T. Boulton, eds. *The Writings and Speeches of Edmund Burke: Volume I: The Early Writings*. Oxford UP, 1997.

Language Matters. Dir. David Grubin. 2014.

Latouche, Serge. *Farewell to Growth*. Polity, 2009.

Miller, Shawn William. *Environmental History of Latin America*. Cambridge UP, 2007.

Mullen, Patrick. "Review: *The Pearl Button*." *Point of View Magazine*, 7 Apr. 2016.

Nichols, Bill. *Introduction to Documentary*. Indiana UP, 2001.

Nixon, Rob. *Slow Violence and the Environmentalism of the Poor*. Harvard UP, 2011.

O'Donnell, Erin. *Legal Rights for Rivers: Competition, Collaboration, and Water Governance*. Routledge, 2018.

Open Pit. Dir. Gianni Converso. 2011.

The Pearl Button. Dir. Patricio Guzmán. 2015.

"Peruvian Subsistence Farmer Asks Court to Continue Case against U.S. Company in Delaware." *Conga Conflict*, 9 Feb. 2018. https://earthrights.org/media/peruvian-subsistence-farmer-asks-court-continue-case-u-s-company-delaware/

The Power of Clean Water. Procter & Gamble and National Geographic, 2018.

Sponsel, Leslie E. *Spiritual Ecology: A Quiet Revolution*. Praeger, 2012.

Trefis Team. "Newmont's Conga Project in Peru Faces Uncertain Fate." *Forbes*, 12 Apr. 2013.

Vaughan-Lee, Llewellyn, ed. *Spiritual Ecology: The Cry of the Earth*. Golden Sufi Center, 2013.

Water and Power: A California Heist. Dir. Marina Zenovich. 2017.

Watermark. Dir. Edward Burtynski and Jennifer Baichwal. 2013.

White Water Black Gold. Dir. David Lavalee. 2011.

6 From Polluted Swan Song to Happy Armadillos

The Cold War's Slow Violence in Nicaragua

Jacob G. Price

During the Cold War, North American intervention penetrated economic and environmental spheres through business agreements that benefited the Central American elite as well as U.S. ideological and economic interests at the cost of the environment and of marginalized rural classes. By 1979 in Nicaragua, the new Sandinista government faced massive ecological damage caused by decades of unchecked industrial extraction during the Somocista regime. Published in the years leading up to the 1979 revolution, Pablo Antonio Cuadra's *Cantos de Cifar y del mar dulce* (1971) serves as an environmental and poetic history that recounts the fragile character of the health of Nicaragua's rural lake ecosystems during the decades of Somocista rule. Thirteen years later, Ernesto Cardenal's first publication after the Sandinista victory, the *Vuelos de victoria* collection (1984), presented a poetic rendering of the value of nonhumans for the Sandinista government and the Revolution. Both these collections showcase the ecological concerns arising from the constant pollution by the Nicaraguan environment over several decades, thereby calling them to the attention of the general public. Slow violence's separation from its point of emergence, due to its dispersion over time and place, often makes it difficult to detect and combat (Nixon 9). However, poetry from the 1970s, in conjunction with literature published in Sandinista Nicaragua, documents the casualties of slow violence perpetuated by capital fiction so that both human and nonhuman casualties do not remain untallied and unremembered. In particular, Cuadra and Cardenal show their commitment to a sense of place and local ecologies. In this chapter, I utilize Nixon's concept of slow violence, in conjunction with Erika Beckman's notion of capital fictions, to elucidate how the poetry of Cuadra and Cardenal historicizes slow violence's impact on the landscape and human/nonhuman networks.[1] The symptoms of slow violence percolate in a variety of ways, from the decrease of a species' population over time to sickness in humans and nonhumans. In the literary case of Cuadra's poetry, slow violence becomes visible when the writer describes how the accumulation of toxins crosses the threshold of tolerance within an ecosystem and cannot then be filtered out or disposed of. Consequently, a large portion of the landscape becomes significantly altered, or the majority of a population in that ecosystem becomes sick, dies, or disappears. The authors' interest in developing a sense of place crafts the locus from which capital fictions, slow violence, and Nicaraguan environmental history all converge within a literary history that critiques the ecology of Cold War capitalism. When read in the context of slow violence, Cuadra's poetry—despite its lack of distinctly environmentalist rhetoric—illustrates the consequences of what happens when the amount of contaminants surpasses the environment's ability to bear pollution. Cardenal's book, when read after Cuadra's, then exemplifies how environmental injustices became politicized in the Sandinista government.

Erika Beckman's book *Capital Fictions* examines the role of capitalism and its fictions of modernity in the first half of the twentieth century through a historical analysis focusing on ecological devastation caused before the Cold War. Cuadra's and Cardenal's poetry illustrate how the relationship between capitalism and the environment then changed in the second half of the twentieth century. For Beckman, capitalism's fictions have significant repercussions on the environment, as the hypercommoditized logic of capitalism alienates humans from nonhumans in order to increase profit (viii–ix). In the poems of Cuadra and Cardenal, the ideological foundations of capitalism that modern ontologies have accepted are challenged. Both poets' works blame the Somocista government's contradictory stance toward watersheds and lakes that allowed them to serve simultaneously as both the main source of water for urban populations and also the dumping grounds for toxic waste. As we shall see, Cuadra's poetry elaborates on the suffering of Nicaragua's lakes by presenting landscape that has surpassed the safe threshold of toxins, resulting in the death of waterfowl. Later, after the Sandinista revolution, Cardenal's poems then challenge the relationship between nature and the market, recasting the human/nature dichotomy with a concern for the well-being of nonhumans and even their inclusion in the revolutionary government. Both poets' works allude to the hypercommoditized logic of neoliberalism and the damaging effects of capitalism. In addition, they also trace an environmental history of Nicaragua that creates space for the agency of nonhumans to be translated into literature. The 1970s and 1980s in Nicaragua are pivotal in literary and environmental histories; their ecologically conscious publications debunk the myths of modernity that capitalism assures.

Pablo Antonio Cuadra's *Cantos de Cifar y del mar dulce* ("Songs of Cifar and the Sweet Sea") is a collection of poetry that chronicles the livelihood of a rural community situated on a fictionalized version of Nicaragua's great lakes. The book, published in 1971, is often compared to the *Odyssey* because it reads like an oral history of the sailor Cifar. Poems are loosely arranged in chronological order, and each of them traces a passage from the history of Cifar and his many adventures out on the lake. The collection details the life and health of lake ecosystems before the Nicaraguan Institute of Natural Resources and the Environment (IRENA) revealed the degree of their pollution and poetically renders how slow violence caused by toxic dumping from extractivist capitalist industries accumulated.[2] The following two poems that I analyze from *Cantos de Cifar* mark the initial moments in the collection when outside influences threaten the ecosystem's health.

The poem "El caballo ahogado" (The Drowned Horse) marks a decisive turn in the book's narrative, shifting the daily routine of rural human and nonhuman life on the lake through the sailors' masculine imagery on the water. Rather than critiquing the Cold War ideological separation of humanity from nature, the majority of the poems reflect the rural community's unique relationship with the environment due to the lake community's relative isolation from global or national influence. Several poems in the collection challenge the human and nonhuman relationships presented as well as the traditional modern dichotomy of culture versus nature; in the poem "El caballo ahogado," this separation dissolves when the ecological connection between the rural community with outside ecosystems is compromised by the arrival of a dead horse:

Después de la borrasca	After the storm
en el oscuro silencio	in the dark silence
miraron sobre las aguas	they saw on the water
flotando	floating
el caballo muerto (131, lines 1–5)	the dead horse[3]

The first stanza ends with this image of a dead horse, which is unusual for the volume, given the collection's adventurous character and sailing motifs. The sailors recognize how strange it is as they watch the horse floating in the storm water, thus illustrating their ability to discern any disturbance of the ecosystem. While writers are in general equipped to reveal slow violence in localized communities when they produce literature with a sense of place, here the introduction of the dead horse challenges the illusion of isolation that a sense of place unwittingly creates. The drowned horse is then also evoked in the final stanza as a metaphor that foreshadows the fate of the humans and nonhumans that live on the lake:

Sintieron	They felt
como un extraño	a strange
presagio	omen
y vieron	and they saw
una corona	a garland
de gaviotas blancas	of white gulls
en el viento. (131, lines 17–23)	in the wind.

The indentation in the third line tilts like a camera, moving the reader's attention the same way as the sailors', first looking at the horse and then the birds in the sky. The poetic gaze's shift is accentuated further by a line that describes the horse's eye as "abierto,/fijo su asombro en el cielo" (131) ("open,/fixing its gaze on the sky" 99). This point of view becomes the vision of death between the horse and the birds, and the horse's line of sight connects the two species as if the horse were actually watching the birds. The omen of the horse corpse brings with it the foreshadowing of the destruction of the lake as a result of an outside influence, just as invasive species are able to alter whole ecologies and push out endemic ones. In the poem, the horse's cause of death is unknown and is never explained. The horse is also a possible threat of further toxic contamination within the lacustrine setting and all of its animals, as well as the sailors.

"El caballo ahogado" and another poem in the collection, "El cementerio de los pájaros" (Bird Cemetery), invite readers to consider the effects of an ecosystem that cannot filter out the toxins that slow violence—announced by the horse in the earlier poem—brings. In "El cementerio de los pájaros," bird populations suddenly decrease inexplicably. Cuadra begins the poem by narrating the arrival of Cifar at one of the many islands in Nicaragua's great lakes:

Arribé al islote	I arrived at the barren island
enfermo	sick
fatigado el remo	the oar weary
buscando	seeking rest
el descanso de un árbol	under a tree.
no vi tierra	I didn't see earth
sino huesos.	only bones.
De orilla a orilla	From shore to shore
huesos	bones
y esqueletos de aves,	and skeletons of birds,
plumas calcinadas,	calcined feathers
hedor	stench
de muerte,	of death,
moribundos	sea birds,
pájaros marinos (135, lines 1–15)	caws
	of agony (135)

This section of the poem presents evidence that the death of the flock had been unnatural, with the lake's ecosystem disturbed by some outside variable. The enjambment of the adjective *sick* in the second line of the poem is the first indicator that their demise was a consequence of the toxic contamination of the ecosystem. The strategic positioning of the verse *sick*, summarizing an ecological imbalance in a single word, also allows readers to interpret both island and rower as suffering from illness, thereby reinforcing the material relationship between the island, rower, and lake. Given the lake-based ecosystem where the rower and the birds live that is depicted in *Cantos de Cifar*, the cause of severe contamination of the water to the point of death is likely the overuse during the Somocista regime of pesticides imported from the United States. Shawn William Miller's assessment of Somocista Nicaragua indicates that "most rivers and aquifers were contaminated, and Nicaraguans suffered the highest number of pesticide poisonings of any nation per capita, 400 of which resulted in death each year" (208). The majority of the pesticides used during the dictatorship were prohibited for use in the United States because of their profound impact on human and environmental health.

Connecting slow violence with the notion of attritional catastrophes, Jorge Marcone describes how slow violence resembles a type of war of attrition carried out by transnational corporations, sometimes in conjunction with national governments against locals (210). The willful pollution of local environments is emblematic of the war against nonhumans in the context of the Cold War. The promotion of capitalism was the principal ideological front line for the United States during the Cold War, but the development of unregulated industries located in Central America, funded through U.S. businesses, left physical, toxic consequences in the environment that resulted in casualties of war. To embrace damaging capitalist industry in support of businesses that favor U.S. economic practices is to accept environmental destruction.

Cuadra's depiction of environmental destruction is the basis through which we may nuance Nixon's concept of slow violence by focusing on the imagery of the birds in "El cementerio de los pájaros." In most cases of contamination, the first victims are smaller plants and animals that are unable to adapt to the influx of pollution. Their deaths may go largely unnoticed, but they ultimately make slow violence visible to those humans who know how to read the landscape and have a sense of place through a connection with the landscape or nature. The poem also redirects the emphasis on how slow violence affects humans toward nonhumans and the ecosystem more broadly. Rather than focusing on how nonhumans respond to slow violence, Nixon emphasizes how human groups are displaced. "El cementerio de los pájaros" does not challenge environmental injustices that humans faced, but the focus on birds illustrates key moments in the lake environment's history that illustrate how slow violence affects nonhuman populations as well. "El cementerio de los pájaros," although narrated via the human gaze, portrays the environment as a community with its own inherent value, shifting attention toward the declining health of the ecosystem as a whole. Starting from the line "De orilla a orilla," Cifar does not insert his own voice. A third person poetic voice takes over, holistically describing the devastation of the island and its avian inhabitants. The landscape evoked in the poem is a symptom of the accumulated waste from decades of toxic dumping during the Somocista regime that decimated human and nonhuman populations alike. When the poetic voice describes the island as sick, this is not metaphorical but literal.

The final stanza of the poem also suggests that this sickness has infected the rower. Upon seeing the island, which the poetic "I" has dubbed "[a] cemetery/for song," the rower decides to turn away:

Con débil brazo moví los remos y di la espalda al cementerio del canto. (135, lines 25–29)	With weak arms I moved the oars and turned my back on the cemetery of song. (135)

To suddenly describe a sailor, particularly the titular Cifar, as weak is surprising, since most of the previous poems present the seamen as masculine figures with interminable strength and the will to face danger and risk. *Cantos de Cifar* positions masculinity against the catastrophic perils of nature in images of storms and threatening nonhumans in order to celebrate men and strength over the natural world. The use of the adjective *weak* in the present poem may imply a sense of defeat when faced with such awful desolation. However, when the "weak" arms are considered in the context of the ambiguous use of the adjective *sick* at the beginning of the poem, this final stanza relies on synecdoche and implies a material and nonliterary sense of sickness that overcomes Cifar. Just as the dead horse foreshadows the death of the birds, the death of the birds in this poem foreshadows the fate of the sailors on the lake.

Francisco Lasarte comments, in his article on the collection, that "something else which threatens the Mar Dulce's idyllic nature is the arrival of 'civilization' and its attendant evils, social and economic injustice" (185). The story and fate of Cifar testify also to the environmental injustice that modernity produces in the environment through the accumulation of toxins; the unchecked toxicity eventually impacts the human group that participates in the lake's ecosystem. After a violent storm seizes the boat in which he is riding, the poem "Pescador" ("Fisherman") announces the fate of Cifar: "Un remo flotante/sobre las aguas/fue tu solo epitafio" (137) ("An oar floating/on the waters/was your only epigraph" 109). Although Cifar's death ultimately is caused by the contamination of the lake and other areas upstream, his passing symbolizes the death of the community's livelihood. Slow violence is often found in poor, rural, and nonwhite regions, and rarely is action taken, even after enough evidence has surfaced to indicate that outside pollution is poisoning the area. The poem's subtle depiction of environmental injustice unmasks the capital fiction of prosperity through modernization. With contamination and other outside influences disrupting the lake's human and nonhuman community, Cifar's death and the conclusion of the volume of poetry exemplify how rural Nicaraguan ecosystems suffered from the effects of slow violence. *Cantos de Cifar y del mar dulce* showcases how authors with a sense of place write about environmental issues and articulate injustices such as slow violence even without the vocabulary offered by the environmental turn.

Cuadra's collection illustrates ecological concern at a critical moment in Nicaragua's rural history. The latent worry for environmental health prevalent in Cuadra's book also underscores Steven F. White's reading of Cuadra's earlier collection *Poemas nicaragüenses* (1950). White suggests that Cuadra writes about the environment as an "organic imagined community" (35). For White, Cuadra writes about nature as a kind of environment that is never a backdrop but instead a living ecosystem in which humans and nonhumans interact with each other. Drawing from Benedict Anderson's famous notion of the "imagined community," White sees in Cuadra's writing a community that is bound together by its material connection and relationship, even on a molecular level (Price 29). This literary theme that White names translates well into *Cantos de Cifar y del mar dulce* as the sense of place that Cuadra exhibits through imagery and narrative. However, the principle difference between *Poemas nicaragüenses* and *Cantos de Cifar y del mar dulce* in the "organic imagined community" is the treatment of pollution. In *Cantos*, pollution disrupts the organic imagined community by decimating different nonhuman populations. The poems toward the end of Cuadra's later work that center on the death of Nicaraguan lake culture and lake-based species act as a swan song for rural Nicaragua. The image of the island that is home to the dead and dying birds functions metonymically as a symbol that Cuadra relies on in order to communicate poetically what is happening to the ecosystem of Nicaraguan lakes at large.

The Sandinista Revolution that ushered in a socialist government during the 1980s in Nicaragua promised to rectify all injustices, including environmental ones. Beneficiaries included people and characters like Cifar, as well as afflicted nonhumans such as the flock of birds from *Cantos*. The revolutionary values that the Sandinista government held were promoted through a literary revival of left-leaning and environmentalist literature that equaled projects in other noncapitalist countries during the Cold War.[4] While serving as Minister of Culture in the new Sandinista government in 1983, Ernesto Cardenal published *Vuelos de victoria* (Flights of Victory). The book pays homage to the Sandinista victory and celebrates achievements made in the government's first four years. The collection covers the same political goals that Ortega had himself outlined in his speech, but the specific poem "Nueva ecología" (New Ecology) distills the Sandinista government's appreciation for the environment and ecological recovery after the revolution. This poem lists all of the environmental damage permitted by the Somocista regime during its rule and ends by advocating for human and nonhuman liberation from oppression through policies enacted by the Sandinista government. While Cardenal's poem directs most of its attention to the contamination of rivers and other bodies of water, it also documents much of the fauna that was decimated during Somocismo. "Nueva ecología" examines the ecology as a whole, emphasizing that in order to rescue Nicaragua as an entire ecosystem from ecological damage, the country must accept that capitalism's promise of wealth and modernity are chimerical. For Cardenal, poetry becomes a medium for responding to the toxic legacy of capitalism and its impact on the environment.[5]

Cardenal accentuates the resilience of the natural world by opening the poem with images of animals resurfacing after having earlier disappeared. The poet foregrounds the poem "Nueva ecología" with optimism. In the poem, just as the natural world is recovering under Sandinismo, so too are humans, the economy, and human rights. The image of the first stanza suggests that nonhuman populations were decimated but are still able to survive the persistent slow violence of the Somocista regime:

> En septiembre por san Ubaldo se vinieron más coyotes.
> Más cuajipates, a poco del triunfo,
>   en los ríos, allá por san Ubaldo.
>   En la carretera, más conejos, culumucos . . .
> la población de pájaros se ha triplicado, nos dicen,
>
> (31, lines 1–5)

> In September around San Ubaldo more coyotes came.
> More alligators, soon after the victory,
>   in the rivers, out by San Ubaldo.
>   Along the highway, more rabbits, raccoons . . .
> The bird populations have tripled, we're told,[6]
>
> (17)

The poem presents the revolutionary government's environmental policies as though their impact had been instantaneous. IRENA and other environmental agencies created by the Sandinista government were measuring the environmental damage left by the Somozas. The poetic subject's assertion that the avian population has tripled is less a literary exaggeration and instead more indicative of data sampling of bird populations. Adrian Taylor Kane supports the factual character of the poem by describing Cardenal's historical and scientific accuracy: "The aesthetics of 'Nueva ecología' are derived from the concept of exteriorism, which Cardenal defines as 'objective poetry, narrative and anecdotal, made with elements of real life and with concrete things, with proper names, details, data, statistics, facts and quotations'" (270). The phrasing "nos dicen" denotes a plural poetic voice that is well informed about the environment's recovery. Given the government's creation of IRENA and concern for the environment, it follows that Cardenal's use of exteriorism gives the poem a scientific and governmental valence and that the poetic voice is the collective voice of government officials and scientists gathering data. This aspect of the poem alludes to a greater sense of transparency between the government and its citizens with respect to the state's environmental policies. The poem leverages science as validation for the Sandinista government's environmental intervention, following up on its promise to "recover and preserve natural resources" (United Nations 261). However, despite the objectivity that exteriorism offers, "Nueva ecología" is also limited by its scientific basis, resulting in the poem's assessment of nature through the nature/culture dichotomy. Cardenal inaugurates a poetics based on cultural, scientific, and political understandings of nature influenced by an environmentally driven new regime.

Cardenal's government-sponsored new ecology also addresses the slow violence effected on rivers. The Somocistas' promise of modernity destroyed many rivers' original paths through projects of diversion, construction of dams, or simple transformation into dumpsites for toxic waste:

> Los somocistas también destruían los lagos, ríos, y montañas.
> Desviaban el curso de los ríos para sus fincas.
> . . .
> El Río Grande de Matagalpa, secado, durante la guerra,
> allá por los llanos de Sébaco.
> Dos represas pusieron al Ochomogo,
> y los desechos químicos capitalistas
> caían en el Ochomogo y los pescados andaban como borrachos.
> El río de Boaco con aguas negras.
>
> (31, lines 8–15)

> Somoza's people destroyed the lakes, rivers and mountains, too.
> They altered the course of the rivers for their farms.
> . . .
> The Rio Grande in Matagalpa, all dried up, during the war,
> out by Sebaco Plains.
> They put two dams in the Ochomogo,
> and the capitalist chemical wastes
> spilled into the Ochomogo and the fish swam and as if drunk.
> The Boaco River loaded with sewage water.
>
> (17)

The Ochomogo and the Boaca rivers in this stanza narrate histories embedded within a landscape shaped by slow violence. As rivers cut through landscape, they create histories that are readable as physical changes of the environment. With the addition of pollution, rivers transmit the history of slow violence through a landscape as the contaminants are deposited in their waters and carried in the riverbeds to other destinations. "Nueva ecología" translates IRENA's discovery of toxins in rivers through scientific processes into environmental history documented in exteriorist poetry. Kane concludes that the use of exteriorism in the poem is ultimately a critique of capitalism: "Indeed, the overall effect of the rivers in 'Ecología' is to portray a country despoiled by greed and contaminated by a corrupt form of capitalism" (271). This assessment locates the poem within the ideological debate central to the Cold War. Given the Sandinistas' position in the Cold War and the celebratory character of Cardenal's *Vuelos de victoria*, "Nueva ecología" functions as a rebuttal to capital fiction, detailing with scientific authority why capitalism does more harm than good to the environment. "Nueva ecología" evokes images of polluted rivers and destroyed landscapes in the context of "capitalist chemicals" that indicate the Sandinista government and socialism in general are inherently better and more inclusive of the nonhuman world.

The example of the Rio Grande River in "Nueva ecología" depicts an ecological ruin that has even greater implications within the context of the Cold War. By declaring that the river was dried up "during the war," the poem locates the temporal site of slow violence within the 1970s. By placing the drying up of the Rio Grande river into the chronology of this war, the poem reveals that the war itself had nonhuman casualties and that the Somocistas, before and during the Sandinista Revolution, had been waging a war on nature. The poem's use of exteriorism in these lines about the rivers emblematizes the political measures of the new government in the context of policy change toward nature. The poem goes beyond the Sandinista government's use of science to assess environmental damage by suggesting that the government can even reverse that damage: "(Hay que verlo otra vez bonito y claro cantando hacia el mar)" (32) ("We will see it clear and pretty again singing toward the sea" 18). This line is not exactly exteriorist in the way that Kane describes it; the poetic voice's tone sounds like that of a campaign promise rather than a purely aesthetic statement because it implies that the government will indeed detoxify the rivers. The parentheses that encapsulate the line are jarring as they remove the reader from the content of the poem in order to provide a more directly political tone. Kane's assessment of exteriorism is viable for much of the content of the poem, but this line makes obvious the poem's propagandistic rhetoric.

The critique of the Somocista regime's use of rivers stands in stark contrast to the Sandinista government's agenda of exposing another capital fiction. Capitalism's promise of modernity and national wealth translates human and nonhuman life into potential use value for commodification (Harvey 250), thus defining nature as a network of potential resources accommodated to basic human needs for survival. Capitalism's ecology is a food chain of potential use value under the rubric of "natural resources," which perpetuates a notion of nature endlessly producing all that is necessary for human population growth and stability. The Sandinista government's response to capitalism's ecology is secular in "Nueva ecología" until the last stanzas of the poem. The poetic voice sidesteps its familiar exteriorist style and alludes to socialist Nicaragua as a type of Eden that challenges capitalist ecology. Stephen Henighan, in his appreciation of Cardenal's *El estrecho dudoso* (The Doubtful Strait), looks through the lens of the writer's liberation theology to examine what kind of Eden he envisions for Nicaragua: "The Eden proposed by Cardenal in this Canto is not that of nature in its virgin state, but of settlements developing in a Christian-influenced harmony with nature" (118). This Edenic vision of nature that the poet conceives in *Vuelos de victoria* is indeed derived from Christianity but is secularized in the last stanza of "Nueva ecología":

> Los cusucos andan muy contentos con este gobierno.
> Recuperaremos los bosques, ríos, lagunas.
> Vamos a descontaminar el lago de Managua.
> La liberación no solo la ansiaban los humanos.
> Toda la ecología gemía. La revolución
> es también de lagos, ríos, árboles, animales.
> (32, lines 47–52)

> The armadillos go around very happy with this government.
> We will save the woodlands, rivers, lagoons.
> We're going to decontaminate Lake Managua.
> The humans weren't the only ones who longed for liberation.
> The whole ecology had been moaning. The Revolution
> also belongs to lakes, rivers, trees, animals.
> (19)

"Nueva ecología" thus does not propose that this Eden be a type of theocracy. In contrast to Henighan's analysis of *El estrecho dudoso*, which relies on a religious reading of the environment, "Nueva ecología" instead offers a revolutionary Eden in which humans and nonhumans coexist peacefully under the auspices of the new government's ecological policy. These final lines of the poem predict a harmonious relationship between humanity and nonhumans that is born from a healthy ecosystem, clean from the consequences of slow violence. In "Nueva ecología," the poetic voice assumes a paradisiacal state of being after the Sandinista victory.

However, the implicit message of the poem is that slow violence has previously corrupted this paradise and that the Sandinista government is the only path back to an Edenic Nicaragua. The poem's second line thus reads like a campaign promise to decontaminate Lake Managua and in so doing reinforces the idea of the loss of a Christianesque paradise because of the Somozas. The poetic voice thereby maintains its optimism, suggesting that slow violence can be stopped and reversed when ecological restoration is characterized through the religious notion of Eden. The poem's confidence in Sandinismo hinges on the idealization of an Edenic Nicaragua as a motivating factor for environmental mitigation.

Finally, the last stanza of "Nueva ecología" portrays nonhumans to be consciously able to renegotiate their relationship with humans: "Los cusucos andan muy contentos con este gobierno" (32) ("The armadillos go around very happy with this government" 19). Cardenal's poem does not allow for nonhuman subjectivities to express themselves in the poem without poetic intervention; the poetic voice speaks on behalf of the armadillos and comments on their general well-being. The poem's exteriorism functions as the medium through which the anecdotal observation of animals' and plants' recovery within specific Nicaraguan locations illustrates how decades of slow violence plagued the environment. The poem's description of the "very happy" armadillos is not a case of irresponsible anthropomorphism since the animals are understood as indicators of the environmentalist policies enacted by the Sandinista government. Such environmental policies evolved throughout the 1980s. By the time of publication of "Nueva ecología," the most direct law related to the environment was one that had been created simultaneously with IRENA in 1979; it mandated that the institute carry out a general plan for conservation (Asamblea nacional 1). With new laws in 1981 and beyond, the government specified how the environment would be affected by agricultural and food industry reforms as well as the creation of national parks. It follows that animal and plant life would be satisfied with a human government that does not regularly disrupt population growth through the life-threatening policies that the Somocista government permitted. The armadillo serves as a synecdoche of animal life that has been able to recover since Sandinistan environmental policies were enacted and enforced. Given the lines of the poem that read like political campaign promises, the armadillo is also a figurative expression of nonhuman approval of the new government. In this portrayal of the armadillo, animal life is being spoken for rather than speaking for itself, and the figure of the *cusuco* is put to propagandistic service. The armadillo reads as an ambiguous symbol that represents the Sandinista government's attempt to reconcile nonhuman oppression under Somocismo without conceding true subjectivity to nonhumans.

The way that nonhumans and humans changed their relationship under the revolutionary government's environmental policies becomes central to the poem's environmentalist spirit. "Nueva ecología" takes another approach apart from exteriorism by using the more ambiguous first person plural to include nonhuman voices and agency in the Sandinista revolution. The poet's call to recover the forests, rivers, and lagoons is made in conjunction with the content's armadillos, conveying a dialogic relationship between nonhuman agency and humans in charge of government policies toward the environment (32). This connection between humans and nonhumans admits that humans alone are unable to "save nature," and the poem's speaker recognizes this by asserting in the opening stanza that nature is resilient. The poetic voice is still operating within the human/nature divide, but by using the first person plural pronoun loosely, it includes nonhumans as potential saviors of the environment through a collaborative effort to construct a new, noncapitalist ecology that combats slow violence and the capital fictions of Somocismo and by extension, the Cold War: "La liberación no solo la ansiaban los humanos" (32; "the liberation was not only a human yearning"). Despite the poem's dealing with nonhumans still largely within the human/nature binary, this line of the poem is the principal moment in which nonhuman agency reveals its awareness of Cold War politics that inform the Somocismo and Sandinismo ideological camps and nature's position within that ethical conflict. In addition to the revolution fundamentally changing life for humans, the poem reveals that nonhumans comprehended their situation as oppressed beings and that there existed a possible avenue of escape. Beyond assigning nonhuman awareness, the last full sentence of the poem also makes nonhumans complicit in the Sandinista revolution: "La revolución/es también de lagos, ríos, árboles, animales" (32; "The revolution/is also of the lakes, rivers, trees, animals"). The poetic subject's claim that "Toda la ecología gemía" (32; "The whole ecology had been moaning") recognizes that nonhumans voiced their opinion about the slow violence that had contaminated nature under the Somocista regime. Cardenal's poem opens up the possibility of nonhuman agency, but it never fully commits to that agency to the extent that the human/nature divide would be erased.

Although neither book tackles questions surrounding the separation of culture from nature, Cardenal's poetry from *Vuelos de victoria* unveils the mystery of slow violence that is subtle yet evident in Cuadra's poetry. Both authors' collections work in tandem, showing how slow violence precipitated and reached tipping points of toxicity in the 1970s and then how during the 1980s, with the Sandinista victory, the government promised to decontaminate the land for the sake of humans and nonhumans alike. By foregrounding literature that deals with Nicaraguan landscapes across the 1970s and 1980s, slow violence becomes visible in both literary and environmental history. Both histories are synthesized in the reading of Cuadra's and Cardenal's poetry and create awareness of the slow violence caused by unchecked capitalism in Cold War Nicaragua.

More broadly contributing to the war on nature was the capitalist resource extraction that encouraged the unchecked use of potent pesticides, the rerouting of rivers, and the dumping of toxic waste into bodies of water. Cuadra's and Cardenal's poetry provide testimony and ideological rebuttals of the oppressive economic structure that foreign businesses accelerated. Both works of poetry document the ongoing destruction of the environment and human livelihood, grappling with environmentalist thought that questions the deeply embedded capital fictions that justified modernization and development at the cost of the rural poor and the environment. The environmentalist rhetoric of the poems, while coming from an anthropocentric relationship to nature, recognizes humanity's material relationship with nature and the inherent value of the environment. In Cardenal's *Vuelos de victoria*, "Nueva ecología" confronts detrimental policies that Somocista left in its wake by providing localized ecological conflicts as evidence against the capital fiction of wealth and prosperity that modernity and capitalism promise.

The Cold War, largely defined by two opposing economic and governmental philosophies, was waged on environmental fronts as well as on ideological and physical battlefields. Central American writers point out how the Cold War trickled down into their own environments. These writers show concern for nonhuman life because they witnessed how either capitalism or governmental brutality enforced capitalist practices that actively harmed the environment while at the same time oppressing and killing human beings. They write about nonhumans as having inherent value and as worthy of consideration in politics. By criticizing the reckless pollution enforced by foreign industry and governments with environmentalist attitudes, Central American writers advocate for nonhuman protections and recognize that nonhumans are essential to the well-being of the ecosystem and of humanity. As the Cold War waged a literal war on nature, Central American authors constructed an environmentalist counterstrike that hoped to broker peace for humans and nonhumans alike.

Notes

1. My definition of a sense of place is built from Heise's discussion of the term in her book *Sense of Place and Sense of Planet*. For this chapter, I understand that the poets have a sense of place that does not rely on human isolation in an environment but rather on humanity's knowledge of how an ecosystem operates. This inherent wish to "reconnect" with nature implies that humans are disconnected from nature and that there exists the possibility of reengaging with nature.
2. When checking the condition of Lake Managua, the governmental organization over the environment, IRENA found that lake water contained hazardous levels of toxicity. By the time IRENA had evaluated the water, the U.S. company Pennwalt had deposited about 40 tons of mercury in the same water between 1968 and 1981 (Faber 168–9) and that, during the time of the study, the city of Managua consistently dumped over 70,000 pounds of sewage per day into the lake.

3. The translation of "El caballo ahogado" is mine. The other poems from *Cantos de Cifar y del mar dulce* are from Grace Schulman and Ann McCarthy de Zavala's book, *Songs of Cifar and the Sweet Sea, Selections from "Songs of Cifar, (1967–1977)"* (Cuadra, "Songs").
4. For instance, in his address to the United Nations in 1979, the then leader of Vietnam, Phan Hein, referred to a turn in Latin American politics that echoed his own, including how Nicaragua and other socialist countries were planning to reverse ecological damage and consolidate political and economic independence.
5. I understand Cardenal's poetic answer to Nicaragua's environmental circumstances not only as an example of an ecologically minded writer who transcribes slow violence into literature but also as a salient departure from Beckman's assessment of ecologically conscious works from the first half of the twentieth century. When discussing José Eustasio Rivera's *La vorágine*, Beckman comments on the novel's implicit critique of poetry's "failure . . . to represent the economy in which the novel as a whole inserts itself" (179). The difference between *La vorágine* and "Nueva ecología" is precisely the recognition of capitalism's flaws. Rivera's novel is rooted in capitalism despite the protagonist's attempt to escape it; he continually comes up against capitalism and participates in it. "Nueva ecología" grounds itself in the physical reality of the environment before and after the Sandinista revolution.
6. I am using Jonathan Cohen's translation from his 1987 publication, *From Nicaragua with Love*.

Works Cited

Anderson, Benedict. "Imagined Communities." *Imagined Communities: Reflections on the Origin and Spread of Nationalism*. Verso, 1983.

Asamblea nacional. *Ley orgánica del institute nicaragüense de recursos naturales y del ambiente (IRENA)*, 112, 1979, http://legislacion.asamblea.gob.ni/Normaweb.nsf/164aa15ba012e567062568a2005b564b/a8c512356eb9a4a7062570a60072c215?OpenDocument&Highlight=2.natural.

Beckman, Erika. *Capital Fictions: The Literature of Latin America's Export Age*. U Minnesota P, 2013.

Cardenal, Ernesto. *Vuelos de victoria*. Visor Madrid, 1984.

Cohen, Jonathan, trans. "New Ecology," by Ernesto Cardenal. *From Nicaragua with Love: Poems 1979–1986*. City Lights Books, 1987.

Cuadra, Pablo Antonio. *Cantos de Cifar y del Mar Dulce*. Libro Libre, 1983.

———. *Poemas nicaragüenses*. Editorial Nascimiento, 1934.

———. *Songs of Cifar and the Sweet Sea, Selections from "Songs of Cifar, 1967–1977"*. Translated by Grace Schulman and Ann McCarthy de Zavala. Columbia UP, 1979.

Faber, Daniel. *Environment Under Fire: Imperialism and the Ecological Crisis in Central America*. Monthly Review Press, 1993.

Harvey, David. *Seventeen Contradictions and the End of Capitalism*. Oxford UP, 2014.

Heise, Ursula. *Sense of Place and Sense of Planet: The Environmental Imagination of the Global*. Oxford UP, 2008.

Henighan, Stephen. "An Ordered Eden: The Ideal Administration in Ernesto Cardenal's *El estrecho dudoso*." *Bulletin of Spanish Studies*, vol. 89, no. 1, 2012, pp. 105–24.

Kane, Adrian Taylor. "The Nicaragua Canal and the Shifting Currents of Sandinista: Environmental Policy." *Ecological Crisis and Cultural Representation in Latin America*, edited by Mark Anderson and Zélia M. Bora. Lexington Books, 2016, pp. 269–75.

Lasarte, Francisco. "Cuadra's Mar Dulce." *Essays on Hispanic Literature*, edited by Sylvia Molloy and Luis Fernández Cifuentes. Tamesis Books, 1983.

Marcone, Jorge. "Filming the Emergence of Popular Environmentalism in Latin American: Postcolonialism and Buen Vivir." *Global Ecologies and the Environmental Humanities: Postcolonial Approaches*, edited by Elizabeth DeLoughrey et al. Routledge, 2015.

Miller, Shawn William. *An Environmental History of Latin America*. Cambridge UP, 2007.

Nixon, Rob. *Slow Violence and the Environmentalism of the Poor*. Harvard UP, 2011.

Price, Jacob. *Environment Under the Gun: Literature and Environmentalism in Cold War Central America*. Dissertation, Rutgers U, 2019.

Rivera, José Eustasio. *La vorágine*. Editorial A.B.C., 1924.

United Nations. *General Debate*. 34th session, 13th plenary meeting, agenda item 9, 28 Sep. 1979.

White, Steven F. "Poemas nicaragüenses: Mapa ecopoético de una comunidad imaginada." *El mundo más que humano en la poesía de Pablo Antonio Cuadra: un estudio ecocrítico*. Multimpresos Nicaragüenses, 2002.

Part III

Protracted Degradation and the Slow Violence of Toxicity

7 Collateral Damage

Nature and the Accumulation of Capital in Héctor Aguilar Camín's *El resplandor de la madera* and Jennifer Clement's *Prayers for the Stolen*

Adrian Taylor Kane

Mexican novelist Héctor Aguilar Camín's 1999 novel *El resplandor de la madera* (The Radiance of Wood) traces the feats, misfortunes, and scandals of the Casares family over five generations. From Rosario Casares, one of the original colonizers of the mythical town of Carrizales and a drunken bar owner, to his great-great-grandson Santiago, entangled in the battles of his father's business war in a modern city at the end of the twentieth-century, *El resplandor* is the story of a family in constant pursuit of financial fortune. The novel's broad historical scope allows for the portrayal of business practices from the colonial era to the present. In doing so, *El resplandor* captures the consistent desire for profit as the protagonists' primary motivation and reveals their indifference to the environmental implications of their practices. Jennifer Clement's *Prayers for the Stolen* (2014) is an equally compelling novel whose insights into nature–culture relations in Mexico are similarly profound. Narrated from the perspective of the protagonist Ladydi García Martínez, an adolescent girl who lives in the mountains of Guerrero in a nearly exclusively female community, it is a story of the ways in which the trafficking of women intersects with the violence of Mexican drug cartels. In the present essay, I analyze the inextricably intertwined relationship between nature and culture in both novels, asserting that these two seemingly disparate works contain a point of contact in their representation of nature as collateral damage in the battle for the accumulation of capital. Engaging with the concept elaborated in Rob Nixon's 2011 study *Slow Violence and the Environmentalism of the Poor*, both novels portray systematic forms of slow violence, which he defines as "a violence that occurs gradually and out of sight, a violence of delayed destruction that is dispersed across time and space, an attritional violence that is typically not viewed as violence at all" (2). Following the lead of Karla Armbruster and Kathleen R. Wallace, who have asserted, "An ecocriticism that sees humans as fundamentally part of nature will attend to representations of human cultures in all their diverse interactions with nature rather than focusing only on texts that show humans observing or experiencing nature in wild or rural settings" (12), in my readings of *El resplandor* and *Prayers for the Stolen*, I focus on the representation of nature in novels where the physical environment plays a secondary role to their stories of human tragedy.

In a preliminary *advertencia* to the reader in *El resplandor*, Aguilar Camín notes that his work could potentially be read as two novels, as it alternates between the present and the past in chapters respectively titled Casares and Carrizales. The chapters titled Casares follow Arabic numbers and relate the events of the present-day financial battle of the great-grandson of Rosario Casares, while the Carrizales chapters, ordered by Roman numerals, narrate the rise and fall of the town of Carrizales. The author also advises that the novel will only reach its "plenitud dramática" (9; "dramatic plenitude") by alternating between Casares and Carrizales chapters, thus allowing the two plotlines to converge. This structure allows the author to foreground the primary theme of the process of writing history while simultaneously portraying the gradual development of capitalism and the ensuing slow violence to nature in Mexico over the course of several centuries. Imagery of the natural environment, capitalism, and war are intertwined throughout the history of Carrizales and the Casares family. Nature and culture are not portrayed as opposites in this novel. Rather, they are presented as formative forces that are inextricably linked. The novel particularly portrays the culture of capitalism as a form of ongoing war in which the tropical forests, the most visible nonhuman elements of nature in this novel, become casualties in a battle for financial profit.

The intermingling of the themes of war and capitalism is present from the beginning of Presciliano the Chronicler's account of the history of Carrizales: "En el principio fue el pontón militar junto a la boca del río, y la bulla de los monos saraguatos, chillando tras el mangle y la madera, como si le gritaran a la luna" (29; "In the beginning was the warship next to the river's mouth, and the racket of the howler monkeys screeching across the mangrove and the woods, as if they were shouting at the moon"). This sentence, undoubtedly a biblical allusion, ironically replaces God—the beginning of all things according to Judeo-Christian beliefs—with a naval vessel. The juxtaposition of the ship with images of nature appears to the reader familiar with Genesis to be an intrusion into the myth of earthly paradise. Inserting this symbol of militarism at the beginning of the history of Carrizales is a reminder that Latin America's colonial history began with a war for control over land and access to natural resources for the purposes of extraction. Indeed, as Mark Anderson writes, "[M]uch of the landscape we know as Latin America today is actually the result of catastrophic land cover changes wrought through the dispossession and genocide of millions of indigenous people and the implementation on a massive scale of extractive colonial land management practices" (x). This colonial history is further evoked in the novel as Presciliano continues his account of the foundation of Carrizales by pointing out that the ship was positioned in the muddiest and most stagnant sector of the bay to prevent weapons and other goods from arriving upriver to the indigenous citizens. He concludes the opening paragraph, "así nació el pueblo de Carrizales, a rajatabla, no donde quiso el amor, sino la guerra" (29; "thus, the town of Carrizales was born with precision, not where love wanted it, but war").

The intersection of images of war and nature continues in Presciliano's narration, as he writes:

> En guerra estaban también los animales de la selva, los jaguares y las nauyacas, y las variedades todas de moscos ubicuos, portadores del paludismo y el delirio. Más mataban las selvas que los indios, más animales que hombres hubo que domar y mantener a raya en aquel campamento del origen (31; The animals of the jungle were also at war, the jaguars and the pit vipers, and the varieties of all the ubiquitous mosquitoes, carriers of malaria and delirium. The jungles killed more than the Indians, there were more animals than men to tame and keep at bay in that original camp).

By suggesting that the animals are also battling for survival, this passage represents humans as yet one more inhabitant of the jungle. Indeed, of the 200 sailors who founded Carrizales, not one returns home in good health due to the brutal conditions of the jungle.

The two groups of settlers that follow the sailors to Carrizales are prostitutes and merchants. The latter come in search of the precious woods and *chicle* found in the inland forests and seek to trade with peaceful indigenous groups. The first member of the Casares family to arrive in Carrizales is Rosario, a drunken bar owner from the islands. Enamored of a prostitute named Adelaida, Rosario follows her to Carrizales as she seeks financial fortune among the sailors. Rosario Casares's precocious son Mariano, who has diligently profited from his father's drinking establishment, later follows him to Carrizales in hopes of exploiting the mahogany forests for trade throughout the Caribbean and the Gulf of Mexico. It was Mariano, writes Presciliano the chronicler, that put Carrizales on the map of prosperity by pioneering the rush for mahogany in the region (32).

A pivotal moment in the Casares's history comes when the local bishop advises Mariano's son, Julián, of a business opportunity in the Forest Reserve of Miranda, the neighboring English colony. The bishop recounts his recent trip to Miranda. Likening it to a return to Eden, he marvels at the natural beauty of the land.

> Tenía climas templados" he recalls, "y lagos inmóviles cuya única misión parecía ser la de reflejar el cielo. . . . En el centro de las selvas

> reinaban las divinidades de la ceiba y la caoba, y la miniatura viva del quetzal que Dios había puesto en la tierra, como al colibrí, para probar el gusto de su alma por las cosas pequeñas y la perfección artesanal de su fantasía (181; It had temperate climates . . . and still lakes whose only mission seemed to be to reflect the sky . . . In the center of the jungles reigned the divinities of the *ceiba* and the mahogany trees, and the live miniatures of the quetzal that God had put on the land, just like the hummingbird, to prove his soul's taste for the small things and the artisanal perfection of his fantasy).

Immediately after extolling the beauty of God's creation, however, the bishop notifies Julián that the recently formed joint government between the Catholic Church and the new military dictatorship in Miranda is in need of revenue and may be willing to open their forest reserve for logging. The great irony, of course, is that the bishop, representing a church that holds nature to be a sacred manifestation of God's beauty and might, is willing to destroy divine creation to ensure the survival of his own institution.

When Julián enthusiastically alerts his wife Rosa that "Los bosques de Miranda están para nosotros" (182; "The forests of Miranda are for us"), she replies, "Esas selvas han sido puestas ahí por la mano de Dios. No pueden ser para nadie" (182; "Those jungles have been put there by the hand of God. They cannot be for anyone"). In response, Julián assures her that God created those forests to serve his children. He manipulates Rosa by playing on her fear that the forests will swallow their children and convinces her that the exploitation of the forests will allow them to send their children to study in the city.

Julián's line of argument in this discussion reflects a commonly held perspective in debates surrounding environmental ethics. The notion that nonhuman elements of nature were created by God for the benefit of humankind perhaps helps to explain the bishop's rationalization of the deforestation of the Miranda Forest Reserve. For many environmentalists, however, this is a dangerous attitude. Lynn White Jr. expresses his concern with this position in his essay "The Historical Roots of Our Ecological Crisis," arguing that, "We shall continue to have a worsening ecologic crisis until we reject the Christians' axiom that nature has no reason for existence save to serve man" (14). In the Casares family, the prevailing notion is that humans are above nature in the hierarchy of creation.

Despite the world view of the Casares men, the text repeatedly returns human beings to their place among nature. Perhaps the best example is the wedding of Julián Casares and Rosa Arangio, which is attended by more than just the citizens of Carrizales:

> A unos metros del festejo, en el lindero de la selva, alertas, desconfiados, acudieron también los invitados del monte, los monos de las

> ramas altas y las serpientes de los bejucos, los venados y los armadillos, las lechuzas y los guacamayos, y un ejército macedónico de insectos que chirriaba sobre el crepúsculo como si anunciara que nadie faltaba en el arca de Noé y todos bajaban sus armas para saludar por un momento los pendones humildes de la felicidad.
>
> (178)

> A few meters from the festivities, on the border of the jungle, alert, mistrustful, also came the guests from the forest, the monkeys from the high branches and the serpents from the vines, the deer and armadillos, owls and macaws, and a Macedonian army of insects that chirped at twilight as if announcing that no one was missing from Noah's arc and all were setting down their arms to salute for a moment the humble standards of happiness.

By including animals and insects as participants in the ceremony, the division between nature and culture is blurred. The humans are surrounded from above and below by wildlife, thus dismantling the image of *Homo sapiens* above nature. Along with passages such as this, descriptions of the sounds, sights, and smells of the jungle surrounding the town of Carrizales help to create the sense that humans are, indeed, in and of nature.

A less subtle reminder in the novel that humans do not ultimately dominate nature is the hurricane that devastates Carrizales in the years following Julian's wedding. Several citizens die, one beheaded by flying debris, another crushed by a fallen tree, others buried in mud and drowned in rising tidewaters. Julián's wife and children narrowly escape drowning in their own home, but the town is leveled. When one of Julián's associates in Miranda asks him "¿Cómo quedó el pueblo?" ("How did the town make out?"), he responds "No quedó" (255; "It's gone"). The hurricane destruction presents major financial losses for the entire town. However, the excessive desire for financial wealth is the one constant in Carrizales, particularly among the Casares family. Julián's father Mariano had once said, "Los negocios no son para compartir, son para acumular" (206; "Businesses are not for sharing, they are for accumulating"). When Julián returns from Miranda after the storm, he finds his brother and father ironing paper currency to dry it out. Despite setbacks from the hurricane's damage, this image makes clear that the Casares's battle for accumulation of capital will continue.

The imagery of war, initiated with the military vessel at the beginning of Carrizales's history, is consistent throughout the novel. When Julián departs for Miranda after receiving a three-year concession to log the reserves, Rosa embraces him as if she were sending him off to war (218). Indeed, in their preparations for deforestation, Julián and his associates develop a plan of attack. As Presiciliano recounts, "Mapearon y midieron el terreno, ubicaron las brechas y los pueblos, los montes y las planicies, el terreno todo en el que iban a emprender su saqueo" (219; "They mapped and measured the terrain, located the breaches and the towns, the hills and the plains, all of the terrain on which they were going to carry out their plundering"). The goal of their battle plan is to earn as much profit as possible. The enemy is not the forest but, rather, their potential business competitors. As represented on their military maps, the forest becomes the site where violence to nature slowly unfolds through the process of logging. The violence done to nature is evident here in the metaphor of sacking. Presiciliano's narration continues: "Se hacían a la selva como los descubridores de antes a la mar, para hollarla y fincar en ella por primera vez, para escarbar sus entrañas y marcharse luego" (220; "They took to the jungle like the explorers of old to the sea, to leave their mark and farm it for the first time, to unearth its bowels and be gone"). The image of physical violence to nature in this sentence is highlighted in the verbs *hollar* ("to tread, trample, or leave a sign of one's path") and *escarbar* ("to dig up, unearth, or scratch"). Finally, when Julián and his associate return to the capital of Miranda, the image is of a victorious return from war:

> Así, cuando, al cabo de treinta y seis días de ganar esa batalla, Salvador Induendo y Julián Casares volvieron a Wallaceburgh en su jeep de altos guardafangos, su entrada fue un acontecimiento y ellos dos los varones triunfadores en la modesta iliada de la selva y la más modesta odisea de su regreso a las cantinas.
>
> (221)

> Thus, when, at the end of 36 days of winning the battle, Salvador Induendo and Julián Casares returned to Wallaceburgh in their jeep with high fenders, their entrance was a grand event, and the two men were triumphant in their modest Iliad in the jungle and the more modest Odyssey of their return to the local bars.

Julián had won this initial battle in his war for accumulation, but he ultimately loses the war. His father successfully ousts him from the forest reserves by undermining his contract with the government of Miranda, and Julián never capitalizes on the fortune to be found in the mahogany forests.

Indeed, the curse of lost fortunes plagues the Casares family from the beginning to the end of the novel. In the Casares storyline, set in the modern city, Julián's son, known simply as Casares, acknowledges the cycle of defeat in his family's battle for financial wealth. He states, "Estoy peleando una guerra que perdí hace años. La perdió mi padre, la perdió mi abuelo. Y me encamino a perderla yo, luego de haberla provocado" (191; "I'm fighting a war that I lost years ago. My father lost it as did my grandfather. I'm on the way to losing it too, after having provoked it"). Despite their misfortunes, the men in the Casares family continually reformulate plans to accumulate wealth. The term *El resplandor de la madera* (The Radiance of Wood) thus becomes a metaphor for greed in the novel. Brilliance is a quality that naturally attracts the visual attention of the subject and is often associated with alluring objects such as precious stones or metals. In this sense, *El resplandor* represents value or wealth. Indeed, for Julián Casares and his father, timber is money. Herein lies their weakness. The narrator describes their obsession with wood:

> Nadie podía dejar [la madera]. Había siempre en su ciclo la promesa de una riqueza mayor. Cuando un maderero novel empeñaba las ganancias de una temporada en la persecución de la siguiente, se decía que lo había picado el bicho de la madera, el bicho que devoraba el centro de la caoba, sin que se notara en la superficie, haciendo árboles huecos de apariencias perfectas, fortunas gigantescas de dineros invisibles.
>
> (250)

> Nobody could leave [the timber]. In its cycle there was always the promise of greater wealth. When a new lumberman invested the profits of a season in pursuit of the following, it was said that the timber bug had bitten him, the bug that devoured the center of the mahogany tree, unnoticeable on the surface, making hollow trees with perfect appearances, gigantic fortunes of invisible money.

The alluring quality of wood for Julián and Mariano Casares, as evinced in this quotation, is not its inherently ecological, aesthetic, or spiritual value but rather its potential for accumulating capital. Ironically, wood does not truly shine until it has been polished—that is, until the tree has been killed, transformed, and delivered to the marketplace. Therefore, the term *El resplandor de la madera* could be a literal reference to the shine of the finished mahogany furniture produced from the timber of the Miranda Reserve or an allusion to the potential of the forest to create wealth for those who exploit it. Similar to the insects that hollow out the inside of certain trees in the previous passage, greed consumes and rots the Casares men. Indeed, one could argue that their behavior displays a form of addiction whereby the short-term desire for financial wealth is ultimately destructive. As Barbara Brandt explains, "Money-addicted individuals, organizations, and governing institutions regularly destroy real wealth—in the form of human health, the natural environment, or community well-being—especially over the long term, in order to make money in the short term" (81). Despite the illusion that wealth and well-being have increased, she argues, short-term financial profits obfuscate the destruction of real wealth (82). Due to the powerful allure of financial wealth, in *El resplandor* the Casares men are blind to the value of what Brandt posits is true wealth.

As demonstrated in these examples, Aguilar Camín's novel reminds the reader that humans are in and of nature, not above it. Nevertheless, the Casares's perception of nature, demonstrated in their actions throughout the novel, is undoubtedly shaped by the culture of capitalism and thus prioritizes economic benefit over the well-being of ecosystems. As Jason Moore observes, "[C]apital's dynamism turns on the exhaustion of the very webs of life necessary to sustain accumulation" (quoted in Anderson xvi). In *El resplandor*, the depletion of the mahogany forest is what sustains the Casares's accumulation of wealth. They are at war with their business competitors, and the destruction of the environment is not their ultimate goal. However, in their view, it is merely a means to an end, a necessary step for earning a profit. In militaristic parlance, the term collateral damage is used to describe unintended civilian casualties that result from tactical operations on military targets. Scholars such as Jessica Whyte have also used the term to argue that the economic pressures of neoliberalism inflict incidental harm on individual rights (147). With respect to the portrayal of deforestation in Aguilar Camín's novel, the physical violence to the environment can also be described as collateral damage in the battle for the accumulation of capital. It is through this portrayal that *El resplandor de la madera* offers a critique of the way nature is frequently viewed within a capitalist framework.

In the 15 years between the publication of *El resplandor* and Jennifer Clement's 2014 novel *Prayers for the Stolen*, Mexico experienced two landmark political changes that have shaped its current sociopolitical status. President Vicente Fox's 2000 election marked the end of 70 years of political dominance by the Partido Revolucionario Institucional, and of equal if not greater significance, Fox's successor Felipe Calderón, also from the Partido de Acción Nacional, launched his war on drugs in 2006, which sent the country into a downward spiral of instability caused by a sharp increase in violence perpetrated by competing criminal enterprises. A reading of Aguilar Camin's text in juxtaposition with Jennifer Clement's 2014 novel *Prayers for the Stolen* reveals similarities in both works' portrayal of nature. *Prayers for the Stolen* is a clear and compelling denunciation of gender violence as "a constant violation of the human rights of women and girls" (Olivera 50). At the same time, its portrayal of drug cartels' armed violence against local people and government forces' careless fumigation—ostensibly targeting drug crops but in reality released over normal crops, forests, and Ladydi García Martínez's village—captures the entanglement of human and nonhuman victims in the use of institutional violence. The novel thus reveals the relation between the exploitation of natural resources and the abuse of women that ecofeminists such as Sofía Kearns contend is endemic to patriarchal societies (115).

Throughout the novel, Clement incorporates imagery of the nonhuman natural world that undermines anthropocentric thinking by continually reminding the reader that humans are part of a larger biotic web of organisms (Naess 4). In the opening chapter, for example, Ladydi describes Guerrero: "A hot land of rubber plants, snakes, iguanas, and scorpions, the blond, transparent scorpions, which were hard to see and that kill. Guerrero had more spiders than any place in the world we were sure, and ants. Red ants that made our arms swell up and look like a leg" (4). She later asserts, "In Guerrero the heat, iguanas, spiders, and scorpions ruled" (18). Such descriptions allude to an ecocentric understanding of the world, in which the interests of humans are not privileged over the rest of the natural world. Indeed, the narrator's assertion that "[w]e were messy and born from the jungle so we were like the relatives of papaya trees, iguanas, and butterflies" situates humans as one more element in a particular ecosystem rather than in a hierarchical position of superiority over other organisms (47). Moreover, by foregrounding the nonhuman natural world, such passages foreshadow the important role that nature will play in the novel's social critique.

When army helicopters are heard approaching the village, the animals and insects are portrayed as being attuned to the threat presented by the herbicide that they will drop. Ladydi notices, "As I moved down the hill an army of ants was marching in several lines down the mountain toward the highway. Lizards were moving in the same direction, moving very quickly. The birds above me were also disturbed and flying away" (53). As Rita, the narrator's mother explains, the Mexican army is paid by the drug traffickers not to drop Paraquat on the poppies used to manufacture heroin, "so they drop it wherever else on the mountain, on us! . . . Those army helicopters had to go back to their bases and report that they had dropped the herbicide so they dropped it anywhere they could" (37). The image of the ants, lizards, and birds fleeing the area calls attention to the fact that humans are not the only species affected by the reckless fumigation of the Mexican army, once again suggesting the text's ecocentric orientation. Sadly, Ladydi and the other women on the mountain have become accustomed to the army's routine and recognize the smell of Paraquat (37). The notion of the community's local vegetation and food sources laced with the herbicide used in the government's supposed war on drugs invites the question of the long-term effects of Paraquat on the local people and the ecosystem that they inhabit.

The result of the use of herbicides in *Prayers for the Stolen* is portrayed as a form of slow violence against innocent Mexican citizens. Although the act of dropping herbicide is more sudden and visible than other forms of slow violence that, for example, are catalysts for rising sea levels, air pollution, or lack of access to food, the novel does, indeed, portray the use of Paraquat as a form of delayed destruction that is dispersed over time. Ladydi is aware of the long-term effects of the herbicide, which are captured in the image of "the unusual dry brown patches of jungle that were suffocating from the herbicide dropped by the helicopters" (125).

"That poison," she acknowledges, "would continue to burn through the land for decades" (125).

As Nixon points out with regard to inscribing slow violence textually, "A major challenge is representational: how to devise arresting stories, images, and symbols adequate to the pervasive but elusive violence of delayed effects" (3). In *Prayers for the Stolen*, Clement primarily navigates this challenge by representing the slow violence caused by herbicides through the use of imagery of the human body. Ladydi's friend Maria, whom she later learns is actually her half-sister, is physically defined in the novel by the harelip that she was born with. When the doctors come from Mexico City to perform surgery on Maria's lip as well as on a two-year-old boy who was born with an extra thumb, Ladydi states, "The truth was we knew the cause behind the deformities on our mountain. Everyone knew that the spraying of poisons to kill the crops of marijuana and poppies was harming our people" (21). Ladydi's mother, Rita, also develops a persistent cough that she attributes to the Paraquat. "My body," Rita exclaims, "is the army's damn poppy field" (37).

In one of the novel's pivotal passages, Ladydi frantically runs for the schoolhouse when she hears the helicopters approaching. She makes it inside safely only to realize that her friend Paula is missing. When Paula finally arrives, it is too late, as she is "drenched in the poison" (54). The girls and their teacher take Paula to the bathroom to wash off the Paraquat, but as Ladydi laments, "we knew much of it was inside her already" (55). That night Estefani, Maria, Paula, and Ladydi menstruate for the first time. Estefani's mother surmises that it is "because of the poison triggering something bad inside of us" (56). The girls' sudden and coincidental menstruation, like the physical deformities and the cough, is a way of textually representing the violence inflicted by the government's reckless fumigation through corporeal alteration. Even if the menstruation comes rather suddenly after the poisoning, the fact that the poison has sufficiently infiltrated their bodies to immediately alter their reproductive systems invites the reader to wonder what other gradual, long-term consequences their contact with the Paraquat will have. Moreover, the specific use of imagery of the reproductive cycle alludes to the possibility of the poison affecting their offspring as apparently was the case for María and the boy born with an extra thumb. In this sense, the novel makes "the relative invisibility of slow violence" visible through images of the body associated with infirmity, mutilation, and alteration (Nixon 2). Such corporeal representations support Rebekah Sheldon's contention that "the child continues to circuit sentimental attention under neoliberal regimes of flexible accumulation more interested in extracting and monetizing subindividual capacities than in maintaining and protecting desirable populations" (177). She continues:

> on the one hand, the child, in her innocence and plenitude, promises another generation of species-survival posed as physiological

> self-similarity even as she begs for protection against the many and varied harms of contemporary industrial practices. On the other hand, while the child appears to vouchsafe the future of the species, her connection to the reproduction opens onto the interlocking biological and physical systems whose livelinesses compose us as much as we compose them. That we too are subject to biotechnical control conversely reminds us of all that escapes from and exceeds that control: the specters of mutation, pollution, proliferation, and dehiscence.
>
> (177)

Through textual allusions to the female reproductive cycle and mutilated children, Clement foregrounds slow violence through toxic herbicides as a primary component of her novel's representation of the disastrous state of Mexican society in the age of narco trafficking and its ominous future.

In addition to this gradual violence, the novel is replete with the imagery of sudden human-on-human violence that portrays Mexico as in a state of internal warfare. This representation coincides with Guadalupe Correa-Cabrera's analysis of Mexico as in a state of civil war shaped by the paramilitarization of organized crime, the increase in violent confrontation between criminal syndicates, and the militarization of the Mexican government's strategy to combat organized crime (126). This war, Correa-Cabrera notes, is driven by a purely economic agenda rather than a conflict of ideologies—a point that also emerges in Clement's fictional rendering of the situation. For example, the day that the doctors from Mexico City arrive to perform surgery on María and the boy with the extra thumb, "a convoy of soldiers" accompanies them "so that they would be protected from the drug traffickers' violent confrontations" (15). Three military trucks park outside the clinic, and 12 soldiers stand watch during the operation (20). The need for military protection in these passages implies that even innocent bystanders and citizens who have no affiliation with any drug cartel can become targets of violence at any given moment. Indeed, this is confirmed when Ladydi returns from María's surgery and her mother alerts her that she has followed a group of vultures circling near their house only to discover the body of a 16-year-old boy. Ladydi muses, "In this land one can go out for a walk and find a huge iguana, a papaya tree covered with dozens of large fruits, an enormous anthill, marijuana plants, poppies, or a corpse" (29). The letter *P*, a reference to the Cartel del Pacífico, is carved on the boy's forehead, and a note indicating that the murderers intend to return for Ladydi's friend Paula and two other girls is attached to his shirt. The extreme and brutal violence carried out by the drug cartels in this image and throughout the novel is the result of what Sayek Valencia describes as gore capitalism. For Valencia, gore capitalism "refers to the undisguised and unjustified bloodshed that is the price the Third World pays for adhering to the increasingly demanding logic of capitalism. It also refers to the many instances of dismembering and disembowelment, often tied up with organized crime, gender and the predatory uses of bodies" (19–20).

The violence that is inherent to gore capitalism is underscored in *Prayers for the Stolen* through representations of the environment and the slow violence of toxicity. The Guerrero countryside in general and the poppy fields in particular are portrayed as a battleground. When the girls ignore their mothers' warnings and venture through the jungle into the hills surrounding their village, they come across an enormous field of poppies. As the narrator describes: "We stood before the brilliance of lavender and black as a huge field, a bonfire of poppies appeared before us. The place seemed to be deserted except for a downed army helicopter, a mangled mess of metal skids and blades among the poppies" (37). She goes on to explain that the poppy growers string wires above the crops in order to down helicopters and sometimes shoot them down with their rifles and AK-47s (37). These images of warfare create the effect of portraying the drug cartels as combatants in a struggle for an ever increasing accumulation of capital. The women of the village, repeatedly doused with poisonous herbicides, become long-term casualties of war to ensure that the cartels continue to profit from the illicit drug trade. Others, like Paula and the local beauty salon owner Ruth, are kidnapped and trafficked as sex slaves. Field hands are stolen from the countryside and forced to pick marijuana crops for the cartels. To live in Mexico in the age of narco trafficking, the novel suggests, is to be caught in the midst of an internal war where there is no regard for the lives of Mexican citizens or the ecosystems that they inhabit.

Clement's characterization of drug cartel members and police officers as the forces of criminal capitalism in Mexico portrays an attitude of indifference that fosters both sudden human-on-human violence and slow violence to the environment. Their utter callousness is evoked throughout the novel through constant comparisons to dangerous predators such as the scorpions, spiders, and poisonous snakes that inhabit the jungle. For example, Ladydi laments, "The night belongs to the drug traffickers, the army, and the police just like it belongs to the scorpions" (52). Such comparisons heighten the sense of constant fear in which the women of the village live and underscore that the drug traffickers and corrupt police officers are as devoid of a sense of compassion and ethics as the scorpions, spiders, and snakes that inhabit the jungle. Take, for instance, the following narration by Ladydi:

> I became very still, like when a white, almost transparent scorpion is on the wall above your bed. Still like when you see a snake curled up behind the coffee tin. Still like waiting for the helicopter to dump the burning herbicide all over your body as you run home from school. Still like when you hear an SUV turning off the highway and it almost sounds like a lion, even though you've never heard a lion.
>
> (41)

The comparisons in this passage imply that the women in the novel feel as much like prey when they come into contact with natural predators as they do when the army or the cartel members, who notoriously arrive in their black SUVs with tinted windows, are nearby. In other words, the narco traffickers and their government facilitators are no more compassionate or altruistic than wild predators. Human life is of no value to them if it does not advance their battle for the accumulation of financial wealth.

The same appears to be true of their view of the nonhuman natural world. In addition to the cultivation of poppies and marijuana for the drug trade, the commodification of nature is present in other aspects of the narco culture portrayed in the novel. For example, when Ladydi is hired as a maid at the Acapulco residence of a man she later learns is a powerful cartel member, her coworker Jacaranda informs her that, behind the house on his ranch in northern Mexico, he keeps cages of old lions and tigers that he buys from zoos. The reader learns that one of Paula's jobs when she is stolen from her village is to use the lion and tiger excrement to wrap into drug shipments bound for the United States as a means of repelling drug-sniffing border dogs (73). Other animals are shipped to the narco ranches for the purposes of entertainment. As Jacaranda states:

> Rich people from the United States liked to hunt there. . . . A deer cost you two thousand dollars to kill. . . . The large living room at the ranch house contained a polar bear rug and dozens of deer heads on the walls. The wide, circular stools were made of elephant feet. The lamps were made of deer legs that had been hollowed out with a long drill. . . . Mr. Domingo would go hunting in Africa once a year and ship trunks of dead animals back to the ranch to be stuffed.
>
> (127)

When Paula escapes her captivity and returns home, she similarly reports that her captor had over 200 pairs of boots "made from every kind of animal and reptile that was in Noah's Ark," including one pair that was purportedly made of human flesh (74). Once again, the value of nature in narco culture is portrayed as purely monetary. The lions and tigers are used as part of the supply chain process, and the hunting of other animals is a lucrative side business. The skins and trophies of animals hunted abroad, some of which are endangered, become displays of wealth and power that reflect yet another facet of the endemic violence of narco culture, this time in the form of the gradual extermination of threatened species.

A close reading of the representation of nature in the novel reveals a portrayal of Mexico as in a state of gradually unfolding ecological disaster caused at least partially by the reckless use of herbicides that are ostensibly used by the police and the military to eradicate drug crops but that, in reality, are indiscriminately sprayed over the Mexican countryside and its citizens, inflicting a form of slow violence. As Nixon contends, "[S] lowly unfolding environmental catastrophes present formidable representational obstacles that can hinder our efforts to mobilize and act decisively" (2). Nevertheless, Clement successfully confronts this challenge in her novel through the use of imagery of the human body and warfare that alludes to the long-term effects of the purported war on drugs. The novel's repeated references to the flora and fauna of the mountains of Guerrero remind the reader that humans exist within broader ecosystems but also portray an utter disregard for the well-being of those ecosystems in the narco trafficking culture. In this regard, nature and women become casualties in an ongoing battle for the accumulation of capital. *Prayers for the Stolen* "demonstrates the direct relation that exists between capital and death, between accumulation and unregulated concentration and the sacrifice of poor, mestizo women, devoured where the monetary and symbolic economies, the control of resources and the power of death are articulated" (Segato, quoted in Franco 221). In the end, the inherent violence of the drug cartels, including the trafficking and rape of women and girls, leaves the protagonist's village torn asunder.

El resplandor and *Prayers for the Stolen* resist dualistic nature–culture conceptions by portraying human cultures as embedded within broader ecosystems and by offering critiques of the environmental repercussions of unfettered capitalism. Anderson postulates that capitalism "has generated environmental crisis not merely as a disagreeable by-product of economic growth (capital expansion), but also as a machine for capitalizing on nature itself" (xvi). Both *El resplandor* and *Prayers for the Stolen* portray nature as being gradually consumed and destroyed by the organizational machines of logging companies and drug cartels that are designed to capitalize on nature. Ultimately, Aguilar Camín's and Clement's novels employ the imagery of warfare to represent the mahogany forests and the poppy fields carved out of the jungle as battlegrounds in a war of accumulation.

Works Cited

Aguilar Camín, Héctor. *El resplandor de la madera*. Alfaguara, 1999.

Anderson, Mark. "Introduction." *Ecological Crisis and Representation in Latin America: Ecocritical Perspectives on Art, Film, and Literature*, edited by Anderson and Zélia M. Bora. Lexington Books, 2016, pp. ix–xxxii.

Armbruster, Karla, and Kathleen R. Wallace. "Introduction: Why Go Beyond Nature Writing and Where To?" *Beyond Nature Writing: Expanding the Boundaries of Ecocriticism*. UP of Virginia, 2001, pp. 1–28.

Brandt, Barbara. *Whole Life Economics: Revaluing Daily Life*. New Society Publishers, 1995.

Clement, Jennifer. *Prayers for the Stolen*. Hogarth, 2014.

Correa-Cabrera, Guadalupe. *Los Zetas Inc. Criminal Corporations, Energy, and Civil War in Mexico*. U of Texas P, 2017.

Franco, Jean. *Cruel Modernity*. Duke UP, 2013.

Kearns, Sofía. "Nueva conciencia ecológica en algunos textos femeninos contemporáneos." *Latin American Literary Review*, vol. 34, no. 67, 2006, pp. 111–27.

Naess, Arne. "The Shallow and the Deep, Long-Range Ecology Movement: A Summary." *The Deep Ecology Movement: An Introductory Anthology*, edited by Alan Drengson and Yuichi Inoue. North Atlantic, 1995, pp. 3–9.

Nixon, Rob. *Slow Violence and the Environmentalism of the Poor*. Harvard UP, 2011.

Olivera, Mercedes. "Violencia feminicida: Violence Against Women and Mexico's Structural Crisis." *Terrorizing Women: Feminicide in the Americas*, edited by Rosa Linda Fregoso and Cynthia Bejarano. Duke UP, pp. 49–58.

Sheldon, Rebekah. *The Child to Come: Life After the Human Catastrophe*. U of Minnesota P, 2016.

Valencia Triana, Sayak. "Teoría transfeminista para el análisis de la violencia machista y la reconstrucción no-violenta del tejido social en el México contemporáneo." *Universitas humanística*, vol. 78, no. 78, July–Dec. 2014, pp. 65–88. EBSCO Host.

White, Lynn, Jr. "The Historical Roots of Our Ecological Crisis." *The Ecocriticism Reader: Landmarks in Literary Ecology*, edited by Cheryll Glotfelty and Harold Fromm. U of Georgia P. 1996, pp. 3–14.

Whyte, Jessica. "Human Rights and the Collateral Damage of Neoliberalism." *Theory & Event*, vol. 20, no. 1, 2017, pp. 137–51.

8 Violence, Slow and Explosive

Spectrality, Landscape, and Trauma in Evelio Rosero's *Los ejércitos*

Carlos Gardeazábal Bravo[1]

Evelio Rosero's *Los ejércitos* (2007) (*The Armies*, 2009) opens with a connection to nature: "Y era así: en casa del brasilero las guacamayas reían todo el tiempo" (11) ("And this is how it was: at the Brazilian's house the Macaws laughed all the time" *The Armies* 3). In this initial scene, all the elements evoke what seems to be an Edenic setting. Ismael Pasos, a retired teacher from a rural area, has climbed to the height of a wall that separates his house from that of his neighbor, Geraldina, also known as "the Brazilian," who enjoys being naked with her husband while playing music. Macaws, naked bodies, cats, children, fishes, and fruits are all the focus of Ismael's voyeuristic lust. Yet this setting radically alters. After one of several sieges, the town and all these elements are disturbingly eliminated. Otilia, his wife, is kidnapped by one of the armies, though it is unclear what army it might be as it is not mentioned by name. The garden at the Pasos's house becomes a ditch full of animal corpses and fruits violently exploded, and the wall is destroyed, as is the division between private and public spaces. Geraldina, who had originally been a strong social presence in the town, is killed, and her objectified corpse is raped by members of an unidentified army. Ismael describes how the conflict surrounding the community directly impacts his loved ones at the hands of the besieging armies. He, his wife, and the local body politic face the traumatic consequences of a war in which none of the armed actors respect the lives of defenseless civilians. The fictitious town of San José thus encapsulates the reality of many small towns from Colombia, including the impact of the conflict on the destruction of the environment. The repercussions of years of governmental neglect, neoliberal policies, and environmental destruction are the bleak apocalyptic backdrop of the narration of *Los ejércitos*.

The Colombian conflict has often been associated with the overt violence of guerrilla warfare, massacres, and terrorist attacks. In addition, the transnational drug trafficking that fueled the conflict is continuously identified with urban, media-friendly, and often spectacular displays of violence. This chapter sheds light on how *Los ejércitos* provides literary insight into other forms of attritional or systemic violence, making apparent the entanglements between trauma, spectrality, and slow violence. By engaging with the notion of "spectral topographies," employed by Juliana Martínez in her own analysis of the novel, this essay investigates Rosero's treatment of the collective trauma of a society that has experienced a devastating war for decades with different degrees of intensity and visibility.[2] The novel negotiates the violent history of Colombian society, haunted by the death of 180,000 civilians and the displacement of nearly 6 million persons over the longest internal war of the hemisphere (1960–2016). Moreover, it addresses different kinds of violence in an interwoven relation, showing how event-centered violence and slow violence are usually linked in contexts such as the Colombian conflict.

Martínez's idea of spectral topographies not only conveys invisible violence but also helps to explain, through its links to Derridean hauntology, the intersections of other kinds of violence, trauma, memory, and time disruption in the novel. While the novel shows how the distinctions between event-centered violence and slow violence are blurred, it also displays how instances of other types of attritional violence have similar qualities as those of slow violence.[3] The spectrality present in the novel, which appears gradually in the narrative, helps to make trauma visible through constant counterpoints with corporeality; it destabilizes boundaries, moving from the site of living bodies to the realm of a haunted space and time. The novel addresses the traumatic effects of the Colombian conflict in a way that invites its readers to interlace a corporeal perspective on violence and spectral views on time and landscapes. The narrative bridges such embodiment with the strategic relationship of conflict, land tenure, and ecocide. These are forms of slow violence connected to processes that gradually affect the most vulnerable communities, as is also present in the "ecological conception of justice" proposed by Elizabeth Anker (215) and understood as "an exercise in unconcealment aimed at divulging the self's cohabitation with other beings" (71). *Los ejércitos* provides a powerful depiction of long-term social, political, and environmental effects of the Colombian conflict while offering an oblique representation of attritional destruction through the spectral qualities of its plot. In so doing, it challenges the visible/invisible dichotomy of violence and shows the long-term effects of both visible and invisible forms of violence. The chapter also appeals to the intersections of Rob Nixon's slow violence and Slavoj Žižek's notion of objective violence to analyze how the attritional and long-term consequences of the conflict are represented in *Los ejércitos*.

Conflict and Environment in *Los ejércitos*

As is clear from several of Rosero's interviews where he emphasizes his belief in the power of literature to effect change (Rosero "I hear"; Rosero and Jiménez), *Los ejércitos* is—in addition to its autonomous aesthetic commitments—motivated by an impulse to remind its readers about the humanitarian and environmental crisis taking place in the Colombian conflict zones. The plot of *Los ejércitos* takes place during the final phase of the Colombian conflict, in which at least four armed actors take part: the guerrillas (FARC, ELN), paramilitaries (AUC), drug traffickers, and the armed forces.[4] The civilian population of rural areas, distant from major urban centers of political and financial power, suffered the brunt of these waves of low-intensity conflicts.[5] Parallel to the increasing impact of war in the civilian population, a set of broad neoliberal reforms also took place in the country throughout the 1990s. The conflict produced millions of displaced persons.[6] By 2007, the year of the publication of the novel, the paramilitaries had begun a criticized process of demobilization clouded by impunity. Meanwhile, the country paradoxically maintained a stable economy while nonetheless becoming one of the most unequal and inequitable societies in Latin America.

The confluence of a low-intensity conflict and neoliberalism has had a negative environmental impact. Massive fumigation procedures have forced communities to move to other regions, including nature reserves. Colombia is considered a megadiverse country, home to 10% of the biodiversity of flora and fauna of the world, because of its intersection with different types of ecosystems and a high number of endemic species. Although the deforestation associated with coca plantations has had a significant impact on forest loss, the biodiversity of these areas is highly vulnerable to *all* kinds of human agricultural activity. Extensive cattle farming and large-scale monocropping projects have also negatively affected communities and ecosystems (Quimbayo Ruiz; Taussig; Mol; Maher; Ojeda et al.).

Nixon defines slow violence as the "violence that occurs gradually and out of sight, a violence of delayed destruction that is dispersed across time and space, an attritional violence that is typically not viewed as violence at all" (2). His approach provides a way to "engage a different kind of violence, a violence that is neither spectacular nor instantaneous, but rather incremental and accretive" (2). In his novel, Rosero engages with the narrative and representational challenges that, following Nixon, are posed by the relative invisibility of slow violence. The impact of the conflict and neoliberal practices is shown in landscapes, bodies of water, and the damage done to precarious communities throughout the novel. As I will explain, these effects work as clear examples of slow violence, even though in some cases that invisible violence is implied rather than explicit. The novel represents the various kinds of violence operating in the Colombian context and unveils the depiction of the relationship between humans and other beings. This serves as a first step toward exploring trauma, understood as another attritional consequence of such types of violence in the novel.

Trauma and Nonhuman Bodies

One of the main issues thematized by the novel is the integration of the animals, plants, rivers, and mountains that surround San José in a robust continuum. These relationships have various functions in the narration. Rosero employs these elements to reaffirm the vitalism that clashes and sometimes intersects with varying kinds of violence, which is manifested both by the nature that surrounds Ismael and the town and by the voyeuristic eroticism that characterizes his worldview. That eroticism, present in the way in which Ismael narrates the story, goes hand in hand with the corporeality and the telluric vitalism projected in the landscape, understood as a reaffirmation of life connected with the landscape.

The links between human/animal corporeity are central to the way the novel approaches trauma and attritional violence. After Otilia is kidnapped, Ismael wanders the town looking for her. The protagonist lets us know that he understands the pain of some members of his community—such as his neighbors, close friends, and even the young soldiers who attack the town—while he is apathetic toward the suffering of several others. The pessimism of his narration allows us to engage critically with the actions and emotions of the different members of his community, an exercise that includes both human and nonhuman entities. The emotional embodiment of trauma in animals belonging to the community of San José is a recurrent trope in the plot. Ismael describes a particular empathic encounter after returning once again to his house while trying to find his wife in town: "Oigo el maullido de los gatos sobrevivientes, girando en torno mío. *Otilia desaparecida*, les digo. Los Sobrevivientes hunden en mis ojos los abismos de sus ojos, como si padecieran conmigo. Hacía cuánto no lloraba" (119) ("I hear the surviving cats meowing, circling around me. '*Otilia is missing*,' I tell them. The survivors sink the abysses of their eyes into my eyes, as if they were suffering with me. How long has it been since I cried?" 111). It is not by chance that it is with his cats that he establishes a connection, given the relationship between human and animal beings in the novel. That vision emerges in the association of different forms of life with the community and within the personal world of Ismael Pasos in particular, which becomes palpable when he comes back to his place and finds it almost destroyed: "en el piso brillante de agua tiemblan todavía los peces anaranjados, ¿qué hacer, los recojo?, ¿qué pensará Otilia—me digo insensatamente—cuando encuentre este desorden?" (101) ("on the ground shiny with water the orange fish still quiver; what to do? Pick them up? What will Otilia think—I wonder foolishly—when she finds this mess?" 103). Dead animals and plants that were part of the usual setting in the Pasos's home are highlighted in his account of the destruction caused by *Los ejércitos* during that attack on San José. The Edenic scene with which the novel began is destroyed by the arrival of war's inferno, suffered by both humans and nonhumans alike. War not only destroys fighters and goods, citizens, and soldiers but also life in general and its chances of preservation.

The limits between the private and the public world have been wrecked as well, intersecting space and nonhuman bodies under the impact of visible violence: "Al fondo, el muro que separa mi casa de la del brasilero humea partido por la mitad . . . la mitad del tronco de uno de los naranjos, resquebrajada a lo largo, tiembla aún y vibra como cuerda de arpa" (102) ("At the back, the wall that separates my property from the Brazilian's smokes where it has been blasted in half . . . half the trunk of one of the orange trees, split lengthwise, still trembles and vibrates like a harp" 103–4). These animals and plants also were part of the Pasos family's way of affirming life around them, confronting the necropolitical order of the conflict. These event-centered violent acts inflicted on nonhuman bodies create a counterpoint to the manifestations of slow violence shown in other parts of the novel. Laura García Moreno, quoting Thomas Dumm, convincingly highlights another connection between animals and trauma in the novel, this time less direct. The name of the narrator and the form in which he introduces himself evoke the narrator of *Moby Dick*, another schoolteacher who was also a witness to and the solitary narrator of a disaster where all the members of his community die. Both novels, states García Moreno, are fictitious testimonies of two survivors who have witnessed the destruction of a collective subjected to continuous violence (148). Similarly, the animals in *Los ejércitos* share the fate of the community. The nature of the violence they endure, however, is selective. As happens to other members of the community, some animals are protected by the operating biopolitical disposition.

The novel offers a bleak representation of the biopolitics operating in that violent order. Although this state of exception of the Colombian internal conflict exposes thousands of citizens to extermination, the powers that dispute the sovereignty over the state make certain groups much more vulnerable than others, in the sense defended by Judith Butler in *Precarious Life*. Rosero shows in the novel those affective dispositions at stake in the Colombian conflict. After the situation in the town turns unbearable, we learn that some of its citizens try to get in contact with the central government. "El profesor Lesmes y el alcalde viajaron a Bogotá; sus peticiones para que retiren las trincheras de San José no son escuchadas. Por el contrario, la guerra y la hambruna se acomodan, más que dispuestas" (124) ("Professor Lesmes and the mayor traveled to Bogotá; their requests to remove the trenches of San Jose are not heard. On the contrary, war and famine are accommodated, more than willing" 116). San José is abandoned to its fate by the state, reducing its inhabitants to entities not only more vulnerable to a violent death but also stripped of the most basic integrity, where attritional violence in the form of famine is turned into another tool for destruction. Another example of the representations of biopolitics is the reference to the private zoo of General Palacios: during the growing violence in San José, the animals that he considers the most valuable are evacuated in cargo helicopters (*Los ejércitos* 164; *The Armies* 161). This is how Rosero thematizes the radical selective ordering that the sovereign imposes, a frame of war within which some animal beings are considered more valuable than humans. Rosero exposes, meanwhile, how trauma and different kinds of violence are tied together, regardless of their politics of speed.

Vanishing Binaries: Slow Violence, Haunted Landscapes, and Trauma

Treatment of the landscape in Rosero's work reveals possibilities for the disruption of the binaries behind the idea of slow violence. A closer study unveils the intersections among his crafting of space, trauma, and the politics of speed, which progressively bestows a spectral character to the places the characters inhabit and traverse. The narrative arc that goes from the idyllic paradise to the apocalyptic destruction of the town has landscape and nature as two of its central elements. The vision of violence deployed in *Los ejércitos* includes a wide panorama that includes the strategic relationship between war and land tenure, a central component of the Colombian conflict, including its impact on the environment.

The novel's geographical space challenges the readers aesthetically, but its function is not limited to only that aspect; the landscape turns out to be a space not only of coexistence but also of destruction as it is transformed into a vector of violence. It can also be considered a haunted space. Juliana Martínez explains how Rosero avoids violence as a simple spectacle and, in its place, chooses an approach that accentuates what she calls spectral topographies. Martínez defines such topographies as "spaces haunted by the uncertainty and unrest that violence (re)produces . . . that encourages novel ways of thinking about Colombia's recent history" (118). Spectral topographies, in works such as *Los ejércitos*, "advance an anti-essentialist, unstable, spectral, notion of space from which alternative and less violent aesthetic spaces of representation can emerge" (124). I add to her argument that, through the spectral landscapes presented in the novel, Rosero dissolves the binaries implied by Nixon's concept of slow violence, while addressing destruction and dispossession in the Colombian conflict. As Ruth Heholt argues, haunting, as well as the figure of the specter, is a way of presenting, even in temporal terms, "the violences of the past . . . terrible oppressions, injustices and traumas" (9–10). Heholt explains how specters and the idea of haunting "are inherently deconstructive" (13). For this critic, "haunting breaks down binary distinctions: visible/invisible, present/absent, alive/dead, here/there. Haunting transgresses boundaries as well as binaries" (13). Elaborating upon the work of Nelly Richard, Jo Lobanyi, and Cristina Moreiras, Alberto Ribas-Casasayas and Amanda L. Petersen explore how spectrality can be a strong response to situations of destruction and dispossession: "Spectrality or haunting rises as an aesthetic opposed to conditions or moods generated by military, political, or economic violence in the context of modernity. It is an aesthetic that seeks ways to counteract erasure, silencing, and forgetting that eschews melancholic attachment to loss" ("Introduction" 6). These haunted spaces convey and confront the quandaries of absence and silence, including the kidnapped, the displaced, and the disappeared, as well as the environmental destruction in the Colombian conflict, as shown in the novel.

A naturalized geography of the conflict has often associated, in the different phases of the war, land tenure with the exploitation of natural resources, and this association is represented in the novel as well. Ismael refers to these naturalizations directly:

> Los cientos de hectáreas de coca sembradas en los últimos años alrededor de San José, *la "ubicación estratégica" de nuestro pueblo, como nos definen los entendidos en el periódico*, han hecho de este territorio lo que también los protagonistas del conflicto llaman "el corredor", dominio por el que batallan con uñas y dientes, y que hace que aquí aflore la guerra hasta por los propios poros de todos
>
> (124, my emphasis)

> The hundreds of hectares of coca planted around San José in the last few years, the strategic location of our town, as those in the know classify us in the newspapers, have made of this territory what the protagonists of the war also call "the corridor", dominion over which they fight for tooth and nail, and which causes the war to surface in everyone's very pores.
>
> (127–8)

The agents that have caused such violence seek to take advantage of strategic geographic spaces, impacting the inhabitants of their communities. San José and the many nonliterary Colombian communities affected by the conflict end up transformed into what Ulrich Oslender describes as "geographical spaces of terror," where armed groups violate the fundamental rights of communities to achieve their economic objectives (80). The resultant social trauma can be seen in the indifference and apathy shown by most of the town's settlers, each time becoming more accustomed to the dehumanizing character of the conflict and the destruction of their surrounding environment and thus mirroring the possible indifference of some urban readers of the novel. Trauma, then, runs parallel to slow violence and other consequences of attritional violence in the novel. It is another long-term, attritional, not-always-visible effect of event-centered violence, which thwarts the binary divisions of violence into slow and event-centered, visible and invisible.

If *Los ejércitos* represents different kinds of nonevident and spectacular violence and gestures toward its causes, where political-economic structures of inequality are coresponsible for these events, the novel does so through a critical representation that enables us to understand the way in which these interconnected forms of violence operate within the Colombian conflict. For Slavoj Žižek, subjective violence is perceived as a visible disturbance of the "normal" nonviolent state of things. Objective violence can be invisible, as structural violence, and can be considered the relatively hidden correlate to the highly visible subjective violence. Žižek argues that there are two kinds of objective violence: symbolic and systemic. Symbolic violence can be traced in forms of language or modes of representation (i.e., implicit sexism, racism, discriminatory language); systemic violence stems from the functioning of economic and political systems. A critique of political economy is required to make systemic violence visible, revealing the relation between its almost silent, nonvisible procedures and its consequences (2). On the other hand, to trace objective violence means to find that which is clouded by our convictions or even our sentimental interpretations that tend to create momentary or shallow empathic links with the victims of such violence. Research on objective violence would aim to make visible, as Marinos Pourgouris states while commenting on Žižek, "as much as possible, the systemic structures that might be imperceptible to the perpetrator and the victim alike" (228).

Similar to other types of objective violence, slow violence is mainly invisible and cannot be seen directly in action, as opposed to the explosive and immediate, subjective violence of the news. The effects of slow violence may develop almost imperceptibly for generations, affecting the most vulnerable populations. The attritional repercussions of the chemical war on drugs, affecting the Colombian ecosystems and the health of rural communities located in the adjacent region, are an example of such operations, as are other transnational enterprises highly demanded by global markets. Fostered by neoliberal policies, all these businesses have impacted ecosystems negatively and have placed communities in a position of precarity, while also directly financing some of the armies described in the novel (Sanchez et al.; Maher).

From this standpoint, Rob Nixon's concept of slow violence can be linked with Žižek's notion of systemic/objective violence, going beyond what Nixon calls "the politics of speed" (11),[7] which is presented as an important matter in the novel. Trauma is a product of both types of violence, not only nonvisible violence. Although the fact that trauma is a long-term, invisible effect of violence does not require its cause to be slow violence, their connection entails an exploration of trauma as a product of an attritional and nonvisible damaging process. The concept of slow violence, in fact, is not represented in the novel as isolated from other event-centered and visible forms of violence. Rosero goes beyond these binaries (visible/invisible, slow/event-centered violence), representing environmental destruction and the trauma shared by the community. Even though Nixon holds that slow violence calls attention "not simply to questions of agency, but to broader, more complex descriptive categories of violence enacted slowly over time" (11), a narrative like *Los ejércitos* helps to show how attrition is also represented as an element in trauma narratives in which environmental destruction goes hand in hand with social damage.

In the novel, the almost dry river of the town, a clear example of slow violence, becomes a metaphor for the progressive destruction of San José and its social fabric. While Ismael gets lost, coming out to the edge of town, he sees a place that he hasn't visited in decades, "moteada de inmundicias y basuras—antiguas y recientes" (39) ("strewn with filth and rubbish—some old, some new" 32), and he notices something familiar down the cliff that resembles a sparkling silver ribbon: "El río. Antes, podía ocurrir todo el verano del infierno, y era un torrente. . . . Hoy, disecado por cualquier pálido verano, es un hilillo que serpentea (39) ("The river. It used to rage all through the hellish summer, and it was a torrent. . . . Today, desiccated by a pallid heat, it is a little meandering thread" 32). Ismael remembers when the members of the community used to fish there, but also how naked girls swam in its clear waters, enjoying the seeming privacy of the quiet forest. A lone noise was "el canto de un mochuelo, el canto de mi pecho en lo alto de un naranjo, el corazón del pueblo adolescente viéndolas. Porque había árboles para todos" (40) ("only the song of a small owl, the song of my chest high up in an orange tree, the heart of every adolescent boy in town watching them. Because there were trees for all" 33). Rosero contrasts here environmental destruction with the multiple powers of the river from the past. Decades ago, the river was a site of coexistence, maybe even the origin of Ismael's voyeuristic relation with orange trees. This body of water in the novel could belong to the group described by Ana María Mutis in her study of literary rivers in Colombian literature.[8] According to Mutis, these rivers canalize the deterioration of the environment, the marginalization of the affected communities, as well as the practices of exclusion and social inequality: the vision of nature portrayed recognizes the rivers as integral parts of such communities.

The novel also engages the realm of what Weik von Mossner describes as "the moral dimensions of our empathic engagements in environmental narratives that are concerned with issues of exploitation, abuse, and injustice" (14). The example of the dry river reveals culturally specific contributions from the engagement with such affective ecologies, even if they can be linked to other Latin American societies as well. Rosero incorporates these interrelations between landscape and conflict in the novel, enunciating their destruction from the humanizing perspective of those who experience them, bringing light to an environmental continuum where animals, plants, rivers, and humans face the consequences of attritional violence. The landscape becomes an agent that motivates noninvasive, situated, reflective empathy for the global readership of the novel. *Los ejércitos* promotes, by prompting a reflective exercise in its readers, not only sustainable lifestyles in the Capitalocene but also an alternative sense of literary justice.

Hope, although tainted with spectral tones, is still present in the displaced communities leaving home as we can see in this dialogue between Ismael and Rodrigo Pinto, a peasant who lived in the neighboring front line. Ismael recounts that Pinto "me repite que no se va, como si quisiera convencerse de eso," adding, "otra montaña sería mejor . . . mucho más lejos" (171) ("repeats that he is not leaving, as if he wanted to convince himself. . . . Another mountain would be better . . . further away" 158). The mountains invite one to overcome the present through what seem like utopian projects for humans and animals; Pinto insists:

> ¿Sí ve esa montaña? . . . Allá voy a irme. . . . Tengo un buen machete. Sólo necesito llevar una marrana preñada, un gallo y una gallina, como Noé. . . . Esa montaña puede ser mi vida. . . . Allá lo espero, cuando tenga a su Otilia con usted. Después nos iremos todos, ¿por qué no nos vamos todos?
>
> (171–2)

> You see that mountain? . . . I'm going to go there. . . . I have a good machete. I only need to take a pregnant sow, a cockerel and a hen, like Noah. . . . That mountain could be my life. . . . We'll be expecting you there, when you have your Otilia with you. Then we'll all go, why don't we all go?
>
> (158)

Rosero describes the landscape as a projection of the desires of the characters where space opens up new possibilities to start again. The hope of the displaced thus contrasts with the destruction that is elsewhere projected onto the mountains, where the author mixes the humanitarian drama of displacement, uprooting, and death with the geography of San José. This affective description of the mountains surrounding San José shows Rosero's counternarrative against the naturalization of slow violence and also other invisible or not directly perceptible kinds of violence. This section of the novel is also connected to the presence of haunted spaces in the novel, as it highlights the idea of displacement and invisibility correlated to the landscape.

Attrition and diffusion, then, are central to the antiessentialist account of trauma offered by the novel. As Nixon states, "we need to account for how the temporal dispersion of slow violence affects the way we perceive and respond to a variety of social afflictions—from domestic abuse to posttraumatic stress and, in particular, environmental calamities" (3). Trauma caused by attrition, in the form of either environmental destruction or severe, decades-long violations of socioeconomic rights, is addressed by the novel through the narrative strategy of spectrality that thematizes such temporal dispersions and absences. These dislocations are exemplified in a critical passage where Rosero crafts the description of the mountains to project a mix of attritional and visible violence, this time in the form of trauma. Ismael tries to find Otilia and walks once again to the cliff by the end of town, where the view of the mountains catches his attention: "En la montaña de enfrente, a esta hora del amanecer, se ven como imperecederas las viviendas diseminadas, lejos una de otra, pero unidas en todo caso porque están y estarán siempre en la misma montaña, alta y azul" (61) ("On the mountain across the way, at this time of the morning, the scattered houses look eternal, far from each other, but united anyway because they are and always will be on the same mountain, high and blue" 58). Ismael once imagined himself living there before meeting his wife; these mountains had been inhabited, but nobody lives there anymore: "no hace más de dos años había cerca de noventa familias, y con la presencia de la guerra—el narcotráfico y ejército, guerrilla y paramilitares—sólo permanecen unas dieciséis. Muchos murieron, los más debieron marcharse por fuerza" (61) ("not more than two years ago there were close to ninety families, and what with the war—the drug traffickers and army, guerrillas and paramilitaries—there are only sixteen left. Many died, most of them must have had to leave" 59). Ismael endures a strong emotional reaction to this view and the history behind it: "aparto mis ojos del paisaje porque por primera vez no lo soporto, ha cambiado todo, hoy—pero no como se debe, digo yo, maldita sea" (61) ("I look away from the landscape because for the first time I cannot stand it, everything has changed now—but not the way it should have, damn it" 59).

In the past, the landscape had been a medium for the construction of social bonds thanks to the common enjoyment it allowed. But in the apocalyptic present of the narration, community and environment have been destroyed by the violence that operates in the conflict, both the direct and palpable kind and those forms of violence with barely visible effects—and therefore is not easily suited for the spectacular, media-friendly violence of war. Such exploration contains an invitation to complex constructions of empathy through these elements, starting with the landscape, a space that is shared by the bodies that suffer such violence—or seem to be condemned to perpetuate it. These spatial disruptions, through spectrality, go hand in hand with a temporal one. As Martínez points out, all the references to time in the second part of the novel are tied to a question mark or an expression of doubt (122). Time vanishes, turning into a haunted measure by the end of the novel when Ismael states that it was not possible to guess what time it was (193). Spectrality, then, becomes a narrative mode to articulate the politics of time and speed associated with attritional violence.

Rosero represents the impact of trauma caused by different types of violence as it operates over the community and its surroundings for decades. The novel, however, is not simply a story about PTSD where ecological destruction works in the background. Through its spectral tones, it instead connects history, different forms of attrition, and even particularities of Colombian and Latin American traditions, creating representational forms of environmental and spectacular violence. For Derek Summerfield, the acritical applications of Western theories of trauma in different cultures can become ways of delegitimizing the knowledge that other communities have of themselves and their ways of dealing with grief and pain. For this author, the generalizing diagnoses imposed by members of other cultures may have purposes other than the resolution of these pathologies: those who speak up may do so not in the context of treatment but because they want to present human rights testimony, in particular "in the context of an assault on their culture and ethnic identity" (1456). Craps asserts, commenting upon Summerfield, that "insofar as it negates the need for taking collective action toward systemic change, *the hegemonic trauma discourse can be seen to serve as a political palliative to the socially disempowered*" (28; emphasis mine). Forms of objective and slow violence remain intact while the Western, urban subjects try to extrapolate the idea of individual trauma to sociocultural contexts where its relevance and applicability are not as central. Michael Rothberg elaborates similarly the relations between trauma and environmental destruction:

> Exploitation and ecological devastation can be traumatic—and can certainly lead indirectly to trauma of various sorts—but their essence (also) lies elsewhere. We need better ways of understanding how different forms of suffering and violence may inhabit the same social spaces and we need to understand what such overlap entails for the possibilities of resistance, healing, and social change.
>
> (xvii)

The novel provides a ground to avoid generalizations and open possibilities for new frameworks beyond political palliatives, contributing to a characterization of violence inside narrative trauma that considers spectrality as one of its constitutive elements.

The narration reveals the deterioration of Ismael's mental health. The lucidity of the retired schoolteacher gives way to memory loss, hallucinations, and, finally, the progressive fading of his identity. The traumatic effects of the conflict in Ismael Pasos are thus not limited to the mourning reactions arising from the kidnapping of his wife, and the sources of this deterioration can be traced back to the attritional atmosphere of fear that reigns in the town long before the first attack of the armies. The traces of the trauma are also evident in many other characters: in Hortensia Galindo, who for four years suffered the consequences of the abduction of her husband, Marcos Saldarriaga, or in Geraldina Almida, the exoticized neighbor of Pasos, who enters into a deep depression after the kidnapping of her son Eusebito, along with her husband. When her son returns home, he remains absent and cannot communicate for a long time (*Los ejércitos* 151). The impact of attritional violence is manifested, then, in the different types of trauma suffered by the characters, but it does not stop there; Rosero does not impose a universal standpoint on such trauma for all the embodied communities while representing the consequences of violence in them.

Through a counterpoint of spectrality and corporeity, *Los ejércitos* explores the role of time, quotidian structural violence, and attrition, while it questions how we define subjectivity insofar as it takes the concept of the Western subject to its limits, incorporating nonhuman entities. It shows how subjects act not only in the face of personal losses and disruptions but also while confronting the weight of history. Ismael well exemplifies this point, given the gradual fading of his memory, of his affective world, of his position in the social order—now turned into a spectral one—and of his control over his own body. At various points, different members of the community—among these Celmiro, Hey, the empanadas seller, and even Ismael himself—assume he is dead. For instance, while not being able to hide during the last attack of the armies, Ismael undergoes a transformation: "me finjo muerto, me hago el muerto, estoy muerto, no soy un dormido, es en realidad como si mi propio corazón no palpitara, ni siquiera cierro los ojos: los dejo perfectamente abiertos" (197) ("pretending to be dead, I pretend to be dead, I am dead, I am not asleep, it really is as though my own heart were not beating, I do not even close my eyes: I leave them wide open" 183). By the end of the novel, Ismael becomes a spectral witness of the dissolution to his own identity (Buiting 142).

The consequences of events that took place decades or even centuries earlier are shown as part of the politics of speed operating in the community. In this respect, Rosero represents social trauma by mixing the description of different precarized communities with the historical implications of violence in them. An example of that mix is the description of displaced people in the novel, before the traumatic attack on the church. The members of the community of San José used to see them in the highway, "filas interminables de hombres y niños y mujeres, muchedumbres silenciosas sin pan y sin destino" (116) ("interminable lines of men and children and women, silent crowds with neither bread nor destinations" 109). Their fate seemed remote and foreign, especially when the displaced were members of racialized communities: "Hace años, tres mil indígenas se quedaron un buen tiempo en San José, y debieron irse para no agravar la escasez de alimentos en los albergues improvisados" (117) ("years ago, three thousand indigenous people stayed for a long while in San José, but eventually had to leave due to extreme food shortages in the improvised shelters" 109). In this the moment, Ismael realizes that their time to exit the town has come, and the described scene is saturated with intricate layers of history. As Ribas-Casasayas and Petersen state, "Trauma affects not only the direct victims, but also their descendants, who find themselves haunted in an affective epistemological process of conflictive and incomplete recovery of the past" (*Espectros* 64). These instances of protracted violence end up being diffused by time, similarly to how trauma caused by attrition is normalized, naturalized, and forgotten—as the novel shows.[9] We can infer that the humanitarian crisis described here has not been caused only by the armed actors of the conflict during the recent past of the narration. Although forced displacement can also be the result of neoliberal extractivism (including large agricultural, mining, oil, and hydroelectric projects), this passage sums up violence that has been enacted for centuries. The fact that the first displaced population described in the novel is made up of members from indigenous communities and the second one is composed of the town's surviving inhabitants invites the readers to reflect on the different consequences of historical violence for the racialized populations. The trauma caused by displacement would not be the same for each group. This is how *Los ejércitos* makes visible the relations between trauma and nonimmediate violence; its causes are found not only in extremely violent events but as a consequence of attritional, forgotten, low-intensity conflicts; poverty; symbolic violence; economic exploitation; and environmental destruction throughout history.

Conclusions

Although the novel focuses on both event-centered violence by emphasizing other nonvisible aspects of war that include trauma and the absent through spectrality, as well as slow violence, *Los ejércitos* shows how different kinds of violence and their effects can frequently coexist in an entangled relationship in contexts such as the Colombian conflict, posing new challenges for their representation. As I have shown, the novel challenges the binary relations established by Nixon, displaying how questions of visibility and speed are more fluid, interconnecting the different types of violence. Overall, Rosero's novel advances a unique stance in which the understated violence of the Colombian conflict and its delayed repercussions are made visible, bringing to light the connections among biopolitics, the degradation of the environment, and the traumatic impact engendered on communities under conditions of precarity. In *Los ejércitos*, both the community and the environment suffer the consequences of the conflict as well as the nonvisible politics of other forms of objective violence.

The temporal dispersion and lack of visibility of slow violence impacts the way local and foreign observers perceive and respond to the Colombian conflict. As an antiwar novel that deals with the complexities of the Colombian conflict, *Los ejércitos* confronts not only the toxic aftermath of the war on drugs but also the different types of violence and environmental destruction. Rosero provides a complex portrayal of the politics of speed regarding environmental damages, contrasted with the presence of media in the conflict. He proposes a narrative where human and nonhuman subjects share an environment that suffers the crude effects of the conflict beyond those portrayed via sensationalist media. Through its spectral account of the conflict, the novel confronts the challenge of representing the politics of speed and the intersections of different types of violence, including the consequences of environmental destruction. Hence, *Los ejércitos* offers a strong counternarrative to neoliberal policies in the Global South and the traumatic correlates that accompany such practices. This counternarrative counterposes a spectral, erotic, and inclusive vitalism to a widespread indifference toward death and destruction in a prolonged conflict, embodying the visible violence while connecting it with ways of revealing the systemic, slow, not so visible types. The representation of slow violence works as an indirect tool to address mourning processes and as a textual device to embody both the visible violence and other forms of violence that are more or less hidden, sometimes invisible. The novel thus articulates a social narrative of the conflict that overcomes historical silences, including the consequences of ongoing ecocide, while inviting readers to reflect on new forms of environmental justice, even though they are not shown directly in the apocalyptic background of the novel.

Notes

1. First, thanks go to the editors of this volume, whose incisive comments made this chapter much better. I would like to thank also the different academic communities involved during the different phases of this paper: Professors Guillermo Irizarry, Jacqueline Loss, and Miguel Gomes at UConn; the participants of the seminar "Postcolonialism and Ecocriticism" at NeMLA 2014; and my colleagues at the Junior Faculty Symposium for "Research in the Humanities" at Loyola University Maryland.
2. See Juliana Martínez's chapter "Fog Instead of Land, Spectral Topographies of Disappearance in Colombia's Recent Literature and Film." For a wider discussion of haunted landscapes, see Downing and Heholt. Regarding the concept of specter, see Blanco del Pilar and Peeren.
3. *Los ejércitos* also provides an embodied, vitalistic worldview, which contrasts with the ecocide and social destruction brought about by war and neoliberal practices, a reading that I have supported in other publications. See Gardeazábal Bravo.
4. Since the 1940s, successive waves of conflicts have beaten up the country. After *La Violencia*, a decade-long, nondeclared civil war between liberals and conservatives, mainly caused by local land tenure issues (1948–1958), Colombian politics were directly influenced by the Cold War. As in other South American nations, several guerrilla movements appeared in the country throughout the

1960s. The Cuban-influenced National Liberation Army (ELN) was formed in 1964. In 1966, several self-defense peasant groups located in the peripheral areas of colonization, ideologically influenced by the Communist Party, transformed into the Revolutionary Armed Forces of Colombia (FARC) (cf. Gonzalez). Drug trafficking has had an important role in the conflict as the financial fuel of some of the conflict actors since the 1980s, becoming a source of political destabilization. Drug traffickers contributed to the creation of the first paramilitary groups around 1982, with the help of politicians, military personnel, and businessmen. The phenomenon began to expand across the country and became a means of acquiring large tracts of land. In 1997, these groups came together to form a single command, the United Self-Defense Forces of Colombia (AUC). Their purpose was to force the guerrillas to withdraw from strategic areas and to control geographic areas that would enable them to sell their narcotics in the global market and also buy smuggled weapons and ammunition. These groups operated with the complicity of members of the state armed forces while carrying out massacres, kidnappings, and selective killings among the communities in those areas. The government of Álvaro Uribe Vélez (2002–10) was then infiltrated by these groups, with dozens of congressmen and members of the government coalition prosecuted for their links to such paramilitary groups. This phenomenon is considered by different analysts as a process of co-optation of the state and its resources (Romero and Valencia).

5. "The term low-intensity conflict . . . is employed more or less synonymously with non-international conflict . . . particularly when these have developed into major operations with the likelihood or reality of atrocities being committed against non-combatants. . . . The term low-intensity has no relation to the severity or violence of the conflict. It is a term used to indicate that the conflict is not between recognized states nor that any major power is directly involved" (Green 493). For Harold Pinter, "Low intensity conflict means that thousands of people die but slower than if you dropped a bomb on them in one fell swoop. . . . When the populace has been subdued—or beaten to death—the same thing—and your own friends, the military and the great corporations, sit comfortably in power, you go before the camera and say that democracy has prevailed" (13).
6. The conflict produced 5,712,506 displaced persons between 1985 and 2012, according to the *Basta ya* report published on July 2013 by the Historical Memory Center.
7. The notion of structural violence was originally described by Johann Galtung as "settings within which individuals may do enormous amounts of harm to other human beings without ever intending to do so, just performing their regular duties as a job defined in the structure" (171). I think that the concept of slow violence works as a projection of this and other theories of violence in environmental issues. Nixon provides some clues here: "What I share with Galtung's line of thought is a concern with social justice, hidden agency, and certain forms of violence that are imperceptible. In these terms, for example, we can recognize that the structural violence embodied by a neoliberal order of austerity measures, structural adjustment, rampant deregulation, corporate megamergers, and a widening gulf between rich and poor is a form of covert violence in its own right that is often a catalyst for more recognizably overt violence. . . . The explicitly temporal emphasis of slow violence allows us to keep front and center the representational challenges and imaginative dilemmas *posed not just by imperceptible violence but by imperceptible change whereby violence is decoupled from its original causes by the workings of time*. . . . To talk about slow violence, then, *is to engage directly with our contemporary politics of speed*" (10–11, emphasis mine).

8. See Mutis. The rivers studied there come from José Eustasio Rivera's *La vorágine* (1924), Gabriel García Márquez's *El amor en los tiempos del cólera* (1985), Laura Restrepo's *La novia oscura* (1999), and Héctor Abad Faciolince's *Angosta* (2004).
9. Nixon refers to memory and slow violence as follows: "[T]he temporal distance between short-lived actions and long-lived consequences [has become disaggregated], as gradual casualties are spread across a protracted aftermath, during which the memory and the body count of slow violence are diffused—and defused—by time" (41).

Works Cited

Anker, Elizabeth S. *Fictions of Dignity: Embodying Human Rights in World Literature*. Cornell UP, 2012.

¡Basta ya! Colombia: memorias de guerra y dignidad. Comisión Nacional de Reparación y Reconciliación. Centro Nacional de Memoria Histórica de Bogotá, 2013.

Blanco, Maria del Pilar, and Esther Peeren, eds. *The Spectralities Reader: Ghosts and Haunting in Contemporary Cultural Theory*. Bloomsbury, 2013.

Buiting, Lotte. "An Impossible Witness of *The Armies*." *New Trends in Contemporary Latin American Narrative: Post-National Literatures and the Canon*, edited by Timothy R. Robbins and José Eduardo González. Palgrave Macmillan, 2014, pp. 133–52.

Butler, Judith. *Precarious Life: The Powers of Mourning and Violence*. Verso, 2006.

Craps, Stef. "Beyond Eurocentrism: Trauma Theory in the Global Age." *The Future of Trauma Theory: Contemporary Literary and Cultural Criticism*, edited by Gert Buelens et al. Routledge, 2014, pp. 63–80.

Downing, Niamh, and Ruth Heholt. *Haunted Landscapes: Super-Nature and the Environment*. Rowman & Littlefield, 2016.

Galtung, Johan. "Violence, Peace and Peace Research." *Journal of Peace Research*, vol. 6, no. 3, 1969, pp. 167–91.

García Moreno, Laura. "La risa y el tiempo en *Los ejércitos* de Evelio Rosero." *Revista iberoamericana*, vol. 250, 2015, pp. 141–60.

Gardeazábal Bravo, Carlos. "Derechos humanos y corporeidad en *Los ejércitos* de Evelio Rosero." *Chasqui: revista de literatura latinoamericana*, vol. 46, no. 1, 2017, pp. 139–52.

González, Fernán E. "The Colombian Conflict in Historical Perspective." *Accord: An International Review of Peace Initiatives*, vol. 14, 2004, pp. 10–15.

Green, Leslie C. "Low-Intensity Conflict and the Law." *ILSA Journal of International and Comparative Law*, vol. 3, 1996, pp. 493–521.

Heholt, Ruth. "Introduction: Unstable Landscapes: Affect, Representation and a Multiplicity of Hauntings." *Haunted Landscapes: Super-Nature and the Environment*, edited by Niamh Downing and Ruth Heholt. Rowman & Littlefield, 2016, pp. 1–20.

Maher, David. "Rooted in Violence: Civil War, International Trade and the Expansion of Palm Oil in Colombia." *New Political Economy*, vol. 20, no. 2, 2015, pp. 299–330.

Martínez, Juliana. "Fog Instead of Land, Spectral Topographies of Disappearance in Colombia's Recent Literature and Film." *Espectros, Ghostly Hauntings in Contemporary Transhispanic Narratives*, edited by Alberto Ribas-Casayas and Amanda L. Petersen, Bucknell UP, 2016, pp. 117–31.

Mol, Hanneke. "Agro-Industry Expansion Through 'Strategic Alliances': The Shifting Dynamics of Palm Oil-Related Dispossession." *Environmental Crime in Latin America: The Theft of Nature and the Poisoning of the Land*, edited by David Rodríguez Goyes et al. Palgrave Macmillan, 2017, pp. 163–86.

Mutis, Ana María. "Del río a la cloaca: la corriente de la conciencia ecológica en la literatura colombiana." *Revista de crítica literaria latinoamericana*, vol. 79, 2014, pp. 181–200.

Nixon, Rob. *Slow Violence and the Environmentalism of the Poor*. Harvard UP, 2011.

Ojeda, Diana, et al. *Caminos condenados*. Pontificia Universidad Javeriana/Cohete Cómics, 2016.

Oslender, Ulrich. "Another History of Violence: The Production of 'Geographies of Terror' in Colombia's Pacific Coast Region." *Latin American Perspectives*, vol. 35, no. 5, 2008, pp. 77–102.

Pinter, Harold. *Art, Truth & Politics: The Nobel Lecture*. Faber & Faber, 2006.

Pourgouris, Marinos. "The Phenomenology of Hoods: Some Reflections on the 2008 Violence in Greece." *Journal of Modern Greek Studies*, vol. 28, no. 2, 2010, pp. 225–45.

Quimbayo Ruiz, Germán Andrés. "Crops for Illicit Use and Ecocide." *Drug Policy Briefing*, vol. 28, 2008, pp. 1–20.

Ribas-Casasayas, Alberto, and Amanda L. Petersen. *Espectros: Ghostly Hauntings in Contemporary Transhispanic Narratives*. Bucknell UP, 2016.

———. "Introduction: Theories of the Ghost in a Transhispanic Context." *Espectros: Ghostly Hauntings in Contemporary Transhispanic Narratives*, edited by Alberto Ribas-Casasayas and Amanda L. Peterson. Bucknell UP, 2016, pp. 1–11.

Romero, Mauricio, and León Valencia. *Parapolítica: La ruta de la expansión paramilitar y los acuerdos políticos*. Intermedio, 2007.

Rosero Diago, Evelio. *The Armies*. Translated by Anne McLean, New Directions Books, 2009.

———. "I hear *The Armies* Charging Across the Land." *Calquezine Blog*, Oct. 2009, http://calquezine.blogspot.com/2009/10/i-hear-armies-charging-across-land.html. Accessed 15 July 2018.

———. *Los ejércitos*. Tusquets Editores, 2007.

Rosero Diago, Evelio, and Arturo Jiménez. "Escribo para exorcizar el dolor de la violencia: Evelio Rosero." *La Jornada*, 6 May 2007.

Rothberg, Michael. "Beyond Tancred and Clorinda: Trauma Studies for Implicated Subjects." *The Future of Trauma Theory: Contemporary Literary and Cultural Criticism*, edited by Gert Buelens et al. Routledge, 2014, pp. xi–xviii.

Sánchez León, Nelson Camilo, et al., eds. *Cuentas claras: El papel de la Comisión de la Verdad en la develación de la responsabilidad de empresas en el conflicto armado colombiano*. Dejusticia, 2018.

Summerfield, Derek. "A Critique of Seven Assumptions Behind Psychological Trauma Programmes in War-Affected Areas." *Social Science & Medicine*, vol. 48, 1999, pp. 1449–62.

Taussig, Michael T. *Palma Africana*. U of Chicago P, 2018.

Weik von Mossner, Alexa. *Affective Ecologies: Empathy, Emotion, and Environmental Narrative*. Ohio State UP, 2017.

Žižek, Slavoj. *Violence: Six Sideways Reflections*. Picador, 2008.

9 The Environmentalism of Poor Women of Color in Mayra Santos-Febres's *Nuestra Señora de la Noche*

Charlotte Rogers

In *Nuestra Señora de la Noche* (2006, Our Lady of the Night), Mayra Santos-Febres presents a fictionalized account of the life of Isabel Luberza Oppenheimer, a historical independent woman of color who ran a successful brothel in the San Antón neighborhood of Ponce, Puerto Rico, for much of the twentieth century.[1] The existing scholarship on the novel examines the intersectional fissures and oppressive social hierarchies to which the novel calls attention.[2] While I build upon those readings, this essay seeks to avoid the perils of anthropocentrism by approaching *Nuestra Señora de la Noche* with an awareness of the ways in which the human and the ecological are deeply intertwined.[3] This study accordingly features the intersectionality of race, gender, class, and environment and in doing so aims to contribute to critical ecofeminism as defined by Val Plumwood and popularized by Greta Gaard.[4]

Santos-Febres's version of Isabel Luberza Oppenheimer is spiritually and culturally rooted in the Portuguese River that runs through the impoverished Afro-Puerto Rican community of San Antón in Ponce. Luberza's affinity for the river stems from her childhood labor on its banks alongside other poor women of color and from her allegiance to the Virgin of Charity of El Cobre. This pan-Caribbean religious icon of Cuban origin is often syncretized with the santería orisha Ochún; both figures are associated with water and the protection of women. In the novel, Luberza uses her wealth and influence to preserve the riverine community from military-industrial development schemes imposed by upper-class businesspeople and landowners. Santos-Febres creates a complex protagonist who rebels against attempts by Puerto Rico's white Catholic patriarchy to harness the landscape's resources for the sake of personal profit under the guise of national modernization. In Santos-Febres's vision, Isabel Luberza Oppenheimer's childhood on the shores of the Portuguese River, her Afro-Caribbean ecofeminist spirituality, and her ownership of land on the riverbank form the bedrock of that rebellion, underlying her economic independence from white men, her protection of young, vulnerable women of color, and her environmental stewardship.

Santos-Febres's ecological sensitivity becomes especially apparent when considered in light of other fictional treatments of Luberza Oppenheimer's life by Rosario Ferré and Manuel Ramos Otero, neither of whose short stories features the environment.[5] Santos-Febres's depiction of "Isabel La Negra," the derogatory term by which she was known, stands out because it goes beyond the domestic and folkloric tropes stereotypically associated with women of color in the Caribbean.[6] Instead, Santos-Febres presents Isabel as deeply enmeshed in her nonhuman environment and consistently aligned with powerful aquatic forces: Luberza is born during a hurricane that causes the Portuguese River to flood. As a child, she is part of the vibrant female community of color that washes the clothes of the wealthy on the banks of the river. As an adult, she comes to own a parcel of land on the river's banks on which she builds a brothel, Elizabeth's Dancing Place. In a clear example of ecofeminism, Luberza ultimately becomes a protector of both poor women of color and of the Puerto Rican landscape. *Nuestra Señora* thus goes beyond earlier portrayals of Luberza that limit their pointed critique to the racial and sexual hypocrisies of Ponce and instead contextualizes these elements within the demographic, economic, and ecological changes taking place in Puerto Rico over the course of Luberza's life, from the turn of the twentieth century to the mid-1970s.

Slow Violence and Environmental History in *Nuestra Señora de la Noche*

Over the course of *Nuestra Señora de la Noche*, Santos-Febres traces the evolving relationship between the river and people of color in Ponce. The *longue durée* of the novel is therefore an ideal vehicle to counter what Rob Nixon calls contemporary society's "representational bias against slow violence" (13), understood as accretive aggressions of the powerful against vulnerable environments and peoples. That slow violence is wrought upon the river itself and the neighborhood of San Antón, a poor community settled by formerly enslaved people of color in the late nineteenth century. It is celebrated as the birthplace of the Puerto Rican music and dance forms known as the *bomba* and the *plena*. The novel fictionalizes the way in which Ponce's expansion and urbanization threaten San Antón's unique character. In particular, Operation Bootstrap, known in Spanish as *Operación manos a la obra*, incentivized industrialization at the expense of rural agriculture in the 1950s and 1960s. As a result of Operation Bootstrap, U.S. and Puerto Rican elites bought up small landowners' property at reduced prices, spurring what was known as the Great Migration to the United States and transforming the economy of the island (Funes Monzote 57). Displaced agricultural laborers flocked to the cities, and Ponce grew from a population of 87,500 people in 1930 to 160,000 in 1960 (Natal 45, 66). Upper-class white Puerto Ricans and

U.S. corporations were the primary motivators and beneficiaries of these demographic and economic changes.

In the novel, the push toward industrialization is exemplified by the activities of the powerful businessman Esteban Ferráns, often called "El Enemigo" (The Enemy). As a youth, Ferráns was a client of Luberza's brothel on the shores of the Portuguese River, while in middle age he seeks to divert the river and build massive highways around it to facilitate industrial development in the area, enriching himself in the process. To do so, he must buy Isabel's land, which she received as a gift from her former lover and the father of her only child, Fernando Fornarís. The fact that Luberza decides to give up her newborn son but to keep Fornarís's property indicates the importance of the independence the land signifies for her. For the rest of her life, she is haunted by her early refusal to care for the child, who is raised by the impoverished woman of color María Candelaria with the Fornarís's limited financial support. Although Isabel is tormented by this situation, preserving the land becomes her primary concern. The environmental and social consequences of Ferráns's scheme are implied yet never directly stated in the novel: the eviction of residents of San Antón, including Isabel and the sex workers, as well as the destruction of wildlife habitat and pollution of the river from petrochemical and metallurgic industries.[7] Environmental historian Reinaldo Funes Monzote emphasizes that Puerto Rico was particularly vulnerable to the negative effects of such industries in the era of Operation Bootstrap because "many factories benefitted from comparatively weak environmental protection regulations, leading to serious damage to ecosystems and natural resources" (57). This situation was not merely the result of governmental neglect; as Julio A. Muriente Pérez has shown, in the middle of the twentieth century, Puerto Rico served as a "third world" economic, social, and political laboratory for U.S. scientists and ideologues (77).

These threats loom over the human and riverine ecologies depicted in the novel and are ultimately fulfilled around the time of Luberza's death in 1974; the novel even suggests that her death (an unsolved murder in the historical record) may have been provoked by her refusal to sell her land. Santos-Febres chronicles how San Antón shifts from being a marginalized but vibrant place on the periphery of Ponce to an *arrabal* ("slum") by the 1970s (319; 315). While *Nuestra Señora* features many locations, from Puerto Rico to Philadelphia and Manila, it always returns to Ponce and the Portuguese River, one of several that runs toward the city's port on the south coast of the island. This aspect of the novel reveals the increasing integration of Puerto Rico into the global flows of capital and people in the mid-twentieth century, while simultaneously making visible the micro-local slow violence of modernization and industrial development in San Antón.

When faced with Ferráns's purchase offer, Luberza steadfastly refuses to sell her land. Her resistance to Ferráns and the modernizing violence he represents makes her an emblem of what Nixon calls the "environmentalism of the poor." According to Nixon, "the environmentalism of the poor is frequently catalyzed by resource imperialism inflicted on the global South to maintain the unsustainable consumer appetites of rich-country citizens and, increasingly, of the urban middle classes in the global South itself" (22). The struggle over the Portuguese River in *Nuestra Señora de la Noche* is provoked by conditions similar to those described by Nixon, with the caveat that Puerto Ricans are at once citizens of the global South *and* the "rich-country" United States: the push to channel the river, industrialize its banks, and displace the inhabitants of its shores will benefit both urban local elites and mainland U.S. corporations that produce consumer goods. In this sense, the novel reflects the historical record; as Carmelo Rosario Natal notes in his history of Ponce, debates about the canalization of the Portuguese River began in 1947 and concluded in the second decade of the twenty-first century (32). The wealthy gradually displaced the poor from the river's banks, much as is the case in *Nuestra Señora*. Indeed, Guillermo Irizarry interprets the drama of the novel as arising from the antagonistic opposition between the power of the dominant class and the force of the subaltern, represented by Luberza ("Pasión y muerte" 212). He shows how Santos-Febres features workers' movements, racial discrimination, the militarization of Puerto Rican culture, and the exploitation of women. I seek to add environmental questions to scholarly discussions of the novel: ecofeminism and ecospirituality are major parts of Luberza's resistance to the dominant forces Irizarry mentions.

Luberza's environmentalism in *Nuestra Señora* makes it possible to consider Santos-Febres as an environmental writer-activist, to use Nixon's term, because her work involves the development of "rhetorical alliances that opened up connective avenues between environmental justice and other rights discourses: women's rights, minority rights, tribal rights, property rights, the right to freedom of speech and assembly, and the right to enhanced economic self-sufficiency" (23).[8] Indeed, *Nuestra Señora de la Noche* does not just create rhetorical alliances but actually embodies those ideals in the life of Isabel Luberza, who attains economic independence, obtains and defends her property rights, advocates for women of color, and protects the environment. Luberza offers women physical protection from abusive men and funds their economic independence, providing housing, salary, and child care in exchange for their sex work. Moreover, Santos-Febres's engagement with environmental questions is similar to that of other black diasporic writers of the Americas who, as Sonia Posmentier writes, confront and counter the "environmental alienation resulting from the vexed legacy of the plantation, urbanization, and various forced and free migrations" (2).[9] I expand on Posmentier's focus on Anglophone and Francophone authors to include Hispanophone writers of color in Afro-diasporic hemispheric ecocritical studies. Santos-Febres's text departs from some of those studied by Posmentier in emphasizing the importance of property rights as a path to self-determination for poor women of color.[10] Like them, however, Santos-Febres brings Afro-diasporic spirituality and environmental sensitivity to the fore. In what follows, I analyze Isabel Luberza's environmental relation to the land and the river as seen in the circumstances of her birth, in her childhood community, in her spiritual practice, and in her environmental stewardship as an adult.

Born in a Storm: Hurricane San Ciriaco in *Nuestra Señora de la Noche*

From the moment of Isabel Luberza Oppenheimer's birth, Santos-Febres identifies her with powerful environmental forces that overwhelm human attempts to constrain them. Indeed, meteorological and ecological phenomena structure Isabel's life. While the year of the historical Luberza's birth is often thought to be 1901, Santos-Febres fixes it on August 8, 1899, the day of the infamous San Ciriaco hurricane.[11] Prior to Hurricane María in 2017, San Ciriaco produced the highest number of storm deaths—an estimated 3,000—the island had ever seen; in both hurricanes, the island's colonial status played a major role in exacerbating the effects of the storm.[12] In 1898, Puerto Rico had passed from the possession of Spain to that of the United States following the U.S. invasion of the island in July and the Treaty of Paris in December. As historian Stuart Schwartz writes, the "fragile infrastructure of transportation, health, and communication turned San Ciriaco from a natural hazard into a major human disaster" (318). San Ciriaco devastated the rural population and landscape: the destruction of coffee crops led to famine and unemployment, spurred migration, and accelerated the transition of Puerto Rico's economy toward the monocultivation of sugar cane by large U.S.-owned corporations. As with most environmental disasters, the poor were disproportionately affected (Schwartz 313–14).

In *Nuestra Señora de la Noche*, Santos-Febres makes San Ciriaco the founding trauma of Isabel Luberza's life. Isabel's caregiver Teté Casiana frequently recounts the legend of her birth: "Tú naciste el mismísimo día de la tormenta. Por eso, negrita, es que a ti hay que tenerte respeto. Cuando naciste, se desbordó el Portugués. Tumbó cosechas y casas. Hasta los americanos tuvieron que refugiarse en los zaguanes de ladrillo del pueblo. Hubo hambruna por meses" (49) ("You were born on the exact day of the storm. So, *negrita*, that means you have to be respected. When you were born, the Portuguese overflowed. It flooded houses and harvests. Even the Americans had to seek refuge in the brick carriage houses in town. There was hunger for months" 41). This origins story foreshadows the respect that Isabel will eventually command as a business owner and philanthropic donor. Casiana's narrative also reveals the political situation and the precarious economic circumstances of the community into which Luberza is born. One character calls the proximity of the invasion and the hurricane a "doble embate" (84) ("double blow" 76) that resulted in unemployment, food insecurity, and migration. Casiana's description of how the Americans found refuge in brick buildings while Puerto Ricans were sheltered by flimsy structures of mud and thatched roofs underscores how the island's colonial status produced unequal suffering in the disaster. The political events surrounding her birth illustrate the precariousness of life on the island, particularly for people of color. Most important, however, by aligning her birth with the hurricane, Santos-Febres implies that Isabel is an indomitable force of nature, like the hurricane and the flooding river, who will resist the efforts of both colonial and Puerto Rican elites to control her.

In a broader context, Isabel Luberza's birth on the day of the San Ciriaco storm reflects how the lives of people of color have historically been shaped by the intersections of meteorological, economic, and geopolitical forces. According to Posmentier, hurricanes "play a central role in black diasporic literature . . . because they retrace the motion of the transatlantic slave trade and the violence and loss of the middle passage" (3). Isabel's birth in the storm signals her position in the long continuum of the Afro-diasporic experience, in which hurricanes disproportionately affect people of color because of the archipelago's geography, its history of enslavement, and its frequently dependent economic status. The suffering the storm provokes among the women of color who raise Isabel makes this dynamic clear. Isabel's caregiver Casiana is injured by a falling *ausubo*, or balata tree, during the hurricane. While the new American occupiers are safe from harm in brick buildings, Casiana goes out into the storm to safeguard her pig, the family's only source of wealth. The balata tree falls on her hip, leaving her lame for the rest of her life. Undaunted, Casiana salvages the balata and shapes it to her purposes:

> Teté talló el ausubo hasta que le sangraron las manos. Taburetes, animalitos, cucharones. Tallaba que te tallaba hasta sacarse de encima la única costumbre de su vida, que había sido trabajar, las ganas de volverse loca al verse como se veía, la cara de su marido contando centavos para comer.
>
> (50)

> Teté later carved things from the balata until her fingers bled. Stools, little animals, cooking spoons. Carving and carving until she could shake the only habit that she had known all her life, which was to work, going crazy when thinking about their condition, the face on her husband counting out pennies so they could eat.
>
> (42)

Casiana's reaction to the storm is significant in several ways: disabled, impoverished, and traumatized by the storm, she expiates her suffering through relentless work. She creates new materials out of the very instrument of her injury. These activities of salvage, labor, and invention are a form of creative resistance to the effects of the storm in the face of government neglect.[13] A biblically resonant 40 days after San Ciriaco, the migrant worker María Oppenheimer begins paying Casiana to care for her newborn daughter Isabel. The money Oppenheimer pays Casiana becomes a form of salvation for her in the precarious aftermath of the storm.

After San Ciriaco, the ecological changes Santos-Febres depicts on the banks of the Portuguese River reflect the larger, interrelated economic and environmental trends in Puerto Rico and, to some extent, across the Caribbean. By the early twentieth century, the sugar plantation–based economy that had dominated the Hispanic Caribbean islands' ecosystems since the time of the Haitian Revolution was in crisis (Funes Monzote 47–51). In the aftermath of San Ciriaco, the crisis intensified, as laborers left destitute by the storm came to the cities seeking work. This ecological and economic disaster led to an increasing urbanization of Ponce and the creation of many working-class neighborhoods, including San Antón, populated by former cane workers and other agricultural laborers.

The Portuguese River: An Afro–Puerto Rican Ecological Experience

From the time of Isabel's birth, her life ebbs and flows around the Portuguese River: it floods on the day she is born; it becomes the site of her labor and joy as a child; it is a source of her spiritual strength and sexuality as a young woman; its sounds and breezes permeate the brothel she later builds on its banks. Luberza shows a prescient ecological awareness at a young age. In an early scene in the novel, the child Isabel and her godmother Maruca, who cares for her along with Casiana, spend all day at the river's edge washing clothes for the wealthy families of Ponce. The natural beauty of the river distracts Isabel from her washing: "Pero el río estaba tan fresquito y el sol ardía tanto. Las aguas brillaban como si estuvieran ellas también hechas de pedacitos de sol . . . [Isabel] quería de todas formas meterse al río" (43) ("But the river was so cool and the sun burned so hot. The waters sparkled as if they, too, were made from fragments of the sun . . . she wanted to go back in the river no matter what" 36). Isabel finds joy in her forbidden play in the river and on its banks. Santos-Febres presents the refreshing, shining river's invitation to leisure as the backdrop to the difficult but lyrically described toil of the Afro–Puerto Rican washerwomen:

> Por entre el pasto se podían ver los cuerpos agachados de otras lavanderas; mangas enrolladas hasta el hombro, pañuelos de Madrás en la cabeza para espantar el caliente del sol. Todas eran grandes y oscuras. Todas se eñangotaban haciendo volar las ropas por los aires,

> las aguas por los aires, donde rayos de sol se reventaban en arcos iris imprevistos y en burbujas de color.
>
> (45)

> They could see the bodies of other washerwomen from the high grasses, sleeves rolled up to the shoulders, madras handkerchiefs on their heads to ward off the hot sun. They were all big and dark. All of them were awkwardly crouched, busy at making the clothes fly in the air, the water fly in the air, where rays of light burst into unexpected rainbows.
>
> (37)

In these passages, Santos-Febres presents the female riverine labor as a source of physical and aesthetic pleasure for Isabel, countering slavery's legacy of environmental alienation for people of color. These brief moments of recreation in a childhood marked by scarcity, racism, and work make the river a source of joy and freedom for Isabel.

Moreover, Isabel marvels at the strength and vibrancy of the community of women of color who meet on the banks of the river to do their washing: "De entre otros matorrales salieron Casilda de Merceditas, Carolina de Vista Alegre, Toñín de Contancia, todas con sus líos de ropa recién lavada, olorosas a agua de río y a sol" (45) ("From other bushes emerged Casilda from Merceditas, Carolina from Vista Alegre, and Toñín from Contancia, all with their bundles of clothes freshly laundered, fragrant of river water and sunlight" 37). In featuring the banter and shared experience of the washerwomen, Santos-Febres establishes an Afro–Puerto Rican sense of place grounded in the river and the community of color that lives and works on its banks. Isabel comes to identify the river as a place of kinship, both social and familial. Indeed, it is on the edge of the Portuguese that Isabel sees her birth mother for the first and only time. María Oppenheimer is a dark-skinned washerwoman; she, too, was given up by her own mother when she was 40 days old so that her mother could continue her work as a migrant laborer. Santos-Febres thus imbues the river with a sense of maternal origins, communal continuity, and spiritual reunification for people of color separated by the harsh dictates of the labor system. As a locus of Afro–Puerto Rican community, the Portuguese forms a hydrological and social barrier between the poor neighborhood of San Antón and the wealthy white enclaves on the other side of the river.

Afro-Diasporic Ecospirituality

Luberza's relation to the environment is shaped by her Afro-diasporic ecospirituality in addition to the circumstances of her birth and her early childhood. From the earliest pages of the novel, Isabel invokes the Virgin of Charity of El Cobre, whose medal she was given by another Afro–Puerto Rican woman. Cachita, as the Virgin of Charity is nicknamed, is one of several Marian figures to whom Isabel and other women of color pray in *Nuestra Señora de la Noche*. Cachita is of central symbolic importance, however, because she is aligned with water and storms: the medal Isabel wears around her wrist is engraved with the words: "Resguardo para los viajeros. Refugio para los golpeados por la tormenta y la alta mar" (112) ("Safeguard for travelers. Shelter for those beaten by the storm and high seas" 104). This aquatic affiliation and the syncretization of the Virgin of Charity of El Cobre with santería's Regla de Ocha orisha Ochún connect Isabel to the river and to the Afro-diasporic spiritual tradition.[14] Her faith in Cachita/Ochún serves as a source of strength for her in her struggles against the white, Catholic, Puerto Rican men in their attempts to control her and the river. Moreover, Isabel shares so many qualities with Cachita/Ochún that she seems to be a living manifestation of her. In aligning Isabel with Cachita/Ochún, Santos-Febres implies that Luberza's fortitude stems from her syncretic ecospirituality.

The Virgin of Charity of El Cobre is the patron saint of Cuba, though she is worshipped in other areas of the Greater Caribbean as well.[15] Popular depictions of her apparition in 1612 show the Virgin rescuing two men and an Afro-Cuban boy from drowning in a storm, a trope Santos-Febres features in the novel by having Isabel, like the Virgin, appear in a storm. Indeed, Isabel's similarities to the Virgin of Charity and Ochún extend beyond her relationship to the waters. In Cachita's *camino* (path) as Ochún, she serves as a protector of women of color; during the Special Period in Cuba, Ochún became aligned with prostitution in popular culture (Schmidt 250).[16] Like Cachita and Ochún, Isabel takes special care of impoverished girls and women of color, offering them medical care, protection, and employment in her brothel.[17] Santos-Febres resolves the morally controversial practice of prostituting young girls by emphasizing Luberza's custodial role as madame, transforming her into an Ochún-like figure. Her wealth also aligns her with Ochún, who is known as a great manager of finances (Sánchez-Blake 197 n. 3).

Santos-Febres thus features Isabel's protection of water and women of color in the novel in ways that emphasize her affinity with Cachita/Ochún and those who live on the banks of the Portuguese River. Literary scholar Rosario Méndez Panedas interprets the parallels between Cachita, Ochún, and Isabel as Santos Febres's creation of mirrors in which Isabel sees herself as a marginal yet empowered figure in Puerto Rican society (n. page).[18] What is missing from this insightful reading is how Isabel's spiritual identification with water creates opportunities for environmental awareness and action. Her ecospirituality as a woman of the African diaspora in the Americas is a foundational component of her environmentalism of the poor. Isabel represents more than a single woman of color; in her embodiment of Cachita/Ochún, Santos-Febres proposes an alternative way of considering the Puerto Rican environment, a way in which ecology, spirituality, and community are all interdependent.

"*No se vende*": Isabel Luberza Oppenheimer's Resistance to Slow Violence

Isabel comes to own a parcel of land on the banks of the Portuguese River in 1932, and she chooses to return to the river and build a business there. Her ecological awareness—the fruit of her childhood—her relationship with her community, and her spiritual life shape her architectural vision. Before she builds the property, she considers its orientation in the landscape: "Isabel contempló la disposición del viento y cómo abanicaba sobre sus tierras. Un gran peñasco delimitaba el comienzo de las tierras húmedas donde empezaba la ribera del río. Hacia allá enfilaría las ventanas, para que el frescor de la ribera calmara los cuerpos sudorosos" (280) ("Isabel contemplated the nature of the breezes and how they blew over her lands. A great boulder marked the first wetlands on the riverbank. She would make the windows face in that direction, so that the breezes from the riverbank soothed the sweaty bodies" 273). This attunement to the earth, wind, and water pays homage to her origins; she even builds the brothel's steps out of river rocks (242). Her use of local materials and the careful inclusion of the environment in her endeavor distinguish Isabel from Ponce's elite, who seek to displace her, channel the river, and build highways across it.

The drama of the latter portion of *Nuestra Señora de la Noche* revolves around how the proposed canalization and industrialization of the Portuguese River may have provoked Isabel's assassination. In many ways, this narrative strand coincides with Ponce's history. For decades after San Ciriaco, the Portuguese River continued to flood in the aftermath of storms, often damaging the homes and roadways of the growing San Antón neighborhood. According to Ponce historian Carmelo Rosario Natal, the local government made fitful attempts, between 1947 and the early 1970s, to dredge, widen, and reroute the river and to clean up its banks (32, 66). This activity destroyed the homes of the poorest residents of San Antón who lived along the river's banks (Natal 32). Despite these efforts, the river still flooded regularly, and the government began discussing the privatization of the channel construction in the 1960s, just as depicted in *Nuestra Señora de la Noche* (66). The action of the novel ends in 1974 with Isabel's funeral, the same year in which the U.S. Army Corps of Engineers continued the channeling work and displaced a further 128 families (Natal 128).[19] This frequently unsuccessful history of human intervention into the course of the river forms the backdrop of the novel's conclusion.

In *Nuestra Señora de la Noche*, Santos-Febres sets up a contrast between Isabel's ecological awareness and commitment to the community of San Antón and her antagonist Esteban Ferráns, who views the land strictly as a commodity to be bought, sold, and developed. Ferráns purchases most of the land around the river from the Fornarís family, and only Isabel's parcel stands between Ferráns and his scheme to become wealthy by channeling the river and building highways around it. To him, Isabel's land is "justo al lado del río. No sirve para mucho, pero está ahí en el mismo medio" (145) ("right on the river. It's worthless, but it is there, right in the middle" 139). Ferráns is convinced that his own land is valuable because he has developed it: "Toda esa tierra de la cual nace cemento y piedra, carreteras para que las cosas salgan y entren y circulen" (145) ("All that land where cement and stone grow, roads so that things come and go and circle" 139). Ferráns's utilitarian and exploitative approach to the land seeks only to produce wealth and commodities from it. As Irizarry has written, Isabel symbolizes the subaltern resistance to the modernizing drive implemented by the Puerto Rican and mainland elites.[20] This resistance stems not just from her status as an independent woman of color but also from her ecological knowledge and her property rights.

Both Ferráns and Luberza use the bribery of corrupt officials and strategic donations to the Church as a means of reinforcing their claims to the land, but Isabel emerges as the more successful manipulator of the system, in part because she also barters the sex work of her employees for favors from officials (142). Luberza's power clearly threatens the entrenched race-class hierarchy in Ponce, a situation that Ferráns will not tolerate. He declares "Esta hija de puta me las paga. . . . Si La Negra no vende, que no venda. El encontrará otras maneras de hacerla ceder" (145, 146) ("This bitch will pay . . . If La Negra doesn't sell, let her not sell. He will find other ways to make her concede" 139, 140). Some time after this encounter, Isabel is shot outside her home. The murder goes unsolved amid rumors of drug trafficking and other illicit activities. Santos-Febres insinuates, however, that the culprit is Ferráns, who has had Luberza killed to eliminate her resistance to his economic scheme.

Reflections on Ecological Loss

Santos-Febres depicts the ecological trajectory of San Antón in the novel as a shift from the verdant, vibrant Afro–Puerto Rican neighborhood of Isabel's childhood to an urbanized, industrialized slum in the wake of Operation Bootstrap. Luis Arsenio Fornarís, the son of Isabel's former lover, returns from many years away from the island and is shocked by the changes in San Antón. As he walks through it, Luis Arsenio wistfully recollects the greenery of his childhood in contrast to the paved roads, electrical towers, and commercialization of the region:

> Entresacados del vecindario se alzaban talleres de todo tipo, de hojalatería y pintura, rotulación, tapicería, mecánica general. El camino

principal estaba pavimentado. Mostraba sus postes de alambrado a orilla de las tres calles que dividían el barrio en abanico hacia adentro, hacia los retazos del cañaveral baldío donde recientemente estaban alzando torres de alta tensión que extendían hacia el cielo una cablería oscura y oscilante. Al otro lado de las torres, el río se escondía por el matorral silvestre. Se veía tan ralo sin su caña cercado de cemento. Del otro lado serpenteaba la carretera militar. Se hablaba de planes para ensancharla y convertirla en una gran autopista, empresa de años. Que los Ferráns se atragantaran el negocio. Él no quería ensuciarse las manos así. Pero así era como ahora se lograban las fortunas en estas tierras. Con cemento. Luis Arsenio no podía creer cómo crecía el cemento, comiéndose lo verde que antes fuera el cerco de su infancia. Aquella cosa verde y trashumante que no lo dejaba respirar.

(320)

Here and there in the neighborhood were workshops of all kinds, for tinplating and paint, sign making, upholstering, general mechanics. The main road was paved. There were wire poles on the side of the three roads that divided the town in a fan shape toward the middle, toward the remnants of the vacant sugarcane field where they had recently erected high-tension lines that extended an oscillating and dark tangle of cables toward the sky. On the far side of the towers, the river hid behind the wild vegetation. The town looked so vacant without its sugarcane, fenced in cement. The military road snaked by on the other side. There was talk of plans to widen it and turn it into a great highway, a task that would take years. Let the Ferránses take that on. He didn't want to dirty his hands like that. But that's how fortunes were made in these lands. With cement. Luis Arsenio could not believe how the cement grew, eating up the greenery that constituted the fences of his youth. That green and migrating growth that had not let him breathe.

(315)

This remarkable passage encapsulates the interrelated histories of Puerto Rico's economy, ecology, and culture. Luis Arsenio's lament for the way in which green space cedes ground to a wave of cement touches on several key aspects of Ponce's modernization. The cane that defined the Puerto Rican landscape and economy for centuries has been replaced by commercial enterprises and bisected by a military highway—these two developments symbolize two main forces of U.S. colonialism on the island: the incursion of the U.S. army and the growth of consumerism, both of which profoundly changed the island. Santos-Febres presents these elements as a threat to the riverine ecology that had previously intimidated Luis Arsenio: the electric towers hang menacingly over the river, which is hidden among the littoral shrubs and hemmed in by the pavement. The replacement of sugar cane with cement indicates the ongoing and evolving exploitation of the island's peoples and ecologies by colonial powers, namely the shift from plantation society to site of neoliberal industrial development.[21] Luis Arsenio recognizes that corruption lies at the core of this transition and does not want to be involved in Ferráns's plans to build a highway over it all. Finally, the division of San Antón by paved roads alters its formerly rural, artistically vibrant identity and destroys its green, migrating landscape. For Santos-Febres, Isabel Luberza represented a cultural and environmental bulwark against the forces of industrial development. Her death in the novel removes the final obstacles to the eradication of green space and the transformation of the unique Afro–Puerto Rican community of San Antón. The swelling crowds that attend her funeral can be seen as a gesture of mourning not just for the philanthropic benefactor of women of color but also for an environmental steward.

The long view taken by the novel thus reveals the slow violence that colonialism and industrialization have brought to poor neighborhoods of color in Puerto Rico. It is a quintessential example of what postcolonial ecocritics George Handley and Elizabeth DeLoughrey call the "inextricability of environmental history and empire building" (20). This neo-imperial exploitation continues into the present. By the turn of the twentieth century, Puerto Rico imported 90% of its food, had 94% of its citizens living in urban areas, and had more vehicles per inhabitant than any other U.S. jurisdiction (Funes Monzote 57). This densely populated, car-based society consumes vast amounts of fossil fuels and gets its power from thermoelectric plants. On a visit to Ponce in September 2018, I observed the effects of the channeling of the Portuguese River firsthand. The portion of San Antón on the banks of the river is now cut off from the rest of Ponce, bisected by power lines and commercial strips lined with car dealerships and U.S. fast-food restaurants. The Portuguese River is a small trickle, overgrown with marsh grasses, running beneath a massive highway.

In *Nuestra Señora de la Noche*, Isabel seems to foresee this ecological loss. In the days before her death, she is haunted by dreams of what she has lost in her life: in one sequence, Isabel chases fruitlessly after her abandoned child, who has hidden himself in the bushes at the banks of the river (348). The novel ends on this somber note, implying that Isabel's steadfast protection of the environment and vulnerable Puerto Ricans has been a valiant yet fleeting stand against the overwhelming forces of U.S. military, economic, and environmental colonialism. This conclusion reveals that Mayra Santos-Febres does not take a solely anthropocentric view of twentieth-century Puerto Rico but rather shows the extent to which human and ecological relationships are imbricated. *Nuestra Señora de la Noche* is ultimately a chronicle of the struggles of a poor and marginalized community against racial, gender, economic, and environmental injustice.

Notes

1. Little is known of her early life, but by the middle of the century, the historical Luberza had become the owner of two brothels and a notorious figure in Ponce and throughout the island. Her unsolved murder in 1974 was the topic of much speculation in the press (Irizarry "Pasión y muerte" 208).
2. Most scholarship on Santos-Febres's novel underscores the ways in which Isabel Luberza Oppenheimer challenges the marginalized status of women of color in Puerto Rican society. For example, Jerome Branche argues that Santos-Febres "reassert[s] the centrality of the black woman to the imagined national community of Puerto Rico" (152). Similarly, Helene C. Weldt-Basson argues that Isabel is "used to symbolize a rebellion against both racism and the subjugation of women in Puerto Rico" (1). Rosario Méndez Panedas examines Isabel as a figure of alterity.
3. In maintaining a focus on both humans and the environment in Puerto Rico, I follow Jana Evans Braziel, who has written, particularly with respect to Caribbean literature, "Ecocriticism must not ignore the human, material consequences of environmental degradation . . .: to do so is to ignore the ways in which history and colonial power have denied large segments of humanity its subjectivity and its ego and to perpetuate the deleterious distinction between 'Man' and 'Nature'" (111).
4. While traditional ecofeminism has been slow to recognize the importance of race and other forms of intersectionality, Gaard writes that "critical ecofeminism benefits from past lessons about gender and racial essentialisms, as well as from the more contemporary critical dimensions of economic, posthumanist, and postcolonial analysis. It offers helpful critiques and augmentations to ongoing conversations within environmental justice and sustainability studies discourse" (*Critical* xxiii). Gaard's observations on the race and class divisions between ecofeminists and social environmental justice activists is particularly pertinent in Latin America ("New Directions" 648). As Mary Judith Ress has indicated in her history of ecofeminism in Latin America, the movement emerged in Central and South America from a feminist critique of liberation theology in the 1980s, largely articulated by light-skinned feminist Christian theologists (44). While Ress indicates the importance of indigenous cosmologies, she does not discuss Afro-Latin American ecofeminism or ecospirituality in depth.
5. These stories are Ferré's "Cuando las mujeres quieren a los hombres" ("When Women Love Men") and Ramos Otero's "La última plena que bailó Luberza" (The Last *Plena* That Luberza Danced). Ferré's story imagines an encounter between "Isabel La Negra" and the white Isabel Luberza, whose husband has left his property to them jointly. Ramos Otero's story depicts the last day of Luberza's life.
6. In an interview, Santos-Febres stated: "[m]e molesta mucho la visión de la mujer desde el espacio de lo doméstico o de lo folklórico. . . . Empecé a ver posibilidades de narrar la vida desde muchos puntos de vista de distintas mujeres caribeñas" ("Mayra Santos Febres" 249; "the view of women in the domestic or folkloric sphere bothers me very much . . . I started to see the possibilities of narrating life from the point of view of different Caribbean women). *Nuestra Señora* brings those possibilities to fruition.
7. Environmental historian Reinaldo Funes points out that the industrialization of Puerto Rico began with the food and textile sectors, but by the 1960s, developing industries included highly polluting metallurgy and the production of petrochemicals and machinery (57).
8. In a contemporary indication of her status as an environmental writer-activist, in 2018 Santos-Febres made the Denver-based Festival Ecológico

de las Américas (Ecological Festival of the Americas) a cohost of the literary Festival de la Palabra (Festival of the Word) she organizes every year.

9. Posmentier examines works by Zora Neale Hurston, Édouard Glissant, Claude McKay, Derek Walcott, Gwendolyn Brooks, Lloyd Lovindeer, Kamau Brathwaite, and Bessie Smith, among others.
10. To mention just two of the writers Posmentier examines, Édouard Glissant proposes an aesthetics of the earth that is based in ideas of rupture, while, according to Posmentier, Sylvia Wynter believes that "the West Indian writer must engage in care-taking rather than property-based relation to the earth in order to make an 'indigenous' claim to the land" (Posmentier 6). For Luberza, ownership of the land enables her to protect it. However, this characteristic could also be interpreted as a perpetuation of individualized ownership, in contrast to, for instance, the idyllic portrayal of the sharing of the river's resources during Luberza's childhood. I am grateful to the peer reviewers of this chapter for this perspective.
11. Studies by Irizarry and Branche state her year of birth as 1901, though Elvira Sánchez Blake places it in 1910, a possible typographical error.
12. The similarities between hurricanes San Ciriaco and María are too many to be discussed in depth here, but it is important to note that the continuing colonial status of the island has left it vulnerable to storm damage, as it was in 1898. Like the 2017 hurricane season, the devastation of San Ciriaco both revealed and reinforced the United States's intentions to maintain the island as a marginalized colony. In his study of San Ciriaco, historian Stuart Schwartz concludes that the hurricane "did not cause the political decision to place Puerto Rico in a dependent status (rather than independent), but it did create a context that made that decision easier" (334). For a history of hurricanes in the Caribbean up to 2015, see Schwartz's *Sea of Storms*.
13. While Afro–Puerto Rican resilience should not excuse the state's negligence in failing to care for its people, it nevertheless remains a testament to the ways in which impoverished Puerto Ricans survive in the aftermath of hurricanes. This creative resistance Santos-Febres envisions is an Afro-diasporic literary phenomenon: Posmentier argues that Afro-diasporic writers give voice to a "poetics of survival, repair, and regeneration" (3).
14. It is important to remember that syncretism is not a simple matching or pairing of icons from different spiritualities but rather an "ongoing process of interaction and borrowing between previously distinct religions" (Schmidt 84). In the Santería pantheon, Ochún is the goddess of fresh waters, and according to Schmidt "the apparition of the Virgin of Charity in the Bay of Nipe signified, for practitioners of Santería, the reconciliation of the two divine sisters, Ochún, the oricha of fresh waters, and Yemayá, the oricha with dominion over the sea" (Schmidt 84). Sánchez Blake similarly observes that the Virgin of Charity is frequently syncretized with Ochún and sometimes with Yemayá (197).
15. Mozella Mitchell has explored the worship of Cachita outside of Cuba, noting that in the case of Puerto Rico, Cachita is not as central as in Cuba but that "the Oshun/El Cobre link is honored along with other syncretisms of Catholicism and African beliefs and practices" (144).
16. The Virgin of Charity was often worshipped by those in desperate situations. Many *balseros* (rafters), fleeing Cuba in precarious vessels during the Special Period prayed and visited shrines to the Virgin of Charity (Schmidt 257–8).
17. Ellen Suárez Findlay describes the state regulation and social marginalization of prostitution in Ponce in the early twentieth century. These efforts consisted of an antiprostitution policing that involved forced vaginal examinations with unsterilized equipment, involuntary medical treatment with mercury, surveillance, and

incarceration for targeted women. When Isabel opens her brothel in the 1930s, she flouts the social opprobrium and state control, drawing instead on her wealth and her identification with Cachita and Ochún to protect her.

18. Méndez Panedas mentions her birth at the time of the hurricane, the proximity of the brothel to the Portuguese River, and the role of water in her powers of seduction (n. page). Elvira Sánchez-Blake notes that yellow and blue, which Isabel wears or admires throughout the novel, are identified with Ochún (201).
19. A few years later, residents complained that the river was still flooding, and even more so in the places where the very work done by the Corps of Engineers created new areas of water stagnation and flooding, especially in San Antón (Natal 130).
20. More broadly, Irizarry proposes that Santos-Febres's style and tropes, including her use of river imagery, offer a form of resistance to the violence of the modern state, especially against poor women of color ("Literatura de violencia" 120).
21. The declaration that wealth is made only with cement may be a reference to the historical Cemento Ponce company run by Luis A. Ferré, father of author Rosario Ferré; as Irizarry notes, this allusion and the similarities between the names Ferré and Ferráns point up tension between Santos-Febres's and Ferré's versions of Isabel Luberza ("Pasión y muerte" 212).

Works Cited

Branche, Jerome. "Disrobing Narcissus: Race, Difference, and Dominance (Mayra Santos-Febres's *Nuestra Señora de la Noche* Revisits the Puerto Rican National Allegory)." *Black Writing, Culture, and the State in Latin America*. Vanderbilt UP, 2015, pp. 149–70.

Braziel, Jana Evans. "'Caribbean Genesis': Language, Gardens, Worlds (Jamaica Kincaid, Derek Walcott, Édouard Glissant)." *Caribbean Literature and the Environment: Between Nature and Culture*, edited by Elizabeth M. DeLoughrey, Renée K. Gosson, and George B. Handley. U of Virginia P, 2005, pp. 110–26.

Celis, Nadia V., and Juan Pablo Rivera, eds. *Lección errante: Mayra Santos Febres y el Caribe contemporáneo*. Editorial Isla Negra, 2011.

DeLoughrey, Elizabeth, and George Handley. "Introduction: Towards an Aesthetics of the Earth." *Postcolonial Ecologies: Literatures of the Environment*, edited by Elizabeth DeLoughrey and George Handley, Oxford UP, 2011, pp. 14–41.

Ferré, Rosario. "Cuando las mujeres quieren a los hombres." *Papeles de Pandora*, Vintage Español, 2000, pp. 22–41.

Funes Monzote, Reinaldo. "The Greater Caribbean and the Transformation of Tropicality." *A Living Past: Environmental Histories of Modern Latin America*, edited by John Soluri, Claudia Leal, and José Augusto Pádua. Berghahn Books, 2018, pp. 45–66.

Gaard, Greta. *Critical Ecofeminism*. Lexington Books, 2017.

———. "New Directions in Ecofeminism: Towards a More Feminist Ecocriticism." *Association for the Study of Literature and the Environment*, vol. 17, no. 4, Autumn 2010, pp. 643–65.

Irizarry, Guillermo B. "Literatura de violencia para tiempos de paz: *Nuestra Señora de la Noche* de Mayra Santos-Febres y *The Brief, Wondrous Life of Oscar Wao de Junot Díaz*." *Chasqui: revista de literatura latinoamericana*, vol. 43, no. 2, 2014, pp. 110–22.

———. "Pasión y muerte de la Madama de San Antón: Modernidad, Tortura y Ética en *Nuestra Señora de la Noche*." *Lección errante: Mayra Santos Febres y el Caribe contemporáneo*. Editorial Isla Negra, 2011, pp. 206–25.

Méndez Panedas, Rosario. "El sujeto caribeño o la seducción de la alteridad en *Nuestra Señora de la Noche* de Mayra Santos Febres." *Espéculo: Revista de estudios literarios*. Universidad Complutense de Madrid, Number 43, Nov. 2009, webs.ucm.es/info/especulo/numero43/sucaribe.html.

Mitchell, Mozella. "Africa's Oshun in Cuba, Puerto Rico, and Trinidad and the Multifarious Image of *La Virgen de Caridad del Cobre* (The Virgin of Charity of Cobre): Symbols of Continuity and Change in Caribbean Religions." *Crucial Issues in Caribbean Religions*. Peter Lang, 2006, pp. 140–9.

Muriente Pérez, Julio A. *Ambiente y desarrollo en el Puerto Rico contemporáneo: impacto ambiental de la Operación Manos as a Obra en la región norte de Puerto Rico*. Publicaciones Gaviota, 2007.

Natal, Carmelo Rosario. *Ponce en su historia moderna: 1945–2002*. Secretaria de Cultura y Turismo del Municipio Autónomo de Ponce, 2003.

Nixon, Rob. *Slow Violence and the Environmentalism of the Poor*. Harvard UP, 2011.

Plumwood, Val. *Feminism and the Mastery of Nature*. Routledge, 1993.

Posmentier, Sonya. *Cultivation and Catastrophe: The Lyric Ecology of Modern Black Literature*. Johns Hopkins UP, 2017.

Ramos Otero, Manuel. "La última plena que bailó Luberza." *El cuento de la mujer del mar*. Ediciones Huracán, 1979, pp. 58–64.

Ress, Mary Judith. *Ecofeminism in Latin America*. Orbis Books, 2006.

Sánchez-Blake, Elvira. "De Anamú a *Nuestra Señora de la Noche*: Poética Errante en la obra de Mayra Santos Febres." *Lección errante: Mayra Santos Febres y el Caribe contemporáneo*. Edited by Nadia V. Celis and Juan Pablo Rivera, Editorial Isla Negra, 2011, pp. 187–205.

Santos-Febres, Mayra. *Nuestra Señora de la Noche*. Espasa, 2006.

———. *Our Lady of the Night*. Translated by Ernesto Mestre-Reed, Harper, 2009.

———. "Mayra Santos Febres: El lenguaje de los cuerpos caribeños, conversación con Nadia V. Celis." *Lección errante: Mayra Santos Febres y el Caribe contemporáneo*. Edited by Nadia V. Celis and Juan Pablo Rivera, Editorial Isla Negra, 2011, pp. 247–56.

Schmidt, Jalane D. *Cachita's Streets: The Virgin of Charity, Race & Revolution in Cuba*. Duke UP, 2015.

Schwartz, Stuart. "The Hurricane of San Ciriaco: Disaster, Politics, and Society in Puerto Rico, 1899–1901." *The Hispanic American Historical Review*, vol. 72, no. 3, Aug. 1992, pp. 303–34.

———. *Sea of Storms: A History of Hurricanes in the Greater Caribbean From Columbus to Katrina*. Princeton U P, 2015.

Suárez Findlay, Eileen J. *Imposing Decency: The Politics of Sexuality and Race in Puerto Rico, 1870–1920*. Duke UP, 1999.

Weldt-Basson, Helene C. "The White Male as Narrative Axis in Mayra Santos-Febres's *Nuestra señora de la noche*." *Studies in 20th & 21st Century Literature*, vol. 37, no. 1, 2013, pp. 1–18.

Part IV

Materialities, Performances, and Ecologies of Praxis

10 Slow Violence in a Digital World

Tarahumara Apocalypse and Endogenous Meaning in *Mulaka*

Lauren Woolbright

"Where There Are Trees, Where There Is War"

> When the tree war started, Jikuli felt very uncomfortable,
> he said goodbye to the mountains, the hilltops, and the green valleys through which the river flowed, he left for a place with no trees, and where he could find peace.
>
> (*Mulaka*)

According to Tarahumara lore, the first people, born of Mother Sun and Father Moon, danced to awaken the earth and bring about all other life. But over time, the fighting and hatred bred among earth's peoples and creatures grew too intense, and the gods decided that the only way to cleanse the world would be to destroy it and begin again. This is not a tale brought to the Global North either directly by the Tarahumara themselves or through a book or documentary film but is told in the opening moments of the 2018 video game *Mulaka*. In this game, players take on the role of a Tarahumara shaman, a *Sukurúame* named Mulaka, who must journey across the varied Tarahumara lands, facing terrifying and powerful beings, aiding suffering people, and gathering the support of animal demigods to find passage into *Re'le Muchúwame*, the Underworld, a quest to defeat the beast blamed for the bloody conflicts above ground. This, Mulaka hopes, will convince the gods not to destroy the world after all.

This game, which joins a growing number of "indigenous games" or "World games," is the result of a partnership between the Chihuahua-based game development studio Lienzo and the Tarahumara people, known to themselves as the *Rarámuri*, which means "the runners" in their language. They are famed for their long-distance running mostly through documentaries focused on this aspect of their identity. The outside world, however, has little idea of their lives, their beliefs, or their troubles. The preceding epigraph appears on one of the game's load screens, granting a fleeting glimpse of Tarahumara culture. Nevertheless, its context remains concealed. Jikuli's identity is the first of 26 secrets that players can uncover, and this one can be found in the starting zone, the desert of Samalayuca. When discovered—which can be done only using Mulaka's signature ability, *Sukurúame* Vision—the game reveals that:

> Jikuli, the advice giver, is one of the Rehpa Muchùwame, the ones above. He shares a sacred link with Father Moon and Mother Sun. Disguised as herbage, he guides us, he gives us true sight. After the Great War of the Trees, he left for Samalayuca. This is why if anyone wants to find him, they must make the sacred journey into the great desert: "I'm leaving, from now on I'll no longer live where there are trees, where there is war. I'll look for a place where everything is calm."
>
> (*Mulaka*)

Hidden items like these make the world richer and more alive, and whether they stumble upon them or quest for them deliberately, players will feel a sense of accomplishment when they are rewarded with this deeper information about Tarahumara folklore. But more importantly, *Jikuli*'s words insist on the importance of trees to the Tarahumara and refers to a history of relentless struggle against logging companies that wish to harvest forests on Tarahumara-protected lands. *Jikuli* may have fled to the desert, taking his advice with him, but the Tarahumara refuse to abandon their trees. Environmental activism is a constant imperative in the lives of many Tarahumara—and one that brings violence into their communities.

Again and again, the game *Mulaka* touches on but never directly addresses the real-life struggles of the Tarahumara people while seeking to enshrine the "most exciting" parts of their folklore in this playable landscape (Lienzo website). Distilling an environmentally oriented message from the game relies on players unearthing its endogenous meaning. This term was repurposed from the biological sciences and leveraged by game designer Greg Costikyan in his 2002 article, "I Have No Words and I Must Design," as a crucial aspect of what makes a game a game. The definition of endogenous meaning aims to articulate the meaning that arises from a game's design—its construction in itself—rather than from the content of the game: "a game's structure *creates its own meanings*," he writes (Costikyan 22, emphasis in original). Just as scientists learn to "read" the earth and interpret what it communicates through its rocks, trees, oceans, climate, and creatures, endogenous meaning can be used to "read" game environments for what they might "say."

This chapter will examine *Mulaka*'s representation of the Tarahumara as well as its insistence on the mechanics of violence, with particular consideration of Rob Nixon's concept of slow violence—environmental effects "that [occur] gradually and out of sight . . . an attritional violence that is typically not viewed as violence at all" (2)—which can be read in the game's central threat of global environmental collapse. *Mulaka* comes close to slow violence, paving the way for it but not quite achieving rich engagement with it. One endogenous meaning arising from *Mulaka*—one that undermines its consoling finale—is the irony of using violence to end violence. With the exceptionalism of the Tarahumara hero and his barely salvaged people as evidence, *Mulaka* presents a world profoundly in need of anthro-decentering. In addition, this project will unpack its emergent meaning(s) as a piece of playable ecofiction with attention to its ecorealities, starting with a broad discussion of the role of violence in video games.

Fighting It Out

Violence has been a fixture of digital games from their earliest origins—text adventure games such as *Zork* and *Colossal Cave Adventure*, which concentrate on exploration and puzzle solving but which always eventually come to blows. Partly, this is to be attributed to their predecessors, tabletop wargames and role-playing games like *Dungeons and Dragons*—and even *Chess* and *Go*. Arcade games intensified the tendency toward violence (and marketing toward men) in the fighting game and bullet hell shooter genres (Cassell and Jenkins). Game studies grew partly from voices pushing away from extreme violence and toward diplomacy and diversity in games, and that branch of the discipline has strengthened over the course of the last decade.[1]

One resounding call, particularly from feminist game scholars, challenges games' reliance on the mechanics of violence. Independent game developer Anna Anthropy describes most games as "men shooting men in the face." She continues: "Surely an artistic form that has as much weight in popular culture as the videogame does now have more to offer than such a narrow view of what it is to be human" (*Rise of the Videogame Zinesters* 3). Nonviolent conflict in games can offer more nuanced and ethical engagement with in-game environments and situations and make the choice to engage in violence more meaningful.[2] Avoiding violence entirely may go too far.

Of the call for alternatives to violent conflict in games, Costikyan has this to say: "The desire for 'cooperative games' is the desire for an end to strife. But there can be none. Life is the struggle for survival and growth. There is no end to strife, not this side of the grave. A game without struggle is a game that's dead" (17). "Struggle" does not have to mean violence, but conflict in storytelling is often violent, regardless of the medium. Games tend to have a worse reputation for violence than other media because they are interactive. Violent movies do not demand that their viewers become complicit in the violence; viewers are merely spectators. But games require players to take action and make choices. Whether wielding sword and shield or choosing dialogue options, the act of participation necessarily changes users' engagement with the content, eliciting feelings of guilt and responsibility, which are nonsensical in other media (Isbister 8). Feelings of guilt are at least possible; so much depends upon the game in question—and the player. Making "evil" choices in a game like *Star Wars: Knights of the Old Republic* or *Dragon Age: Origins* feels quite different from mowing down aliens in *Halo* or even crashing through pedestrians in *Grand Theft Auto*. Game developers draw on players' complicated relationship with violence, either nudging them to think carefully about their choices to deepen their investment in a game's narrative or encouraging them to rationalize away or even enjoy acting violently. Gameplay, after all, must be well received by players in order for a game to be commercially successful, whether the content is violent or not.[3]

Digging deeper, Tilo Hartmann argues that players often see violence in games as justifiable. He pools a number of studies of video game depictions of warfare and theorizes:

> Videogames are entertainment products—and they are designed in such a way as to make violence enjoyable. Violent videogames frequently embed moral disengagement cues that effectively frame the violence as justifiable. Accordingly, players tend to enjoy videogame violence and related warfare scenarios (rather than feeling guilty or empathetic distress) because they are morally disengaged while playing.

According to Hartmann, players naturally empathize with game characters but at the same time recognize that games are not real—an idea explored more fully in Jesper Juul's *Half-Real: Video Games Between Real Rules and Fictional Worlds*. One motivation of players who choose violent video games is the ability to do things they cannot in their real lives (Huntemann and Payne). In sum, game designers realize that human beings have strong empathetic inclinations, and in order to make violent stories "fun," they have to invent story-, world-, or character-based reasons for players to put aside their natural empathy and fight their way through a game. Hartmann lists eight tactics enumerated by Bandura (1991, 2002) that work to morally disengage players, some of which include killing for the greater good, attributing blame or greater moral violation to the victim, distance from the corresponding aftermath of the violence, and dehumanizing the enemy (Hartmann). All of these strategies are deployed in *Mulaka*. The literature review that follows contextualizes *Mulaka* within the spectrum of Latinx and Latin American independent (indie) games, and will show its place in the context of representation of slow violence in the field.

Violence in Independent Latin American/Latinx Games

Latin America's real-world violence appears more obliquely in its video games than in games developed in the largest markets—the United States,

Japan, Canada, and Europe. Where games of the Global North tend to use cultures of violence as an opportunity for empowering heroic fantasy, often with no regard whatsoever to the environmental implications—and are more than happy to use Latin American settings to do it (Penix-Tadsen 174)—the selection of indie titles discussed here challenges these norms, aligning themselves more with Nixon's insights regarding slow violence and its repercussions on environments and people. As he writes, "The long dyings—the staggered and staggeringly discounted casualties, both human and ecological that result from war's toxic aftermath or climate change—are underrepresented in strategic planning as well as in human memory" (2–3). Timothy Morton uses the term "hyperobjects" for problems that stretch through space and time and points out how exceptionally difficult they are to represent in media (Morton). In spite of that challenge and the fact that most games do not attempt such depth, the following examples bring slow violence into view in some way.

Colombian Vander Caballero is the head creator and creative director of the Canadian studio Minority Media, developers of the highly esteemed puzzle-platformer game *Papo & Yo*. The landscape of this game is the *favelas* and the negative environmental impacts that all-too-frequently go along with them. Players take on the role of Quico, a young boy whose abusive, alcoholic father's rage pushes him to enter a fantasy world where traversing the *favelas* becomes a puzzle, and this can be accomplished only with the help of Monster, Quico's imagination's rendering of his father. Monster is helpful until he consumes the red frogs that emerge, which cause him to turn violent, and Quico must flee. The violence of the game, while domestic in nature, together with the *favelas*-as-world emphasizes fear and powerlessness and evokes forms of violence in the lives of Latin American impoverished peoples: gang violence, police violence, domestic violence, and the institutionalized violence of the very existence of the *favelas*. In the context of Nixon's book, *Papo & Yo* depicts the human costs of everyday violence, evoking empathy in players and pointing out the interconnectedness of social and—I argue—environmental justice (the *favelas* are an affront to both).

By contrast, Bethesda's *Fallout* series could not be a more extreme example of spectacular violence in games; one of the selling points of the newest game in the franchise—*Fallout 76*—is that players can nuke each other's bases. *Fallout Shelter* is a mobile off-shoot of this franchise, which was contracted out to Behaviour Interactive of Quebec, an independent studio that includes a number of Latinx developers. The game takes a much more subdued approach to the postapocalyptic scenario, focusing on survivors in an underground vault. The game's backdrop of complete nuclear devastation is its connection to slow violence: where the main *Fallout* series involves combat against the twisted humanoids and creatures of the postapocalypse, *Fallout Shelter* allows players to micromanage the mundane concerns of survival: "Congratulations, friend. . . . You will have total authority to create the perfect underground community," the trailer boasts, an ironic accentuation of the player's lack of control over the political conditions that led to bombs dropping in the first place.

The game *Bullet Boy*, in turn, undermines its own depiction of violence through humor and art style and is described as an "action cannon blaster game" on its webpage. Rocketing through a floating game world of shops, wind turbines, and oversized birds, players try to traverse levels by firing their avatar, Bullet Boy, from one cannon to another, earning power-ups and points as they go. If Bullet Boy hits a bird, he gently bounces off, and each level ends not with an impact but with a ceaseless upward blast. The assumed inevitable destruction of the player-as-bullet dampens the game's lighthearted tone somewhat—but only if players pay attention to their embodiment as the ever disposable missile.

Rock of Ages, by Ace Team of Chile, also relies on player interpretations of its endogenous meaning(s). With its Monty Python–esque art style, the game pits players against one another; one player sets up obstacles in the form of troops, war machines, towers, and beasts of war, and after an allotted amount of time, the other player tries to smash through all the defenses to take down their opponent's gate. The game runs the risk of making a farce of violence in that its fun lies in literally squashing opponents with a giant rock rather than dealing with any of the emotional or physical consequences of actual historical warfare. Even with these critiques in mind, the game emphasizes the metaphor of the rock as a representation of war when players learn in the very first cut scene of the game that this is not just a piece of solid mineral material but rather, and no less than, Sisyphus's rock. Where *Bullet Boy* positions players at the micro level of the single bullet, here, the object of destruction is a geological force on a time scale far vaster than the fleeting conflicts of humanity. The game skips hundreds—sometimes thousands—of years between levels, so it has a much longer temporal reach than most games. Interpretation of the rock as both a force of nature and Sisyphean eternal punishment reveals this game's slow violence, even if this rhetoric is well beneath the surface.

Some games benefit from a more realistic depiction of violence. *Borders* is an arcade game art gallery installation created by three Mexican-American college students that simulates the all-too-often deadly experience of crossing the border into the United States. The game displays a skeleton for every player who perished trying to make the crossing in an effort to illustrate the human cost of militarizing the border. Skeletons litter the game's desert, piled up by the dozens. This game's violence is systemic and institutionalized, born of racism, nationalism, and xenophobia that affects the region's immigrants and refugees, thus connecting it to Nixon's argument. The protracted, seemingly endless buildup of this violence (literally in the form of digital bodies, which build their own grisly wall) makes the persistent visibility of those-who-came-before into a devastating digital memorial to all who witness the game.[4]

Building on this pattern of the significance of in-game physicality, *Bleeding Border* by Uruguayan Curse Box Studios is the darkest of the games described here and, like *Rock of Ages*, must be interpreted metaphorically to develop a connection to slow violence. The border here is not a geographical location but the body itself. The game begins with three heavily armed characters entering a building, whereupon they are immediately attacked; two men are killed outright, and the player takes control of the remaining character, a woman, who has one of her hands bitten off by a twisted, humanoid monster, but when it touches her blood, it instantly dies. Struggling to bandage the wound and hide, now armed only with this knowledge, gameplay becomes a gory balance of bandaging the player-character's arm so as not to bleed out and strategically reopening the wound to defeat enemies. The game tracks blood loss, and if the character loses too much, she will die. In short, this game is in itself an economy of blood. From a symbolic standpoint, the game's twist on the survival horror genre takes the player agency of popular game series like *Bioshock*, *Dead Space*, and *Doom* and links tonally with descriptions of the U.S. military use of depleted uranium, cluster bombs, and unexploded mines. Nixon's examination of the grim realities of a human body turning on itself as "uranium replaces calcium" (215) reflects an instance when any ordinary moment could suddenly turn fatal. The disgust players experience in having to reopen their own body to the horrors that assail it in the desperate hope of saving themselves in *Bleeding Border* is only a brief discomfort compared with the realities Nixon describes, but they represent a powerful dismantling of stereotypical video game violence nonetheless.

Mulaka sits in the middle of this spectrum of lighthearted to darkly satiric depictions of violence. It unquestioningly offers fighting as an entertaining mechanic at the same time its narrative raises questions about the repercussions of such violence and the hatred it signals, as though to say that slow violence is condemnable, but fighting may be justifiable at the individual level. Such privileging of the player-character is near-universal in games but, in light of the game's ending, proves problematic here. An analysis of the game's content and of details of its creation, especially the partnership between the Tarahumara and Lienzo, sheds further light on the game's environmental themes.

Making of *Mulaka*

Of all Latin American nations to be depicted in video games, Mexico is the most common, likely because of its proximity to the United States, the world's leader in game development. Mexican landscapes appear infrequently in games, but when they do, they are mostly sites of searches for treasure in jungle ruins, zones involved with the drug trade (Hartmann), or stereotypes of Mexican cultures. Nintendo's popular title *Super Mario*

Odyssey, for instance, includes a Sand Kingdom zone made up of a mixture of colorful, Mexican-style architecture and a Mayan/Aztec-inspired inverted pyramid in the desert. This zone is populated by sombrero-capped skeleton people and features jarabe tapatío (Mexican Hat Dance) music, eliciting mixed feelings from Mexican audiences. While some players were upset about the blatant stereotyping of Mexican culture, others were happy to see Nintendo take enough of an interest in Mexico to represent it at all, which speaks to players' expectations of the treatment of Latin American cultures and stories in games.

When Lienzo began making games, they first avoided projects related to their home nation. Edgar Serrano, director and cofounder of Lienzo, comments on the state of the games industry in Mexico, which he characterizes as "almost non-existent" ("Mining for Lore"). While there are over a hundred game development companies in Mexico, most of them do outsourced jobs for larger American, Canadian, and European game companies. He notes that of the 15 or so companies in Mexico who do original games, most of them focus on mobile and PC games. "We are actually the first studio to manage to release the same product on the three major gaming platforms," Serrano points out. He explains, "Most Latin American game developers seem afraid of showcasing their Latin roots" and seem to be trying to behave more like game companies of other markets: "They feel that if they present something that's clearly done with a Mexican flair to it, it's going to do badly" ("Mining for Lore"). In an interview with *Variety*, Lienzo lead writer Guillermo Vizcaino comments:

> We want to make games that are ours, we want to make games that have an impact—what better way than to start with something that's local, something that's dear to us and something that speaks to the Mexican folklore and also stays away from the clichés. We don't want to do piñatas, and Día de los Muertos and Cinco de Mayo and all of that, and this was the perfect opportunity to do something like that (quoted in Sanchez).

Across various interviews, Serrano outlines his reasoning for choosing the Tarahumara folklore as the subject of a game: few people, even living in the city of Chihuahua, know much about their neighbors who live in the mountains so nearby. Worse, they often have negative perceptions of Tarahumara people when they see them around the city, stating that they are "no-good vagabonds" ("Mining for Lore"). He explains that "this racism and this discrimination, was hurting the culture a lot" ("Mining for Lore"). Lienzo hopes that *Mulaka* helps to remediate the Tarahumara image locally and demonstrate their cultural worth to Mexico—and to the wider world. To this end, Lienzo collaborated closely with local Tarahumara and anthropologists, meeting to identify folktales that would lend themselves to a playable medium. Eventually, Lienzo enlisted

Tarahumara performers to do voice acting and provide traditional music for the game's soundtrack, as well as soliciting interviews for the three-part documentary video series released in advance of the game for the purposes of education as well as promotion. While the series is an effective marketing strategy, highlighting some of the social issues the Tarahumara face, it does not address any of their environmental struggles.

Like many regions in Latin America, the Sierra Madre mountain range has suffered devastating environmental losses due to deforestation and adverse effects of climate change. Logging companies blatantly disregard regulations barring them from cutting the old growth forests of the Sierras, and ever worsening drought has afflicted the region for over a decade, destroying Tarahumara crops and forcing the community to become more dependent on the outside world to survive. They cannot rely on government agencies to protect their rights. In her examination of Tarahumara indigenous rights with regard to their forest resources, Silvia Romero explains the struggles the group faces in seeking legal support protecting their forests. She writes, "It highlights illegal logging, wood theft and opposition in some ejidos to the commercial exploitation of the forest. . . . The argument of the Tarahumara is that they are culturally assumed as guardians of the forest, because their daily life depends on it" (Romero 252). Even if they do seek jobs nearby, they find it difficult to obtain employment outside of their traditional agricultural techniques and suffer systemic racism from nonindigenous Mexicans (Romero 236). Vizcaino discusses the early struggles Lienzo had making contact with the Tarahumara due to drug activity in the area:

> Back then, it was harder to get to the community because there was a lot of violence, like drug-dealing violence going on in the state and there were a lot of warnings too about going into the Sierra, the mountain-range because you heard stories about people getting kidnapped or straight-up shot. We started working with some NGOs, non-governmental organizations that did stuff for the Tarahumara. . . . [T]he first trips were done with them, in their trucks and they took us into their communities because the drug dealers don't really mess with them because they know they're doing good for their community.
>
> (quoted in Sanchez)

Danger from drug traffickers is only one of the ill effects of proximity to outsiders for the Tarahumara; environmental activism poses risks as well. On January 15, 2017, Tarahumara community leader Isidro Baldenegro Lopez was shot and killed, and his death has been associated with his stance against illegal logging. His is not the first loss suffered by the Tarahumara; in 1983, Baldenegro's own father was killed for standing up to the logging companies, and Baldenegro, then only a child, witnessed his murder. He nevertheless pursued his father's legacy, winning the Goldman Environmental Prize in 2005. One report of his death adds, "Investigative journalism group Global Witness reports that Latin America had the highest murder rate of environmental activists in 2015, compared to other regions," a staggering 122 of the 185 assassinations that year ("Award-Winning Mexico").

Mulaka does not directly deal with any of these problems because the game is set in a time before colonization. Given the developers' desire to represent an indigenous culture that is still in existence rather than one lost to history, as Serrano notes games so often do, citing the popularity of Viking and Samurai cultures ("Mining for Lore"), it seems odd to choose a distant-past version of the Tarahumara. The nonplayer characters' (NPCs') clothing, dwellings, and daily activities are pastoral and verge on an econostalgia for a simpler human existence free from modern technological encumbrance. The NPCs' lives are deeply dependent on the functionality of the natural world, as their frequent pleas for Mulaka's assistance attest. In this way, *Mulaka* represents indigenous people in a way that nonindigenous players expect to see them, which is disheartening, given that native essentialism is a kind of stereotyping that can be even more harmful than other forms of racism (Wildcat 36). This precolonial context nonetheless maintains the othering of wilderness spaces common to literature and other media by emphasizing supernatural corrupting influences there and locating the animal demigods' shrines in tucked away zones.

Nevertheless, the game demonstrates that a lore-steeped narrative drawn directly from real life is a viable way to enact indigenous ways of knowing, giving players a chance to explore values and customs in a culture not their own. Further reason for the value of a game as cultural preservation, Tarahumara themselves are losing grip on their traditions because their young people are growing up more connected with the modern world than with their heritage. They are less likely now than ever to speak the Tarahumara language or know the stories of their own culture's oral tradition, and their culture could eventually slip away completely ("Mulaka—The Game"). In addition to the folklore that constitutes its narrative, Lienzo was able to incorporate Tarahumara ideas into core game mechanics such as the three-soul health system: the Tarahumara believe that men have three souls (and women, four), so Mulaka's health appears as three glowing, white diamonds in the game's interface. Damage taken blackens each diamond in turn, and once one goes fully dark, the game animates a shimmering ghost of Mulaka being ripped from his body. Mulaka can heal in combat by performing a dance to summon his lost souls back into his body.

Mulaka is one way to raise awareness of Tarahumara culture through a medium that may resonate with a younger audience, meeting them where they are, so to speak. Serrano comments, "I also think that it's kind of naïve to expect people to get into these cultures if you don't present them in the right medium, which—I think it's video games" ("Mining for Lore"). The game's ability to appeal to an international audience is encouraging; *Mulaka* has been well received in spite of some clunkiness of gameplay, and this bodes well for the World game genre in the future.[5]

Mulaka's Toxic World

Part of the appeal of World games thus far has been their stunning visuals and how gameplay communicates meaning just as much as the narrative does. *Mulaka*'s colorful palette is visually striking, but the world is also deadly, and over and over again the game presents the world's aggression and toxicity as unnatural, altered by outside forces. The demigod Cho'Mari, appearing to Mulaka as a moss-draped elk at the end of the game's starting zone, chooses Mulaka for the quest of preventing the apocalypse after watching him set the desert lands back into balance. Cho'Mari informs Mulaka that Terégori, the Lord of Death, is befouling the world above, and he beseeches Mulaka to seek the support of the other demigods—the woodpecker, the bear, the snake, and the puma—to gain their powers, enough strength to defeat Terégori. The player encounters evidence of environmental corruption across the regions in the game from the struggles of the people to the dangerous creatures inhabiting the lands. At times it is unclear whether the monsters have always been this way or if they are a product of Terégori's corruption, but some are clearly described as such. For example, in the jungles of Hueráach, Mulaka finds purple mushrooms called Wekogi Nori; the game explains: "Unlike other enemies, Wekogi Mori's attacks are not directly physical. This mushroom will create an expanding toxic fog that inflicts damage over time. There is no clear record of these creatures existing in the past. It seems like a new species, evolving in demand to the necessities of modern day survival" (*Mulaka*). The ecological understanding in the language used here could just as easily refer to modern Chihuahua, and while on the surface it is meant to indicate Terégori's influence, it also conveys an analogy between the digital world and the real one.

While some of the low-level monsters seem to be acting in accordance with expectations for their actions, larger enemies have darker roots, a trait that underpins their aggressive behavior. Like many cultures across the globe, the Tarahumara believe that the afterlife is an inverted version of life and that the spirit world is ever present in our daily lives. Many of the monstrous creatures Mulaka encounters are the product of imbalance between the two worlds. Wa'ruara Watakari (the emperor bullfrog), for example, is described this way: "The Guardian of the mystical lake. . . . It generally stays away from people and other creatures, living an isolated life far from the other dwellers of these lands. While generally tranquil, recently, the beast has shown unnaturally hostile behavior" (*Mulaka*). Another example is Ganó, the most ancient and formidable of the Ganoko (giants made of stone, earth, or plants), who has been corrupted by the evil he sought to imprison by guarding the gate to the Underworld. Once defeated, Ganó opens the way for Mulaka, who journeys into the Underworld and faces Terégori, who appears as a wolf-shaped creature half-submerged in an oily pool that is also his fur, which sometimes parts to reveal sharpened bones and a bulbous red heart beneath.

Once Mulaka defeats Terégori, however, all is not well. Mulaka learns that the gods intend to go forward with the destruction of the world, in spite of his efforts. Mother Sun and Father Moon themselves deliver the brutal truth:

> You have assumed that Terégori, the cursed one, was spreading foulness and corruption around the world. . . . The truth is that it is not because of Terégori that people's hearts became cold and dark. It is because life became corrupted and creatures turned against each other that Terégori came to be in the first place. He blossoms from that corruption, he is indeed born from foulness and hate, not as a cause, but as a consequence. You can defeat him as many times as

Figure 10.1 Terégori Esqueltico, the Lord of Death, as he appears in a cutscene (in-game cinematic passage during which the player does not have control of the player-character).

> you want, he will just be born again, fueled by the corruption of the world. . . . At this point, a new start is the only solution. A new start where we use what we have learned to make a better world.
>
> (*Mulaka*)

Terégori is a literal embodiment of human hatred, violence, and abuse of the creatures with whom they share the world. Consideration of the game's endogenous meaning here invites us to conclude that the looming apocalypse is therefore human caused. Having reached the tipping point, the gods have no choice but to act, regardless of the admirable efforts of one man. Our interconnectedness with all life entails that we cannot remove ourselves from the consequences of our actions, and in Mulaka's case, the violent conflicts and related environmental degradation arising from human greed mirror our own.

The subtext of this moment is significant because of what Mulaka has done to get here, namely killing dozens of creatures and taking on the powers of the animal demigods. Progressing through the game requires both. Mechanics communicate values and hierarchies and, when focused around a character's prowess in combat, reinforces messages sanctioning that physical force is admirable and that killing is not only an option but also yields rewards. *Mulaka* mediates these moral lessons somewhat with a focus on exploration of the world to find the stones that unlock each subsequent zone, but the impetus to proceed is always to defeat the next enemy (always an animal or elemental being) in combat; such a practice is hardly based on an environmental ethic.

Furthermore, ending the game on this dispiriting note is a surprise. Games may sometimes wrap up their narratives leaving behind loose ends or lingering questions. Sometimes the hero dies in the process, but games do not usually have the hero fail in his quest. The consequences of failure feel more real when they cannot be circumvented—and especially when players cannot just try again to change the outcome. In this regard, *Mulaka*'s ending is profoundly environmental; it takes a standard narrative arc, which sets up players to expect success, and refuses to deliver a tidy conclusion. Players will not be able to save the world with the strength of their combat skills—not this time. The game does grant Mulaka the consolation prize of saving his people; however, while the Tarahumara may survive the apocalypse, the aftermath will be a desolate rebeginning that they will have to face alone, leaving players with a mixture of regret and hope as the credits roll. The danger here is the myth of the "new start": many films and other media[6] on this subject soften the blow of climate cataclysm by positioning viewers with the survivors, but the vast majority of us will not be so lucky—the poor certainly will not. We will not be gifted with a pristine new world from the gods. To consider this direction gives us a false sense of satisfaction when we should be feeling disturbed to the point of taking action.

The *Seeló* and Animal-Being

A further core tension of the violence in *Mulaka* is the smaller-scale, yet no less significant conflict between the Tarahumara people and the Seeló, giant mantis-men, and further tension arises from Mulaka's own similarity to them. Mulaka's ability to shape-shift seems to bring him into closer relationship with the nonhuman natural world; however, this connection does not go beyond the functional properties of these creatures: leaping (puma), flight (woodpecker), strength (bear), and swimming (river snake). There is no depth to play-as-animal in this game. The game also does not try to reconcile the killing of Seeló with Mulaka's transformational abilities, but the endogenous meaning of those abilities—communicating that animals are powerful protective forces with admirable traits that humans can learn to emulate—clashes sharply with the vicious Mantismen. Mulaka's powers are evidence of his privileged position as Sukurúame: only he is capable of winning over the demigods, always through the core mechanics of violence, always justified via one of Hartmann's eight ways to engender moral distance in players. Enemies are rendered monstrous by their descriptions in the game, which pop up each time the player encounters a new one. These text descriptions explain the enemy's nature, often providing direct links to Tarahumara mythology, and give hints at how to defeat them. Most of the killing is justified by the explanation that Terégori has driven them mad and made them dangerous when they naturally would not be. This is only a partial justification, however, for the Seeló.

The Seeló are part of Tarahumara mythology. Enrique Servín, the cultural anthropologist with whom Lienzo worked, remarks that the Seeló are likely a symbolic representation of the Tepehuanes, with whom the Tarahumara had conflicts ("Mulaka—The Mythological Creatures"). Christophe Giudicelli describes these conflicts in detail in an article, explaining that the Spanish worked to turn the two peoples against one another for their own purposes of colonization and religious conversion. The two groups do differ in their worldviews: the Tarahumara worship the sun and moon, while Tepehuanes do not. However, it seems that much of the conflict between them was instigated by their colonizers as the Tepehuanes were brutally subjugated by the Spanish while the Tarahumara were granted peace treaties and lands. This history deeply problematizes the monstrous nature of the Seeló, if they do indeed represent the Tepehuanes.

Developer Alan Márquez comments on their design, thinking for the Seeló: "Because they are the only humanoid enemies, you get to feel hand-to-hand combat more. So they are, like, civilized, but when it comes to walking and jumping, they are very animalistic" ("Mulaka—The Mythological Creatures"). The game itself frames the Seeló leader, Wa'ruara Gu'wi, as an antagonist whom it is perfectly justifiable to kill:

> After rising through the ranks of the Mantis-men forces, this Seeló has led their armies through what has been considered the golden years of the Mantis reign. Though, the passing of time has clearly taken a toll on the mantis leader's state of mind. The once mighty, wise and sound general is now sickly, unpredictable and severely bipolar. The mantis attack on the city of Paquime came out of nowhere, no provocation no real reason to invade. Even amongst the army's troops themselves, a deep sensation of confusion could be felt. Although some might question Wa'ruara Gu'wi's sanity, he is still a formidable fighter and a very dangerous foe to face. He will try to use his expertise with magic to try and vanquish his enemy.
>
> (*Mulaka*)

While the description expresses respect for the general, it sets out a clear rationale for the unavoidable violence that must ensue between him and Mulaka, which the player needs to hear in order to accept this course of action from a spiritual leader like the Sukurúame. After Mulaka defeats the Mantis-men, E'láwi, the woodpecker demigod, tells him that the Seeló have acted in this irregular way because "they know the world's destruction is looming near and they are rampant, blinded by rage" (*Mulaka*). This layers on further justification for murdering them while at the same time inviting pity for them. A greater spiritual quest for Mulaka may have been an attempt to free them from this rage as he later does with Ganó, but the option does not exist.

Perhaps the most unsettling aspect of the perceived necessity of destroying the Mantis-men is their dehumanized representation in the game. Lienzo was being true to Tarahumara lore when they chose to include

Figure 10.2 Wa'ruara Gu'wi, the Seeló boss in the town of Paquime.

the Mantis-folk as they did, but by the game's end, the revelation that the world will be destroyed in spite of Mulaka's efforts calls into question the human–animal relationship that had seemed to be well established by his acquisition of shape-shifting powers from the demigods as well as the gods' perspectives on violence. After the Rehpa Muchúwame explain the true nature of Terégori, they say: "It is because of you that we now see the value of the Tarahumara people. And that is why we have decided to allow your people to transcend into the new world. A world where they shall thrive, for as long as they can follow your footsteps" (*Mulaka*). This Noah's Ark scenario means that everyone the player helped along their journey—the only everyday people the game depicts, all unambiguously good, victims of the Seeló and Terégori—will be saved, but the Mantis-people, the creatures, monsters, and beautiful landscapes will be destroyed utterly. With that, any remnant of the hatred between the peoples (homogeneity is the only path to peace?) and any environmental devastation they might wreak via warfare will be erased, gone from existence as well as from Mulaka's conscience—but perhaps not from the player's—and therein lies *Mulaka*'s greatest potential as a text engaged with slow violence: its potential impact on players.

Conclusions

Mulaka is part of a larger and ongoing conversation among originary nations about the world's social and environmental crises. In *Red Alert! Saving the Planet with Indigenous Knowledge*, Daniel Wildcat of the Euchee people—part of the Muskogee nation in the American Midwest—cautions us to consider that "[t]he peoples on the planet who have had the least to do with the climate crisis and yet are the most vulnerable to its destructive impacts should rightly lead these discussions" (60). Wildcat goes on, "Because indigenous people have paid attention to our Mother Earth, it is important to listen to what we can share with humankind. These knowledges are bound in unique lifeways—customs, habits, behaviors, material and symbolic features of culture emergent from the land and sea" (Wildcat 17), which, I would argue, can be especially effective when expressed through narrative in interactive media like games. Games "show, don't tell," as it were, and putting the responsibility for meaningful choices in players' hands is a powerful move, as proven by neuropsychologists, who in one study have found stronger similarities between playing games and doing physical activities than between consumption of games and watching movies (Isbister 3). *Mulaka* could be a crucial piece in the rhetoric of indigeneity to communicate the dire situation we are in.

With endogenous meaning foregrounded, successful games like Mulaka could pave the way for more explicit interactive media engagements with elements of slow violence that could help illustrate Nixon's connections between postcolonial and environmental themes as well. Consumers in the

Global North seem infatuated with the idea of human capacity to overcome the worst imaginable scenario, which is likely linked to our need for hope in dire circumstances, to see ourselves as victors over the worst nature has to offer (even if its worst comes from our own actions), including full-scale global catastrophes like climate change. Even our postapocalyptic media tend to ignore the lingering devastations of slow violence on the surviving population. But why not, for instance, use mechanics that richly simulate the effects of nuclear radiation and dirty bombs, as some survival games have done?[7] Why not place players in environments ravaged by "[climate] change, the thawing cryosphere, toxic drift, biomagnification, deforestation, the radioactive aftermath of wars, acidifying oceans" (Nixon 2)? Why not tell stories of characters who must become ecological detectives,[8] determining the cause of disease and death in their postapocalyptic communities and enacting solutions to improve life for all? And why not engage with stories from outside the Global North to illustrate these ideas, rather than sticking to the American canon of games? Nixon indicates that scholars benefit from being more inclusive of marginalized voices of the Global South (235); he writes, "There is much to be said for a bioregional approach. . . . However . . . [all] too frequently, we are left with an environmental vision that remains inside a spiritualized and naturalized national frame" (238). It would behoove us, therefore, to think globally in our games as well, as addressing slow violence in any form that requires us to do.

Mulaka's bittersweet ending is powerful rhetoric in our cultural imagination, but it fails to fully embrace the risky, unsatisfying tactic of denying audiences the "reset button;" the world may end, but the player-character and his people live on. Still, the game opens important questions of how game structures can be leveraged to create meaning for players. It is all too easy to write off the depth here as mere entertainment, but the endogenous environmental narrative carries weight—if players are open to it. According to Tarahumara lore, the first people danced to awaken the earth and bring about all other life. And we may be able to alter the course of slow violence, if we pay close attention to the endogenous meaning arising from games like this—and from the earth itself.

> Mulaka's efforts have proven that humanity has courage. But sometimes it is wiser to simply start again.
>
> (*Mulaka*)

Notes

1. Controversy over depiction of gender in games dates from the 1980s. See Cassell and Jenkins (*From Barbie to Mortal Kombat*, 1998), for example, and the post-#GamerGate explosion of books on this topic, one of which, by Kafai, Richard, and Tynes, directly updates Cassell and Jenkins's text (*Diversifying Barbie and Mortal Kombat*, 2016). Coverage of this subject in scholarship and journalistic media is vast.

2. See, for example, Anthropy's games *Dys4ia*, *Triad*, and *Queers in Love at the End of the World*.
3. The controversial game *Postal*, for example, positions the player as a disgruntled postal service employee who goes on a shooting spree. The game was poorly received, not because of its level of violence (although it is often cited for this) but because of its average gameplay.
4. Greg Ulmer's book *Electronic Monuments* explores the concept of public and personal memorialization in digital media including concepts of peripherals—attaching digital artifacts to existing monuments, for example, with the use of augmented reality technology—and testimonials—websites that explain the monument and all of its various elements. Since the book's release in 2005, numerous examples have arisen such as memorial Facebook pages and memorialization for fallen soldiers in Iraq (www.ndsu.edu/pubweb/~kbrooks/memorials.html).
5. The first World Game was arguably the widely acclaimed *Never Alone*, made by Upper One Games in partnership with the Inupiaq people of Alaska in 2014.
6. Examples of this trope include the climate-themed apocalypse films *2012* and *The Day After Tomorrow*. It is also common—though much darker—in zombie media. For zombie media, such rebuilding may take hold briefly, but it nearly always comes apart.
7. For example, *Metro 2066* and, to a lesser extent, games in the *Fallout* series.
8. Sara L. Crosby's term for several of Edgar Allen Poe's protagonists and a healthier engagement with ecohorror than the ecophobia common in American media (520).

Works Cited

Alvarez, Gonzalo. *Borders*, 2018, https://gonzzink.com/portfolio/borders/.

Anthropy, Anna. *Rise of the Videogame Zinesters*. Seven Stories P, 2012.

———. Games and Zines by Anna Anthropy, https://witch.io.

"Award-Winning Mexico Indigenous Environmental Activist Murdered." *teleSUR*, 17 Jan. 2017, www.telesurenglish.net.

Barasch, Alex. "'Mulaka:' Struggle of Exploring Tarahumara Lore in a Video Game." *Variety*, 10 Apr. 2018.

Bleeding Border. Curse Box Studios, 2015, www.curseboxstudios.com.

Borderlands 2. 2K Games, 2012, https://borderlands.com/en-US/.

Bullet Boy. Pomelo Games, 2015, www.pomelogames.com.

Cassell, Justine, and Henry Jenkins. *From Barbie to Mortal Kombat: Gender and Computer Games*. MIT Press, 1998.

Costikyan, Greg. "I Have No Words and I Must Design." *Proceedings of Computer Games and Digital Cultures Conference*, edited by Frans Mäyrä. Tampere UP, 2002.

Cronon, William. *Uncommon Ground: Rethinking the Human Place in Nature*. Norton, 1996.

Crosby, Sara L. "Beyond Ecophilia: Edgar Allan Poe and the American Tradition of Ecohorror." *Interdisciplinary Studies in Literature and Environment*, vol. 21, no. 3, Summer 2014, pp. 513–25.

Fallout Shelter. Behaviour Interactive, 2015, www.bhvr.com.

Giudicelli, Christophe. "Un cierre de fronteras . . . taxonómico. tepehuanes y tarahumara después de la guerra de los tepehuanes (1616–1631)." *Mundos Nuevos*, 18 Mar. 2008, https://journals.openedition.org/nuevomundo/25913.

Hartmann, Tilo. "The 'Moral Disengagement in Videogames' Model." *Game Studies*, vol. 17, no. 2, Dec. 2017, http://gamestudies.org/1702/articles/hartmann.
Huntemann, Nina, and Matthew Thomas Payne. *Joystick Soldiers: The Politics of Play in Military Video Games*. Routledge, 2010.
Isbister, Katherine. *How Games Move Us: Emotion by Design*. MIT Press, 2017.
Juul, Jesper. *Half-Real: Video Games Between Real Rules and Fictional Worlds*. MIT Press, 2011.
Kafai, Yasmin, Gabriela T. Richard, and Brendesha M. Tynes. *Diversifying Barbie & Mortal Kombat: Intersectional Perspectives and Inclusive Designs in Gaming*. ETC Press, 2016.
Lienzo Mx. "Mulaka: The Game." *YouTube*, 19 Dec. 2017, www.youtube.com/watch?v=-omk0r3rzs0.
———. "Mulaka: The Mythological Creatures." *YouTube*, 5 Dec. 2017, www.youtube.com/watch?v=mSLfNMG-Pog.
———. "Mulaka: The Tarahumara Culture." *YouTube*, 21 Nov. 2017, www.youtube.com/watch?v=986XvM3IQNg.
"Mining for Lore: A Chat with Edgar Serrano on *Mulaka* and the Tarahumara." *Not Your Mama's Gamer*. Hosts Samantha Blackmon and Alisha Karabinus. Episode 170, 30 Mar. 2018, www.nymgamer.com/?page_id=53.
Morton, Timothy. *Hyperobjects: Philosophy and Ecology after the End of the World*. U of Minnesota P, 2013.
Mulaka. Lienzo Studios, 2018, www.lienzo.mx/mulaka/.
Never Alone. Upper One Games, 2014, http://neveralonegame.com.
Nixon, Rob. *Slow Violence and the Environmentalism of the Poor*. Harvard UP, 2011.
Papo & Yo. Minority Media, 2013, https://weareminority.com.
Penix-Tadsen, Phillip. "Latin American Ludology: Why We Should Take Video Games Seriously (and When We Shouldn't)." *Latin American Research Review*, vol. 48, no. 1, Spring 2013.
Rock of Ages. Ace Team, 2011, www.atlus.com/rockofages/.
Romero, Silvia Jacquelina Ramírez. "Derechos indígenas y redes de mediación políltica en la Tarahumara: Los actores sociales y su interrelación en el conflicto territorial de Pino Gordo por la tierra y el bosque." Doctoral thesis, 2007, http://repositorio.flacsoandes.edu.ec/bitstream/10469/1330/1/TFLACSO-02-2007JRR.pdf.
Sanchez, Antonio. "Tell the Story, Play the Myth: How Video Games Translate Folk Tales." *Variety*, 29 Aug. 2018.
Sicart, Miguel. *Beyond Choices: The Design of Ethical Gameplay*. MIT Press, 2013.
Star Wars: Knights of the Old Republic. BioWare, 2003, www.starwars.com/gamesapps/knights-of-the-old-republic.
Super Mario Odyssey. Nintendo, 2017, https://supermario.nintendo.com.
Ulmer, Gregory. *Electronic Monuments*. U of Minnesota P, 2005.
Wildcat, Daniel R. *Red Alert! Saving the Planet with Indigenous Knowledge*. Fulcrum Publishing, 2009.

11 Slow Violence in the Scientific Ecosystem

Decolonial Ecocriticism on Science in the Global South

Thaiane Oliveira

According to Rob Nixon, slow violence is "a violence that occurs gradually and out of sight, a violence of delayed destruction that is dispersed across time and space, an attritional violence that is typically not viewed as violence at all" (2). In his book, Nixon then proceeds to elucidate how political and literary forms of resistance to slow violence give voice to the environmentalism of the poor, whereby communities—mainly from the so-called Global South—mobilize to demand environmental justice and contain the destruction of their environment. These socioenvironmental mobilizations are, for the most part, invisibilized by the media, but as Nixon notes, they constitute an attempt by those in the Global South to withstand the environmental violence that, in the name of progress, both puts the immediate survival of people in the ecosystem at risk and also allows the North to exert control over the South (4). Such a perspective is similar to what Joan Martinez-Alier refers to as "ecologism of the impoverished" (13), in which the wealth of developed nations entails the exploitation of the natural resources of poor or developing countries.

The intent to dominate the Global South is not limited to the exploitation of its natural resources. Veiled racism in the epistemic construction of the central North and the peripheral South reifies modes of thinking that naturalize the dominance of the former over the latter and also occludes the scientific circulation of knowledge as a strategy to keep the region in a peripheral condition. According to Walter Mignolo, it is widely believed that "the first world has knowledge, the third world has culture; Native Americans have wisdom, Anglo Americans have science" ("Epistemic Disobedience" 2). Calling on epistemic disobedience to offer transformations (revolutions) rather than relying on reforms promulgated by civil disobedience, this reflection about the discursive domination that divides the center and the periphery relates to Aníbal Quijano in his project to replace distorted paradigms of knowledge. Quijano proposes to disarm the coloniality of thought, that mode of thinking and understanding of the world according to structures of rationality brought by the foundational principles of modernity's project (548). Similarly, Antonio García Gutiérrez proposes a declassification of the world that challenges coloniality.

For García Gutiérrez, declassification is an open system that installs logical pluralism at the core of processes of enunciation through a variety of metacognitive tools problematizing the dichotomies that structure inherited epistemologies.

This chapter addresses the scientific systems of the countries of the Global South from a decolonial ecocritical standpoint. By understanding the scientific dynamics and their power relations as a form of ecosystem, I will bring together key ecological concepts such as resistance, resilience, and networks in order to understand the transformation of the scientific ecosystem in the Global South. Focusing on Brazilian scientific production concerning climate change, as well as its relationships with neoliberalism, research agendas, and slow violence within the circuits of science, I propose epistemological alternatives for understanding the role of the research institutions of the Global South from an ecocritical perspective, starting with a declassification of the world (García Gutiérrez 10).[1] Slow violence in the context of my study has two main forms: the first is inflicted on science in general through campaigns that promote "fake science" and contribute to fostering a "culture of doubt" in the epistemic and political spheres; the second is situated within the field of science and is exerted through modes of thinking and frameworks of collaboration that foster an academic imperialism that oftentimes omits and disqualifies research undertaken in the Global South.

The declassification exercise, as a form of epistemic disobedience, consists of identifying hierarchical, classificatory orderings and their dependency on neoliberal logics, while also embracing the value of the plurality of knowledges within the scientific ecosystem. The analogy between forms of exchange in biological systems and the scientific system, understood as an ecosystem itself, is chosen here to express mutual interdependencies and synergies in the creation of knowledge, given that the production of new knowledge is nourished by the prior existence of relevant sets of knowledge and their recombination (Mazloumian 1). The analogy of the scientific ecosystem helps illuminate changes in incentives and scientific requirements, as well as relations of inclusion and exclusion. Scientific systems function similarly to biological ecosystems, which adapt to selective pressures through networks and plural interactional exchanges (Smaldino and McElreath 13). An examination through the analogy of ecosystems helps elucidate the involvement of actors within a very comprehensive network of interdependencies.

Among the factors that influence the complex ecosystem of science—and fake science—are the stakeholders identified by Nixon as agents in the process of manufacturing and sustaining a culture of doubt surrounding the science of slow violence (40). Lobbyists, political advisors, media plutocrats, right-wing think tanks, fake citizen groups on Facebook, academic reviewers of climate science written by scientists who are not familiar with the reality in the region, and pseudoscientific websites all contribute to the creation of this culture of doubt that postpones the making of policies aimed at helping to control the long-term effects of climate change. The many stakeholders promoting a culture of doubt constitute a significant element within scientific ecosystems based on networks of economic, political, and social groups that in turn directly affect scientific and technological production. Such a culture of doubt has become increasingly evident in the policies proposed by leading government representatives. A case in point is Brazil, where current President Jair Bolsonaro considers the climate agenda a left-wing political plot and has moved to cut 95% of the budget for combatting climate change. Brazil was scheduled to host the Climate Conference in 2019; however, budget cuts and lack of government interest in the conference agenda has led to the country's withdrawal from organizing the event. In the first 100 days of his government, Bolsonaro relaxed environmental crime laws, eased pesticide regulations, and pronounced himself against the redemarcation of indigenous lands.[2]

A more capacious understanding of science, therefore, is one recognizing that the scientific ecosystem not only consists of scientists, research and educational institutions, and development agencies but also encompasses participants who do not visibly belong to the circuits of science production yet directly affect research agendas. Additionally, conceiving of science as itself forming a scientific environment can shed light on the slow, dispersed, inconspicuous violence effected by the diversity of participants within scientific systems in the Global South. It is helpful, then, to integrate an ecological element into the epistemology and circulation of scientific knowledge in order to consolidate those types of concepts described by Nixon as connecting different academic disciplines (31). Within such a context, the production of knowledge based on a "scientific truth" is never a disinterested search for knowledge; rather, it is a process that demands questioning of the models of visibility imposed on the circulation of scientific knowledge in contemporary times. Scientific truth emerges as a discursive exercise that, while under the guise of a rational social construction based on rigorous methods and instruments, rests on predetermined notions of objectivity for its validation. These preestablished norms hold the power to exclude certain stakeholders in order to maintain a monopoly of control over what does and does not pass as objective scientific knowledge within these larger ecologies of knowledge (Santos 35). In order to remove the harmful influence of these hierarchies on knowledge, culture, and human existence, we must accordingly seek to understand the cognitive patterns and material structures that sustain them. Inherited notions that label regions according to stages of "worldism" and allow wealthier countries in the Global North to chart the direction of research, funding, and the circulation of scientific knowledge serve as examples of factors that have long fostered a chasm between hemispheric binaries.

The Agendas of Classification and Declassification of the Global South

The hierarchical ordering of the scientific ecosystem rests on the classification of the world according to particular, inherited notions of order, which remain unnoticed in fora dedicated to the promotion of new knowledge. One of the main traits for classification of the region appeals to ordinal terms such as *Third World*, a notion fashioned by analogy to that of the Third State, which was used in the French prerevolutionary system to describe those who did not belong to either the clergy (the First State) or nobility (the Second State). Such subjects, including artisans, peasants, and the bourgeoisie, supported and were subservient to society and the state. Thus, the definition of the Third World, coined by French demographer Alfred Sauvy in 1952, grouped nations that had belonged to what European powers had conceived as colonial (or, later on, as neocolonized) peripheries. The shared economic model thus involved an extractive system based on abundant natural resources produced for export. However, after the fall of the Berlin Wall, the collapse of the Soviet Union, and consequently the sudden crisis of socialism in Eastern Europe and the economic opening of China, capitalism became the predominant global economic system. Since that time, the division between First, Second, and Third Worlds has "no longer had theoretical or operational consistency, as Second World countries (socialists) were becoming Market Democracies" (Visentini 7). As an alternative to a post–Cold War label of *Third World*, scholars in world-system theory then coined the notion of the Global South, thereby implementing a strategy aimed at depoliticizing the meaning of Third-Worldism. A second conceptualization conceives the term to encompass a deterritorialized geography of what is considered an externality according to capitalism. This framework accounts for subjugated peoples in wealthier countries and acknowledges that there are Souths in the geographic North as well as Norths in the geographic South (Mahler). This deterritorialized connotation of the term, in turn, allows for a third meaning that includes the imaginaries of resistance of transnational political subjects under extant global capitalism. A division of the world between South and North, center and periphery, however, prevents us from thinking in terms of relations, instead of oppositions within the classification process and further reinforces the abyssal division between colonizers and colonized acknowledged by Santos ("Beyond Abyssal Thinking: From Global Lines to Ecologies of Knowledges").

Publishing networks dedicated to the dissemination of scientific research are an example of such naturalized and unnoticed epistemic colonization of the South. These fora restrict and limit the space dedicated to scholarly production on and by scientists from the Global South in journals of wide circulation. These publications reproduce models established by institutions of higher education in countries considered to be at the center of these ecosystems (Larivière et al.) and uphold their standards as the sole measure of validity to evaluate what constitutes science. Scientific committees lack representation of evaluators from peripheral and semiperipheral countries (Dhanani and Jones), and scholars from wealthier centers assume positions of authority, imposing methodological models that foster dependencies on infrastructures and systems of measurement in order to evaluate science (Haustein). In this scenario, publishing companies, guided by these one-sided standards, dominate the market of scientific knowledge and dictate limited, exclusive parameters for impact assessment, quality, and scientific legitimation based on the geopolitics of knowledge (Mignolo, *Histórias*). Often, external support and pressures from industry interests influence new paths for research and determine the agendas of inquiry. In this neoliberal-governed scenario, circuits of validation create a chasm between ostensibly valid and invalid knowledge, occluding power disputes. That is, a scheme is created that delineates a binary of "quality science" versus peripheral science (or even pseudo science), in which what is not measurable by the models of scientific evaluation remains virtually invisible.

This difficulty reflects the dispute in which some scientific fields predominate over others according to their scientific authority—in particular, the natural sciences over the social sciences, with the latter understood as a field of interpretation. In addition to these disputes among fields, there is even a whole rationale for the researcher's "locus of speech," established by a system of colonial prestige based on multiple hierarchies: an ordering of agents or institutions in which Anglo-American universities are ranked according to criteria that respond to the realities of these institutions; a hierarchy of problems, domains, or methods, with research agendas prioritized by global realities to the detriment of local interests; and a linguistic hierarchy in which English is upheld as the "universal language." Also, innumerable forms of epistemic violence conceal colonial logic through metaphorical masks that silence and make unfeasible other forms of knowledge production from countries of the Global South. Grada Kilomba notes that "we are not dealing with a peaceful coexistence of words, but with a violent hierarchy that defines those who can speak" (301). Similarly, there is the issue of what can be said and where it can be expressed. An inquiry into the global scientific ecosystem can expose such epistemic violence imposed by the scientific agenda of the North onto the South, a violence that takes shape through negotiations and impositions on the visibility of scientific accomplishments and via reinforcing biases of backwardness, eccentricity, and precariousness.

Boaventura de Sousa Santos reminds us that these distinctions based on regimes of visibilities and invisibilities reify "abyssal cartographic lines" once used to demarcate a divide between the Old and New Worlds in colonial times and continuing to exist in the structures that inform the exclusionary logics of modern Western thinking. In the scientific ecosystem, this abyss "consists of granting modern science the monopoly of the universal distinction between what is true and what is false to the detriment of two alternative bodies knowledge" (Santos 47). According to this Portuguese scholar, it is necessary to democratize and decolonize knowledge, while acknowledging the importance of multiple epistemologies as well. To this end, it is necessary to admit that knowledge is a tool that must serve beyond the traditional spaces of knowledge production, starting from the notions of open access, sharing, and public commitment—all key elements for science within Latin America (Vessuri et al.).

Understanding knowledge production from a decolonial perspective entails thinking from the languages and categories of thought not included in the foundations of Western thought (Mignolo, "Epistemic Disobedience" 18). As Santiago Castro-Gómez reminds us, decolonizing the Latin American university means introducing decolonial thinking through the incorporation of transdisciplinarity and complex thinking that allows a cognitive exchange between Western science and other forms of knowledge production (281). The decolonization of higher education, therefore, is not a "reversion of the colonial moment by the postcolonial" (Colaço and Damázio 8) but rather a position of continuous struggle for a more open, plural, and participatory university. The decolonization of science entails the unveiling of the logic of coloniality as well as the disassembly of processes that contribute to the reproduction of the colonial matrix of power. Decolonial ecocriticism contributes to exposing oppressive logics in order to destabilize normative, Western-oriented ideas about environmental awareness.

Negative connotations and the reinforcement of cognitive patterns on the basis of pernicious depictions permeate the tone and nature of the relationship between nations (Visentini) through the codification of notions of difference between developed and central countries and the peripheral and semiperipheral ones (52). A quick survey of research agendas involving the Global South yields an abundance of terms like *precariousness*, *disability*, *malnutrition*, *marginalization*, and *poverty*, alongside words such as *social justice*, *human rights*, and *empowerment*, in the context of a discourse against *dependency*. These terms are commonly associated with projects that ostensibly promote transformative social change financed by European and U.S. institutions, and their emphasis calls attention to the negative framing around the reality of the region.

The promotion of this agenda of precariousness by central countries is one of the outcomes of this form of epistemic domination as it informs the academic imperialism (Alatas 29) that in turn fuels arguments justifying economic imperialism. According to Achille Mbembe, "knowledge has become a commodity. Therefore, these are new ways in which markets, states and higher education interrelate, with new implications for obscuring the boundaries between these spheres" (40). It is thus necessary to understand the sphere of knowledge production as historical and situational concepts constitutive of scientific knowledge itself, a knowledge based on the subalternization of social groups and the displacement of alternative epistemologies. On the other hand, a classification based on a pluridiversity that fosters multiple interpretations and avoids reducing reality-imposed models of validity and simplistic dualisms would address Santos's call to find paths for the region to escape the binary geopolitical thinking of North/South (38).

Currently, academic imperialism is manifested in the preeminence of epistemic neocolonialism, where pseudo universalist arguments inform explanations that are imposed as global answers to all inquiries. This one-size-fits-all model is reproduced in theories, categories, and methodologies developed exclusively to address the needs of Western societies. But these structures of thought reject and ignore the specificity and complexities of their own countries. Bypassing acknowledgment of this blind spot, they position themselves discursively as an ersatz for modernity, and they uphold progress and wealth as univocal, essential characteristics of the central countries in a world system structured upon the basis of rigid, false binary formulas. The South meanwhile is reduced to totalizing notions of precarity and poverty.

In the words of Ngugi wa Thiong'o, "we must coldly and consciously look at what imperialism has done for us and our view of ourselves in the universe" (88) and propose other epistemological approaches to the spaces of knowledge production. Accordingly, the decolonial project seeks to reinvent such spaces as ones used for the production of differences by subaltern subjects. According to Bernadino-Costa and Grosfoguel, "re-reading authors who have been silenced by the academy does not only mean witnessing the effects of colonial domination; it means having to register multiple voices, actions and dreams that fight against marginalization, discrimination and inequality, while seeking social transformation" (21). Nevertheless, even this can be an ineffectual strategy when one considers the whole set of dominating elements in the scientific ecosystem as they function in an integrated and interdependent way (Oliveira 191). These strategies also echo the arguments that often justify the financing of studies about the Global South. These contexts reveal more spaces for inquiry about the elaboration of epistemologies of difference and their place in the study of the "South," and the sole emphasis on differences results in a counterproductive exercise that, in the end, reinforces the abyssal division in the classification of the world. How, then, might we think of the division of the world by an alternative theory that allows us to think of the differences and also unveil the discursive strategies of domination?

Decolonizing Ecology and Ecologies of Scientific Practice: Resilience, Resistance, and Networks of Knowledge Classification

One step for both moving beyond thinking in dichotomies and also better understanding the interdependence of stakeholders in supporting and advancing particular notions of global knowledge entails, as proposed by García Gutiérrez, considering the way in which metaphors inform how we think about agendas and initiatives that emerge in the production of knowledge. Metaphors are more than poetic ornaments; they are strategic cognitive devices that contribute to the interpretation of new knowledge (Scolari) and serve to create new scientific paradigms (Kuhn). Furthermore, they represent a "practice at stake in the claim to impose, in the name of epistemology or the sociology of science, the legitimate definition of the most legitimate form of science—the science of nature" (Bourdieu 9). And as noted by Santos, metaphors also enable us to take a more comprehensive view toward what we know, as well as what we do not know. Decolonial ecocriticism proposes destabilizing normative and Western-centered ideas that conceptualize the environment and its complex systems (Hume and Rahimtoola), and deploying the metaphor of scientific ecology (Oliveira) allows us to engage concepts such as *resilience*, *resistance*, and *network* to interpret the scientific environment in new ways. The ecological metaphor goes beyond the epistemological resource of translating symptoms or disputes between areas; notions of resilience and ecological resistance employed to understand the global scientific ecosystem reveal the imposition of a scientific agenda from the North.

Two concepts originating from the fields of engineering and psychology that are currently being used in political movements in the Global South to characterize ecological models are the terms *resilience* and *resistance*. However, neither term represents an approach that challenges inherited models for scientific systems, but each is an agent of stability. Within an ecological framework, *resilience* refers to an ecosystem's ability to respond to a disturbance, to sustain damage, to recover, and to adapt rapidly. The notion gained prominence in 1970 through the work of renowned Canadian researcher C. S. Holling, who described it as the persistence of natural systems in the face of changes in ecosystem variables, due to natural or anthropogenic causes (6). These perturbations and disturbances can include stochastic events such as fires and floods, as well as human activities such as deforestation, soil fractionation for oil extraction, and the spraying of pesticide into the soil. *Resistance*, in turn, is understood as a system's ability to maintain its structure and operation in the face of a disturbance, with the ability to bear or tolerate something being included in the various meanings of the term. In clinical psychiatric studies, for instance, resistance appears as a force against any attempt to break the isolation established by the repression of a set of representations, preventing the success of the treatment.

The application of these concepts to the global scientific ecosystem suggests that initiatives of resilience from countries of the so-called Global South seek stability within the contexts of scientific circulation dominated by countries from the center. Such initiatives of resilience are accompanied by models of scientific subservience, in which the Butler syndrome manifests itself in semiperipheral countries when said nations serve as landlords by ignoring their own caste (Jauretche 14; Bennett 88). Partnerships of Latin American governments and scientific oligopolies and even financing publishers whose presence in these same countries is underrepresented are some examples of models of subservience and resilience that disturb projects of egalitarian, global scientific circulation.

Initiatives such as Sci-Hub and Libgen foster practices supporting open-ended processes to break the monetization of knowledge, in these cases by functioning as search engines that infiltrate paywall barriers. Both engines also exemplify the necessity of declassification that has been dividing the Global South and North, with their portals being established in Kazakhstan and Russia, countries that are located on the fringes of the global scientific circulation circuit. However, despite opening the path to knowledge through copyright infringement, these search engines still contribute to the citational capital of these publishing sources. That is, the engines break the access codes, but they do not break the consolidated structure around citation and visibility for central countries, which results in the invisibility of the scientific production of peripheral countries. Instead, the citations of these closed articles are increased, thus reinforcing the discourse of quality certification by which these companies have positioned themselves centrally within the global scientific marketplace.

If, then, neither resilience nor resistance seeks to break established paradigms, what is the possible alternative for decolonizing this scientific ecosystem? Could it be revolution? In ecology, despite being anchored in the precepts of the quest for stability in the face of climate change, the concept of sustainability emerges as a discourse for the "green revolution," an ecological revolution that echoes a series of criticisms related to the social and economic spheres, as discussed by researcher José Roberto Moreira (39). The notion of sustainable development is understood as development that meets the needs of the present without jeopardizing the needs of future generations, a concept widely disseminated by the Brundtland Report, *Our Common Future*, published in 1987 (41). But despite the predominance of this understanding about the concept of sustainability, it is not yet a finished concept and reflects certain geopolitical disputes within its interpretation. According to Moreira, the discourse on sustainability, which was shaped by a neoliberal agenda with the opening of markets in the 1990s, was forced to confront clashes in the global order from two critical theoretical-interpretative perspectives (45). The former understood sustainability as emphasizing environmental issues more present in the capitalist countries of the North and in wealthier social strata. This perspective tended to propose a new relationship between human beings and nature, in both its technical and its existential dimensions, proposing an unobtainable and utopian balance in the proper use of the planet's natural resources. The individual dimension is of greater relevance in this type of northern approach, ignoring the degree to which the implementation of sustainability is actually feasible without solving underlying structural problems. Conversely, a second perspective, in which the dimension of social equity becomes more evident, came mainly from peripheral countries in the South and from the poorer strata of capitalist societies. The concept of revolution—as exemplified by such discourse around sustainability as a "green revolution"—thus clearly involves a heterogeneous field of disputes in which neoliberalism becomes an astute stakeholder in the co-optation of the production of global ecological narratives.

Consequently—and in alignment with Thomas Kuhn's views on revolutionary sciences—in order to overcome the stability produced by a paradigm, it is necessary for subjects to acknowledge crisis and risk and to propose alternative analytical models to the obvious and unquestioned assumptions on which the established paradigm is based. There is a dynamic of dispute and interdependence between "innovative" and "conservative" stakeholders. Revolutions are not made in the pursuit of stability; therefore, resilience and resistance do not lead to revolutions. Revolutions arise to propose alternative models to an already established system. What alternative models are these? Can these revolutions arise from the epistemological bosom that nurtures the abyssal division between North and South? From an ecocritical perspective, network classification[3] addresses latent, interdependent relations and acknowledges the networks of this complex system. A decolonial perspective brings to light the occluded colonial matrix of power and the knowledge that informs the articulation of these structures.

The Case of Brazil: Heterogeneity, Nesting, and Modularity

Since the publication of the Fourth Report of the United Nations Intergovernmental Panel for Climate Change (IPCC), also known as Climate Change 2007, a race has been underway in the search for technical and scientific paradigms to meet the goals of global sustainability and to address the growing concerns on the social and environmental impacts of climate change within a global research agenda. According to Fabrício Neves and João Vicente Costa Lima, "pode-se dizer . . . que, nessas 'agends quentes' da pesquisa, a ciência produzida pelos países centrais reproduziu padrões de pesquisa e resultados que restauraram sua própria posição de liderança nas fronteiras do conhecimento" (273; "it can . . . be said that, in these 'hot agenda' of research, science produced by central countries has reproduced patterns of research and results that restore their own leading position in the frontiers of knowledge"). Therefore, the precedence of preestablished "hot topics" of research increases a country's material and symbolic capital through systems of production of science, technology, and innovation, as articles and patents constitute important landmarks for future research. In addition, they allow for scientific nation branding, in which a national brand becomes attractive for scientific competitiveness and fosters migratory and information flows for international cooperation.

In the first decade of the twenty-first century, Brazil, a strong strategic competitor on the international stage with a significant role in environmental issues due to its biodiversity, attempted to align its research focus toward national and international trends of research on climate change. Investments in international sectors drove budget resources and public policy, prompting commitments from international agencies at meetings, ranging from Eco-92 to the United Nations Conference on Sustainable Development, known as Rio + 20, held in 2012. The main focus was to reduce deforestation of the Amazon rainforest and to construct the first national program to reduce CO_2 emissions caused by deforestation and forest degradation. Concerned with its public image, the STI (Science, Technology and Innovation) made many investments in these areas. The 2008 Climate Change Plan included extremely relevant public consultations under the name of *Sectorial Dialogues*. Through fora such as the Third National Conference on the Environment and the meetings of the Brazilian Forum on Climate Change—a milestone for this position within the global agenda—a strategy was put into place in order to establish a dialogue with central countries through a bioenvironmental model in which biological areas and the Earth were the main focus (Glänzel et al.).

As seen in Figure 11.1, there was a significant increase in scientific scholarship the year after the Climate Change Plan was published in 2008; it

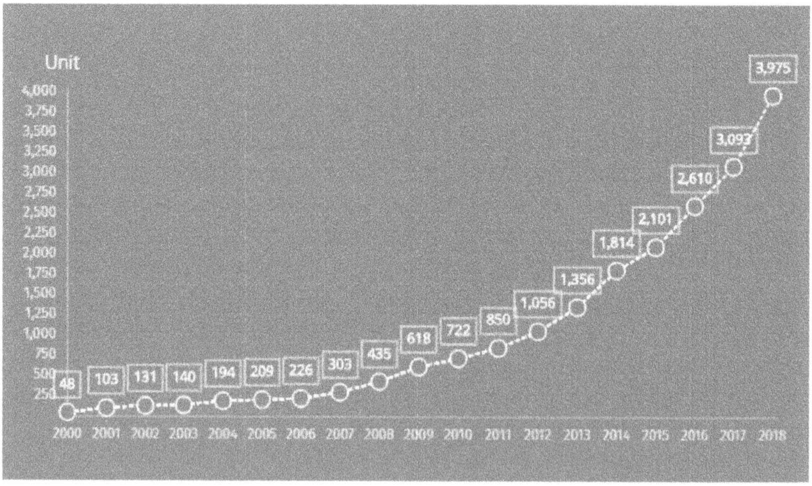

Figure 11.1 The entirety of the Brazilian scientific output on climate change.

Source: Data extracted from the Dimensions Platform, with an excerpt on Brazilian scientific production on climate change (titles and abstracts). By Thaiane Oliveira.

provided feedback for ongoing investment and helped consolidate research agendas with relevant encouragement for studies in diverse areas. In addition to the Brazilian Coordination for the Improvement of Higher Education Personnel (CAPES) and the National Council for Scientific and Technological Development (CNPq), as well as local agencies such as FAPESP in São Paulo and FAPERJ in Rio de Janeiro, international agencies invested in the joint understanding of science as a common good and contributed to the political and scientific capital through network-based knowledge production. The United Kingdom's National Environment Research Council (NERC), the United States' National Science Foundation (NSF) and National Aeronautics and Space Administration (NASA), Portugal's Foundation for Science and Technology (FCT), the Australian Research Council (ARC), and France's National Agency for Research (ANR) have been among the main investors in climate change research in Brazil, along with the European Union. All these countries occupy key positions in a network of climate change research in coauthorship with Brazilian agencies.

The network of relations of coauthorship shows how the structures of colonization of knowledge remain in place materialized by dense relations with certain stakeholders. The analysis of the co-authorship network graph (Figure 11.2) sheds light on the spheres of influence and their contexts. The current Americanization of Brazilian scientific systems can be understood as a neocolonizing endeavor in which control over a territory is not limited to territorial space but exercised through commercial, cultural, and scientific influence—hence the centrality of the United

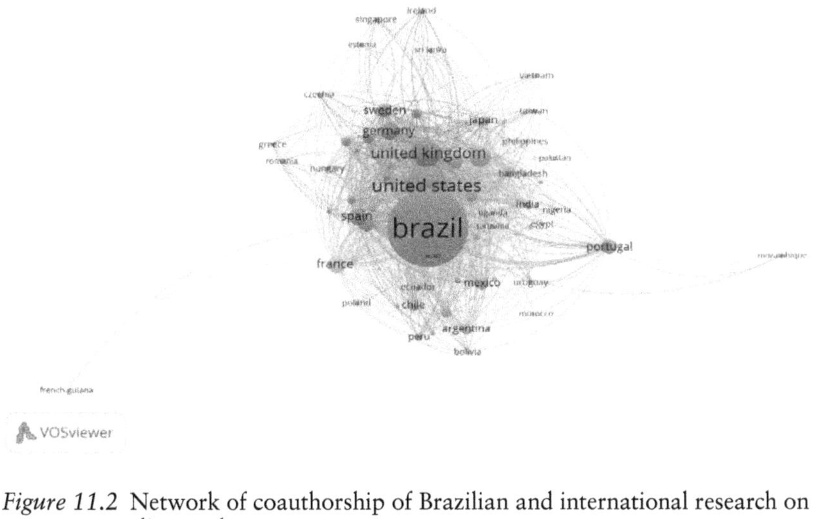

Figure 11.2 Network of coauthorship of Brazilian and international research on climate change.

Source: Dimensions and VOSviewer.

States in Figure 11.2. Brazil's link with other Latin American countries is marked by its proximity to the United States. The figure also displays a strong interaction with the United Kingdom, another member of an oligopoly of scientific publishing that has dominated the market of scientific production for more than a century (Larivière et al. 11), directly affecting the spaces of visibility in scientific circulation (Sousa and Oliveira 84). The next cluster features a set of European stakeholders that marks their relationship with Brazil.

The ecologies of research in coedited projects are characterized by heterogeneous interactions organized through nesting and modularity. *Heterogeneity* consists of the presence of species that present a greater number of connections. The analysis of social networks includes species that have a high reputation and hence increased influence within their network. The scientific ecosystem includes countries that hold positions of centrality within a determined network. Examples are the colonial countries, which connect in the Americas (the United States) and Europe (United Kingdom and Germany). The concept of nesting in ecology refers to the pattern of exchanges by which interspecies (specialists and generalists) interact (Gaiarsa). In this dynamic, the latter (generalists) have major connections with the stakeholders of the former group (specialists). Applied to the global scientific ecosystem, generalists propose universal epistemologies to account for phenomena that meet their own logic, thus dismissing the emergence of new paradigms that may revolutionize previously established paradigms. In addition to their high degree of input, the connections they have with other stakeholders in the cluster—through research in coauthorship and funding, for example—turn these generalists into dominant stakeholders in the network. The United States network has the characteristic of linking other Latin American stakeholders as participatory stakeholders in the process of developing international research; the links between the United States and Brazil are much stronger than those between Brazil and other Latin American countries. Some of these Latin American connections must pass through the United States network, even though this requires a detour and longer distances. Additionally, the study of modularity allows us to identify the clusters that are formed out of the process of interactional dynamics. For example, Portugal emerges in a cluster of few connections but with dense edges. These fringes relate to other colonized countries because of their linguistic proximity. Such is the case of Mozambique, where the colonial mark following the subtraction of native languages has persisted over the centuries and manifests itself today in new guises.

While the production of scientific and technological knowledge was previously concentrated in three regions of the world by the end of the twentieth century—Japan, the United States, and parts of Europe—the new millennium brought with it new geopolitical dynamics. The growth of the so-called emerging countries changed the flow of investment and production in science, technology, and innovation (Cassiolato and Vitorino 23) and contributed to the emergence of the BRICS (Brazil, Russia, India, China, and South Africa) and the MIST (Mexico, Indonesia, South Korea, and Turkey) blocs as world powers. It also fostered the economic growth of China and the Asian Tigers (Korea, Taiwan, China, Hong Kong, and Singapore), as well as New Asian Tigers (Malaysia, Vietnam, Thailand, Indonesia, and the Philippines).[4] In this scenario, countries are appearing in the spaces of the circulation of science, despite restrictive editorial practices of the scientific oligopoly reinforcing the centrality of the hegemonic countries. However, the category of countries of the Global South transcends a single geographical delimitation and conveys the idea of an expanded locality that defies world (dis-)order (Levander and Mignolo 8; Resende and Thies 10). This cluster to the right of the center of the graph is composed of countries of the so-called Global South, with the addition of Australia, and marks a new reconfiguration in science in this new millennium. It breaks away from the invisibilities imposed during three centuries of colonial science (Sousa and Oliveira 86) by proposing new challenges for the scientific movement through South–South cooperation. In the field of ecology are indications that the effects of disturbances are propagated more slowly in modular networks than in nonmodular networks. The identification of a modular South–South cooperation network allows us the strength not to seek stability but rather to seek ruptures within a scientific ecosystem in which predators have dominated, feeding off the invisibility of countries outside the central axis.

Final Remarks

Network classification (redecslassificação) is a proposition that focuses on key concepts of the analysis of social and ecological networks to unveil the violent dynamics of global scientific neoliberalism. This critical approach is in line with decolonial ecocriticism, as it understands the role of networks as a major analytical tool for understanding interactions with a group of stakeholders, while also proposing to denounce how colonial logics continue to be reflected in scientific production and how subservience models remain reproduced in the global scientific ecosystem. Despite the inclusion of stakeholders as funders and researchers in the discussion, as noted by Nixon, numerous participants are propagating a culture of doubt around climate change (39). And these agents have been increasingly central, even occupying important positions of political representation, as in the case of the Brazilian president Bolsonaro and a large part of his team, who attribute progressive advances to a left-wing agenda and therefore deem them as initiatives to be defeated.

The relationships among notions of trust, truth, and objectivity are characteristic of the institutions of modernity. For example, scientific truth has consolidated itself around the assumption that scientific knowledge is the only possible knowledge. These presuppositions lead to the suppression of local knowledge and the destruction of alternative epistemological foundations extant today. According to Luiz Signates, science is currently undergoing a series of crises, among them a crisis of truth (143). For him, the crisis of truth is produced by a postmodern understanding of scientific knowledge as the only process of inquiry allowed to mediate the many representations of reality, a process that inevitably leads to alternative facts—the material most shared by the culture of doubt. Climate change is part of the elaboration of a culture of doubt. This affects both citizens who have lost trust in the institutions as well as the deliberative public sphere itself. The case of Brazil serves as an example where the process has yielded an agenda of environmental and social retrogression.

One of the characteristics of the countries of the so-called Global South, besides their own local knowledge, is the wealth of natural resources and the attendant potential for exploitation by dominant countries. According to United Nations reports, countries of the Global South will lack access to basic sanitation by 2030 (United Nations 16). These countries will be the ones most affected by climate change. Such warnings bear little meaning when the culture of doubt is triggered by political disputes over the Latin American agenda. In Brazil, Bolsonaro's government has announced its intention to leave the Paris Accord; the country will also no longer host the next round of COP-25, the UN-sponsored talks on climate. President Bolsonaro has vowed to open the rainforest to commercial exploitation despite evidence about the short-term, worldwide consequences of deforestation.

Some sectors of Latin American economies are, of course, more vulnerable than others to having their assets devalued or rendered unusable by climate change and are therefore greatly affected by the policies adopted to mitigate this loss. For instance, large reserves of fossil fuels cannot be exploited if Latin America intends to comply with the Paris Accord, which requires that almost 50% of the region's oil and gas reserves and 75% of its coal reserves remain buried. But this and other examples show the need to go beyond models that advocate simply for a new relationship between human beings and nature centered only on notions of stability. These models have a utopian dimension when we consider the political guidelines and power disputes over climate change. As stated by Ngugi wa Thiong'o, one must look consciously at what colonial imperialism has done for the countries of the so-called Global South (88). To do so, it is necessary to propose alternative epistemologies to reenvision spaces of knowledge production from perspectives that understand the relationships among multiple stakeholders. It is also necessary to acknowledge the violence ingrained within the configuration of certain sectors themselves. The proposal of decolonial ecocriticism is important for this goal, as it facilitates proposing analytical models at the level of the entire ecosystem while considering the very political disputes raised by challenging issues such as climate change, which is considered to be one of the greatest crises for humankind. And if it is a concern for all of us, social equity ought to be at the heart of the discussion beyond classifications of the world and beyond utopian models of sustainability and stability or dependency and insertion. We must think of epistemic revolutions before neoliberalism turns the planet into an inhospitable place for all of us, whether we are from the Northern or Southern Hemisphere.

Notes

1. The proposal here follows in the footsteps of García Gutiérrez by retracing the history of classification of the scientific ecosystem towards a reconstruction of the concept of declassification.
2. This redemarcation is an important initiative to foster the environmental preservation of these territories in the fight against *grileiros*, people who falsify documentation for illegal possession of indigenous lands.
3. The junction of the prefixes *re* ("repeat") and *des* (a prefix of negation) forms the word *redes* ("networks") in Portuguese. This linguistic peculiarity has become an important epistemological construct for studies on the ecosystem in the contemporary world (Gaiarsa). It is an exercise of declassification, proposed by Antônio García Gutiérrez, which marks the search for a reclassification from an ecosystemic and ecocritical view based on networks of interdependent relationships.
4. These are countries characterized by strong economic and social development, low dependence on raw materials, and industrialization intensified by economic openings for transnational markets.

Works Cited

Alatas, Syed Hussein. "Intellectual Imperialism: Definition, Traits, and Problems." *Asian Journal of Social Science*, vol. 28, no. 1, 2000, pp. 23–45.

Bennett, Karen. "The 'Butler' Syndrome: Academic Culture on the Semiperiphery." *Revista Canaria de Estudios Ingleses*, vol. 69, 2014, pp. 155–71.

Bernardino-Costa, Joaze, and Ramón Grosfoguel. "Decolonialidade e perspectiva negra." *Sociedade e Estado*, vol. 31, no. 1, 2016, pp. 15–24.

Bourdieu, Pierre. "Le champ scientifique." *Actes de la recherche en sciences sociales*, vol. 2, no. 2, 1976, pp. 88–104.

Brundtland, G. H., Khalid, M., Agnelli, S., Al-Athel, S., & Chidzero, B. (1987). *Our Common Future*. New York.

Cassiolato, José Eduardo, and Virgínia Vitorino, eds. *BRICS and Development Alternatives: Innovation Systems and Policies*—Vol. 1. Anthem Press, 2009.

Castro-Gómez, Santiago, and Desiree A. Martin. "The Social Sciences, Epistemic Violence, and the Problem of the Invention of the Other." *Nepantla*, vol. 3, no. 2, 2002, pp. 269–85.

Colaço, Thais Luzia, and Eloise da Silveira Petter Damázio. *Novas perspectivas para a antropologia jurídica na América Latina: o direito e o pensamento decolonial—Volume IV*. FUNJAB, 2012.

Dhanani, Alpa, and Michael John Jones. "Editorial Boards of Accounting Journals: Gender Diversity and Internationalisation." *Accounting, Auditing & Accountability Journal*, vol. 30, no. 5, 2017, pp. 1008–40.

Gaiarsa, Marilia Palumbo. *Efeitos em cascata em redes mutualistas*. Diss. Universidade de São Paulo, 2017.

García Gutiérrez, Antonio. "Declassification in Knowledge Organization: A Post-Epistemological Essay." *Transinformaçao*, vol. 23, no. 1, 2011, pp. 5–14.

Glänzel, Wolfgang, et al. "Science in Brazil: Part 1: A Macro-Level Comparative Study." *Scientometrics*, vol. 67, no. 1, 2006, pp. 67–86.

Haustein, Stefanie. "Grand Challenges in Altmetrics: Heterogeneity, Data Quality and Dependencies." *Scientometrics*, vol. 108, no. 1, 2016, pp. 413–23.

Holling, Crawford S. "Resilience and Stability of Ecological Systems." *Annual Review of Ecology and Systematics*, vol. 4, no. 1, 1973, pp. 1–23.

Hume, Angela, and Samia Rahimtoola. "Introduction: Queering Ecopoetics." *ISLE: Interdisciplinary Studies in Literature and Environment*, vol. 25, no. 1, 2018, pp. 134–49.

Jauretche, Arturo. *La colonización pedagógica*. Universidad Nacional del Nordeste, 1967.

Kilomba, Grada. "Africans in the Academia: Diversity in Adversity." *Kritische Migrationsforschung*, 2012, pp. 299–304.

Kuhn, Thomas. *Revoluções Científicas*. Perspectiva, 1978.

Larivière, Vincent, et al. "The Oligopoly of Academic Publishers in the Digital Era." *PloS One*, vol. 10, no. 6, 2015, pp. 1–15.

Levander, Caroline, and Walter Mignolo. "Introduction: The Global South and World Dis/order." *The Global South*, vol. 5, no. 1, 2011, pp. 1–11.

Mahler, Anne Garland. *From the Tricontinental to the Global South: Race, Radicalism, and Transnational Solidarity*. Duke UP, 2018.

Martínez-Alier, Joan. *The Environmentalism of the Poor: A Study of Ecological Conflicts and Valuation*. Edward Elgar, 2003.

Mazloumian, Amin, et al. "Global Multi-Level Analysis of the 'Scientific Food Web'." *Scientific Reports*, vol. 3, article 1167, 2013, pp. 1–5.

Mbembe, Achille. "Decolonizing the University: New Directions." *Arts and Humanities in Higher Education*, vol. 15, no. 1, 2016, pp. 29–45.

Mignolo, Walter D. "Epistemic Disobedience, Independent Thought and Decolonial Freedom." *Theory, Culture & Society*, vol. 26, no. 7–8, 2009, pp. 159–81.

———. *Histórias Locais/Projetos Globais: Colonialidade, saberes subalternos e pensamento liminar*. Editora UFMG, 2003.

Moreira, Roberto José. "Críticas ambientalistas à revolução verde." *Estudos Sociedade e Agricultura*, vol. 15, 2000, pp. 39–52.

Neves, Fabrício Monteiro, and João Vicente Costa Lima. "As mudanças climáticas e a transformação das agendas de pesquisa [Climate Change and the Transformation of Research Agendas]." *Liinc em Revista*, vol. 8, no. 1, 2012, pp. 268–82.

Nixon, Rob. *Slow Violence and the Environmentalism of the Poor*. Harvard UP, 2011.

Oliveira, Thaiane. "As políticas científicas na era do conhecimento: uma análise de conjuntura sobre o ecossistema científico global." *Perspectivas em Ciência da Informação*, vol. 24, no. 1, 2019, pp. 191–215.

Quijano, Aníbal. "Coloniality of Power, Ethnocentrism, and Latin America." *Nepantla*, vol. 1, no. 3, 2000, pp. 533–80.

Resende, Fernando, and Sebastian Thies. "Temporalidades enredadas no Sul Global." *Revista Contracampo*, vol. 36, no. 3, 2017, pp. 2–19.

Santos, Boaventura de Sousa. "Beyond Abyssal Thinking: From Global Lines to Ecologies of Knowledges." *Review*, vol. 30, no. 1, 2007, pp. 45–89.

Scolari, Carlos A. "Media Ecology: Exploring the Metaphor to Expand the Theory." *Communication Theory*, vol. 22, no. 2, 2012, pp. 204–25.

Signates, Luiz. "Epistemologia e comunicabilidade: As crises das ciências, ante a perspectiva da centralidade do conceito de comunicação." *Comunicação & Informação*, vol. 15, no. 2, 2012, pp. 133–48.

Smaldino, Paul E., and Richard McElreath. "The Natural Selection of Bad Science." *Royal Society Open Science*, vol. 3, no. 9, 2016, royalsocietypublishing.org/doi/10.1098/rsos.160384.

Sousa, Lumárya Souza de, and Thaiane Oliveira. "Cartographies of Action Research: Searching for Displacements of the Epistemology of the South." *Comunicação e Sociedade*, vol. 33, 2018, pp. 83–106.

Thiong'o Ngugi wa. *Decolonising the Mind: The Politics of Language in African Literature*. East African Publishers, 1992.

United Nations. *Sustainable Development Goal 6 Synthesis Report on Water and Sanitation*. United Nations Publications, 2018.

Vessuri, Hebe, et al. "Excellence or Quality? Impact of the Current Competition Regime on Science and Scientific Publishing in Latin America and Its Implications for Development." *Current Sociology*, vol. 62, no. 5, 2014, pp. 647–65.

Visentini, Paulo Fagundes. *O caótico século XXI*. Alta Books Editora, 2015.

12 Bodies, Transparent Matter, and Immateriality

Compagnie Käfig's Ecodance Performances

Ilka Kressner

For Jesús Alonso-Regalado[1]

Rob Nixon highlights invisibility and temporal dilation as main challenges for artists engaged in conceptualizing and portraying slow violence and ecological vulnerability. The media he discusses in *Slow Violence and the Environmentalism of the Poor* are literature, film, and political activism. Taking up Nixon's question, "How do we both make slow violence visible yet also challenge the privileging of the visible?" (15), this essay examines dance as an art form to engage environmental issues and communicate unsettling slow violence, without exclusively relying on the visible sense. It explores two performances by the dance troupe Compagnie Käfig: *Agwa* (2008; Water) and *Pixel* (2014), both created by the founder of the company, Franco-Algerian choreographer Mourad Merzouki, and a group of 11 male dancers from Brazil, the cast of *Agwa*, and nine male and one female dancers from Brazil, North Africa, and France, the cast of *Pixel*.[2] *Agwa* was created in Brazil; it was the result of Merzouki's first collaboration with the dancers, who then became members of his troupe based in France (Merzouki "Uncaged," n. page).[3]

The aesthetic emphasis of Compagnie Käfig lies on creative synergies of styles. In contrast to most art forms shaped by individual artists, contemporary dance situates itself within transnational aesthetic contexts. Dancers, choreographers, and musicians are frequently crossing national and cultural boundaries to participate in ensembles, and troupes are traveling to perform, teach, and engage in joint ventures while participating at international festivals. This nomad lifestyle is certainly the case for the dancers of *Agwa* and *Pixel*: both pieces have been performed each about 350 times in 17 (*Agwa*) and 19 (*Pixel*) countries (Compagnie Käfig, web). Within this transnational artistic context, my motivation for selecting the performances as examples of a Latin American exploration of ecofictions and ecorealities is based on the fact that about two-thirds of the artists involved in the making and performing of both pieces are from Brazil and have had their initial artistic training in the country. In addition to elements of samba, bossa nova, hip-hop, ballet, modern dance, and other street and folk dances, the main aesthetic inspiration of both pieces is the dance of capoeira. Born out of the experience of uprooting and slavery, capoeira cultivates a specific conception of a shared space that allows for the expression of the voice of the marginalized.[4]

Alongside the choreographic emphasis on the physicality and athleticism of the human body, both performances include transparent and immaterial things that seem negligible at first sight: piles of plastic cups in the case of *Agwa* and swarms of dots of white light in *Pixel*. My essay explores the connections and conversions between the accentuated physicality of the human body, and the transparence of plastic and immateriality of pixels. By doing so, it invites the rethinking of notions of presence and materiality from an ecocritical perspective, grounded in performance studies. Given the fact of the copious and constant presence of plastic cups and pixels, it also examines the challenges of creating and referencing metaphors of intrusion and interference in the context of a dance performance.

The art form of dance, as a sequence of bodies moving in space, reflects ecological debates from a decidedly material standpoint or, more precisely, point in motion. Meaning is conceived and shown to be necessarily grounded in bodily experience and related to its surroundings (Johnson 12). In Alexa Weik von Mossner's words, it is a result of our being "embodied, embedded and affective" (22). In contrast to, for example, an ecocritical novel about the experience of living near a polluted river written by one single author, the presentation of material and electronic disturbances and possible threats in the corporeal art form of dance creates works that are being experienced jointly by the dancers and their audiences in a shared here and now. In the two pieces by Merzouki, the audience's understanding of human interaction with things and the metaphorical valence of plastic and pixels builds up associatively, through reiterations of gestures and movements by individuals and groups of dancers in changing formations, and the presence and agency of plastic cups, flowing water, sonic and electromagnetic waves, and moving pixels.

This focus on materiality and the more-than-human in *Agwa* and *Pixel* alludes to temporal forms that may differ from the established notion of a single, neutral chronology. Dance scholar Mark Franko termed such a time of bodily gestures and interfaces of bodies (animated and inanimate) during a performance an "intertemporal" time (8). If, following Franko, a dance performance, with its concentration on an intense and shared presence, signals an "intertemporal relationality" of dancers' gestures "that militates against progressive change, characterized by modernism" (8), the emphasis on gestures that connect the animated with the inanimate becomes a means of critique of this aesthetic of modernism and, in extension, the (chrono-)logic dictate of modernization. I propose to read the two performances as counter- and postmodern reflections on the fragility of ecosystems, conceived as sites of dynamic and synchronous interactions of human and nonhuman agents (Rigby 4–5; DeLoughrey and Handley 16).

Criticism on performance and ecocriticism, in both the Anglophone and Hispanophone scholarly contexts, is only nascent: *Readings in Performance and Ecology* (2012), edited by Wendy Arons and Theresa J. May, is to date the only book-length ecocritical approach to performance, with most of the essays analyzing modern dance. This only recent seizing of the topic may surprise, given dance's deep material and spatial implications, its connections to the "more-than-human world" (Arons and May 2), and, in the contexts of capoeira and hip-hop, the commitment to speak out and challenge a status quo perceived as unjust. In Latin America and in Brazil in particular, dance has been playing a foremost social role in transforming cultural attitudes about gender, race, and sexuality, oftentimes broaching situations of bias and injustice.[5] Capoeira in particular, the dance developed by the slaves of African descent in sixteenth-century Brazil that is now performed on the stages of mainstream dance venues, surfaces questions of power related to exclusion and the challenge to make systemic violence visible in dance. Capoeira, this "fortuitous play, serious struggle, game, sport, *malicía*, dance, ritual, musical performance, theater, drama, philosophy and life" (Merrell 43), is a powerful interactive voice to challenge the politics of racial and social exclusion. Alongside these themes, contemporary dance is currently exploring ecocritical aesthetics and sensibilities by highlighting an epistemology based on our shared materiality and experience of place.

Environmental concerns, in particular human–nonhuman interactions, are expressed in manifold guises in contemporary performances. Puerto Rican–born choreographer Merián Soto takes her dancers and audiences outside: *Three Branch Song* (2006) is performed in plain air. Dancers are balancing wooden sticks and branches on their shoulders, backs, and heads and move in slow motion, thus performing a metonymic balancing act of our fragile human interaction with nature. Madrid-based Camille Hanson's *The Sacrifice of Giants* (2016) combines dance with background images and videos of marine landscapes and specific props, among those, a large net, to illustrate the story of illegal hunting and slaughtering of whales and dolphins. In Hanson's piece, the dancers emulate first the motions of sea mammals swimming and are then trapped in nets, which are lifted up in the air and hanging some ten feet above the stage. The performance culminates in the dancers' mimicking of the final convulsions of the moribund mammals. Other performances visualize environmental contamination through mimicked metamorphoses of the bodies of the dancers into beings of monstrous shapes, through movements, costumes, and special effects.

Merzouki's two works are less descriptive of and focus less on specific forms of violence imposed on certain species. Instead of connecting the dances to real cases of pollution—the heavy metal contamination of the Tietê River running through Sao Paulo state, which was a widespread media presence during the early 2000s and a possible stimulus for the choreographer and dancers to create *Agwa* (da Silva, Ivone et al. 105)—the critical power of both pieces unfolds in a more allusive and ambiguous staging of interferences within existing ecospheres. *Agwa* references gestures of using, throwing away, but also collecting plastic: dancers drink from the cups, play with them, throw them around on stage, sweep them toward the side of the stage with their feet and hands, collect and rearrange them. *Pixel* emulates the effect of electromagnetic impulses on and interferences with the human body. Both pieces resemble ecosystems in the sense of biological organizations, "composed of all the organisms found in a particular physical environment, interacting with it and with each other" ("ecosystem," *OED*). The conditionality of animated and inanimate matter challenges what Emilio Santiago Muíño describes as one of modernity's "mitos constituyentes . . . [e]l titanismo explícito. . . el dominio del hombre sobre las cosas . . . espejismo de independencia de las sociedades respecto a los ecosistemas circundantes" (33–4; "founding myths . . . the explicit titanic power . . . man's domination over matter . . . the illusion of independence of the societies from the surrounding ecosystems"). In *Agwa* and *Pixel*, the human body is much less in charge of objects but interacts with matter in progressive, contextual, and open-ended ways.

Unlike most other contemporary performances of comparable lengths, neither *Agwa* nor *Pixel* have intermissions. Both pieces build rich arcs of suspense that develop during the 40 (*Agwa*) and 70 (*Pixel*) minutes of the performances, driven by the sound tracks, which were commissioned and developed in collaboration with the dancers, choreographer, and cochoreographers. In conjunction with the music, props, and subtle light arrangements that guide and react to the movement on stage, the dancers create situations that explore changing junctures of agency, (self-)control, and identities, including partially prosthetic (human–nonhuman) and swarm identities.[6] The interplay of entities of "vibrant matter," in the sense of Jane Bennett's coining, is in both performances generated mainly out of muscular energy, synthetized products based on petroleum (plastic cups, ponchos, umbrellas, and shoe soles), electronic impulses (pixels, light, and sound waves), and water.

Agwa and *Pixel* invite highly diverse critical approaches, among which I propose a material ecocritical framework. This perspective reverses the vector of classical thought from the deliberations of the thinking mind toward a treatment of matter. It argues that such a reversal of perception may originate not only in thought but also in praxis, such as our everyday handling of cups and interaction with electromagnetic waves. In "Stories Come to Matter," the introduction to *Material Ecocriticism*, Serenella Iovino and Serpil Oppermann highlight that:

> [b]odies, both human and nonhuman, provide an eloquent example of the way matter can be read as a text. Being the "middle place" where matter enmeshes in the discursive forces of politics, society,

> technology, biology, bodies care compounds of flesh, elemental properties, and symbolic imaginaries. Whether performing their narratives as statues in a square, teachers in a classroom, plankton in the ocean, fossils trapped in a stone wall, or chickens in industrial factory farms, bodies are living texts that recount *naturalcultural*[7] stories.
>
> (6; emphasis in original)

The authors conclude that the recognition of the creative force of agency "not only insinuates new conceptions of nature, life, and materiality, but also relocates the human in a larger material-semiotic 'collective'" (6). The focus on the reality of interdependency, the exemplified benefits of forming alliances with other bodies (animated or not), and the awareness of a shared use of space make *Agwa* and *Pixel* fitting examples to help sharpen ethical sensibilities within a material ecocritical framework.[8]

The aesthetics of both *Agwa* and *Pixel* is marked by athleticism and fast-paced movements, often accompanied by drums and accentuated by the dancers' stomping. As is characteristic of all works by Merzouki, the two pieces are highly energetic. The soundtracks by musician AS'N (*Agwa*) and sound artist Armand Armar (*Pixel*) include traditional capoeira instruments, most prominently the berimbau, a single-string percussion instrument, with instruments of Western origin, such as guitars, pianos, saxophones, xylophones, and the human voice. The sounds are further fused with techno and electronic music. From the onset, Merzouki's choreographies are built on rapid and fluid changes between solos, duos, trios, small groups of four to five dancers, and the entire cast. These changing assemblies contrast with the traditional configuration of two dancers engaged in capoeira movements and that of one single performer of hip-hop, surrounded by a circle of observers. The costumes resemble street clothes: dancers wear shorts or jogging pants and go either shirtless or wear T-shirts or sweaters in earth tones. Most of them dance in Converse sneakers.

The dancers come from a wide variety of performing disciplines that include capoeira, acrobatics, and street theater. About half of them are self-taught (Gonçalves do Nascimento Leitão and Faxola Franco, interview). Reviewers of the two performances oftentimes highlight the combination of rawness and refinement, point out the raucousness of individual talents, and the dancers' athleticism (Molzahn for *Chicagotribune*, Morgan for *Critical Dance*). According to Debra Cash's review in *Artfuse*:

> Merzouki's confident direction brooks no challenge about where the audience should focus its attention. The dancers are up, they're down, they're center stage, they're in the deftly deployed spotlight. Some of the men lie on their stomachs and drum on the floor with their palms. . . . [They] are gifted performers and major hunks—scene

> after scene, they take a moment to convey "have you noticed I've taken my shirt off again?"
>
> (Cash)

Put aside the partially exotifying undertones of this passage that also hint at the reviewer's likely training in more traditional dance forms, Cash's observation conveys an admiration of the dancers' bravado and gifted performativity. The objects with which they share the stage are seldom mentioned in reviews and, if so, rather negatively, as props that limit their expressiveness. Critic Chelsea Thomas describes in *Dance Informa* that, "[w]hile the choreography is mostly stimulating and the idea is one-of-a-kind . . . [t]he meticulous use of the cups overshadows the dancing at times, and with such strong and passionate dancers, it seems a shame to see them contained" (Thomas). In contrast to such a position, I argue that plastic cups and pixels are not mere things that limit the creative exploration of human movement but instead vital elements of the performances of *Agwa* and *Pixel* that articulate a reflection on the human body's interaction with material and immaterial forces. The stage is a shared artistic and political space that explores the impact of invisible and transparent objects and flows of energy on, related to, and within the human body.[9] My conception of a political space follows Judith Butler's notion of a space that allows for a deep "worldly" connectedness and offers the potential of agency based on shared materiality (26–9).[10]

An example of the biotic-abiotic nexus and of awareness of the human body's participation in a broader material-semiotic process that goes beyond the individual bodily limits is a sequence in *Agwa*, in which six dancers, at this moment wearing rain boots, position water cups next to one another. Their aim is to arrange ten lines of about 15 cups each that run parallel to one another, from the front edge of the stage to its back. The assembling of these cups is performed first individually, resulting in only partially parallel lines. After the dancers observe one another's skimpy attempts at setting up straight lines, they redo the assembly, now in sync with one another, guided by the music, correcting and helping one another. While their movements are individual, they now all follow the rhythm of the music, by accentuating each strong first beat with a vertical movement of their limbs and/or torsos, such as lifting an arm, leg, standing up straight after a squat or jumping into a handstand (Figure 12.1).

The result of this partially individual, partially collaborative arranging is a painstakingly accurate setup of ten parallel lines made of plastic cups that divide the stage into long areas of the same size. During that time, a seventh dancer, seemingly ignoring his colleagues' endeavors, has been running and leaping across the stage, without touching any of the cups. After the assembling of the lines of cups has been finished, the six dancers recede to the back of the stage, while the single dancer begins to cross the stage from left to right and back again doing a series of somersaults.

Figure 12.1 *Agwa* (2014).
Source: # 16110503–10BF, reprinted with permission by photographer Benoîte Fanton.

It is only now that the audience realizes that the group's assembling of cups also served as a preparation for the artist doing somersaults, where the division of space had to be carefully calculated to allow him perform. Here, the collaborative setting up of a grid structure is motivated by a single dancer's body shape and idiosyncratic way of motion. This sequence illustrates how individual prowess can take place only as the result of a close collaboration within and with a shared space. Plastic is at first a literal hassle, a hurdle; then it becomes a figurative stepping stone and structuring device that brings together a joint effort toward the performing of acrobatic leaps.

The plastic cup as the emblem of our throwaway society is in *Agwa* multiplied by the hundreds. It becomes a stubborn material reality to be reckoned with by the performers, who carry stacks of cups in multiple forms, fill them, arrange them on the floor, and move them back and forth so that the assemblages of cups subdivide the stage, form barriers and circles that condition, restrict but also highlight their motion. At the end of the presentation, all dancers are drinking the water. By doing so they directly reference our human need of water to live and move, in addition to, more indirectly, the masses of water required to produce plastic. Then they throw the empty cups up in the air in a joint gesture, before throwing themselves on the floor, while the cups fall down on them (Figure 12.2).

Figure 12.2 *Agwa*, final moments.
Source: # 16110503–25BF, reprinted with permission by photographer Benoîte Fanton.

The performance finishes with the dancers and cups lying on the ground, illuminated by spotlights positioned at about knee height. Now, plastic, with its multicolored light reflections, has become visually more intriguing than the human bodies. *Agwa*'s focus on plastic makes our throwaway practices, based on an out-of-sight-and-out-of-mind attitude toward waste, present and tangible. In their transparent plasticity and tenacious presence, the hundreds of cups act on, condition, and at times objectify the human body, as in the case of the final scene. Their presence on scene is a powerful visualization that plastic's disposability and easy disappearance is a dangerous illusion, both optical and epistemological.

Plastic is likely the most touched and least noticed human-made material of our everyday lives. In its omnipresence in a vast variety of shapes, it has altered our perception of reality. According to Jeffrey L. Meikle, "The very stuff of existence in the late twentieth century is synthesized in chemical refineries from petroleum, a universal medium of exchange" (1). Compared with nineteenth-century artifacts, such as iron and steel, plastic seems light, ephemeral, easily replaceable, and disposable (Meikle 2). Its malleability dazzles and challenges our attempts at its conceptualization. As Roland Barthes notes in an essay included in *Mythologies*, plastic is nothing without a context. Yet at the same time, it gestures at dissolving context (Barthes 171; Meikle 3). In its luring embodiment of the idea of an infinite transformation of matter, plastic is context bound and produced for the here and now, the moment when we grab a cup of water to satisfy our thirst. But it is also decontextualizing in its embodiment of the idea of limitless transformation. For Barthes, plastic is less an object in itself than the trace of a movement (171). We might add, with a less exclusive emphasis on the past, that it is also the indicator of a future conversion of matter. It may return, after a renewed polymerization, in the soles of our shoes, grocery bags, and cell phones. More likely, if it belongs to the 91% of plastic waste that is not being recycled worldwide (Geyer, Jambeck, and Laender Law 1), it reappears in the form of invisibly small pieces floating in the air, sea, and drinking water, enters our bodies, and, in an even more toxic concentration, enters those of our cohabitants in the Global South.[11]

While for many early enthusiasts, plastic embodied the human yearning to mold and control nature, "shape the stuff of existence at a fundamental chemical level, . . . imbue it with properties, textures and colors unknown to earlier generations, . . . mold from it objects and environments unknown to prior civilizations" (Meikle 9), plastic's slow decomposition has turned it into a metaphor of human hubris and ignorance of the profound connectedness on a planetary level.[12] In recent years, among the most repeatedly visualized motifs of plastic's contamination of our habitat are those of heaps of plastic floating on the ocean and of sea animals trapped in plastic nets or chunks. The presence of these images in the visual repertoire of contemporary mass media is much needed to communicate to a broad audience the pernicious after-effects of our wasting of plastic.[13] Still, the visually shocking heaps of plastic litter covering a surface area of 1.6 million square kilometers in the Pacific (Lebreton et al.; Moore and Phillips 57)—a form of an "instant spectacle," following Nixon (6)—refer to a much less insidious form of the impact of waste on our biosphere, compared to the slow violence of micro particles that are too small to be photographed or felt by touch, whose pernicious effects range from the transnational to cellular levels. The potential problem of those photographs is their locating of the environmental threat at a distant region; hence they risk diminishing the images' potential to spark critical engagement. Nixon takes seriously the challenge of scale and asks:

> How can we imaginatively and strategically render visible vast force fields of interconnectedness against the attenuating effects of temporal and geographical distance? . . . Wendell Berry has warned against the potentially debilitating effects of such large-scale approaches . . .

> I would argue, however, that although advocating personal environmental responsibility is essential, to shrink solutions to the level of the private and the small is evasive, even if it does constructively enhance one's sense of agency.
>
> (38–9)

Merzouki and his dancers, I argue, approach the dilemma of scale (too large and abstract to be tackled individually versus too private to reach a social consensus) through an intricate linking of singular with joint interactions with plastic that takes place during the shared space-time of a performance. *Agwa*'s portrayals of individual experiences with plastic include those moments when specific dancers pick up a cup and interact with or react to it individually, in direct causal relations. These singular gestures, set in strong spotlights, are embedded within the broader social network (and body work) of the performance: a few seconds after the aforementioned individual scene, other dancers, in a gesture emulating the collecting of garbage and sweeping up of piles of cups with their hands and feet, take a cup that had been used by another dancer, rearrange it on the floor, and pass it along to others. Patterns of individual consumption are broken up, brought into a social realm and sphere of a performative reuse of the object in new, artistic contexts.

Following dance scholar Cristina Rosa's description of dancing bodies as entities that are "capable of articulating ideas as bodily writing" (69), the 11 performers in *Agwa* write with their bodies and plastic cups, ponchos, and rain boots; their new "lettering" is thus the result of a mesh of bodies and matter in motion that communicates alternative engagements with plastic by taking it out of the ordinary context and redefining its use, which is a form of imaginative recycling. This becoming aware of the presence of the all-too-present things and the shared urgency of their ethical use is portrayed to be inextricably social and relational; it necessarily builds up over time and through an inventive use of space.

Agwa's main intertextual (or more precisely interperformative) inspiration regarding its specific practice of space is the *jogo* (game) of capoeira: "[t]he *jogo* is the action in which two individuals play inside a real or imaginary circle called *roda* . . . . The Capoeirista is someone who moves unpredictably without preset pattern of actions. He uses the space in its full dimension" (Almeida 16, 52). The use of a clearly marked space in its three spatial in addition to the spatiotemporal dimensions in the *roda*, or in *Agwa*, the circle of light on an otherwise dark stage, emphasizes a shared here and now and invites the performers and audience to explore this intense presence via repetitions, variations, and modifications of gestures and situations. The sound waves are of crucial importance in this context, as their traveling through space, both illuminated and dark, connect the stage with the space of the viewers. This combination of a single action (the dancers) with a shared active contemplation (by the audience) links the singular use of plastic with a joint reflection about our contemporary practices and attitudes toward waste.[14]

Like few other materials, plastic connects the transnational with the local, present with past, and natural environment with artifice. Just as natural rubber, one of the key materials from which the use of plastic has evolved, plastic, too, is a type of polymer. While early plastics were derived from biological material, it is today mostly produced synthetically or semisynthetically. In the Brazilian context, the reference to rubber has become a metonym for exploitation and (neo)colonialism, as the earliest profitable natural plastic was *caoutchouc* derived from trees that are native to the Amazonas Basin (*Hevea brasilensis*).[15] The Latin American rubber boom, dating roughly from the 1860s to early 1900s (Barham and Coomes 142), was a direct "perpetuation of colonial terror" (Beckman 170), which saw mass enslavements and killings of indigenous workers and "free" *seringueiros* (rubber tappers) by the rubber men.[16] The exploitation had deep and long-lasting implications for the region's ecosystems: erosion, changed water cycles, reduced biodiversity and soil fertility have been realities for those living in regions marked by excessive rubber exploitation (Cotter et al. 111). The ultimate cause of the indigenous peoples' and nature's suffering was situated at a spatial distance: "the industrialized world's appetite for rubber" (Tully 86). Its aim was to increase the mobility of others, as it was rubber that would revolutionize land transport "against that other great communications marvel of the nineteenth century, the railroad" (Tully 22). Today, more than a century after the decline of the natural rubber boom, the synthetic rubber industry has still not yet been able to manufacture a product as good as natural rubber (Tully 357). The violent exploitation of rubber continues, but so does the resistance to environmental rapine: in Brazil, Chico Mendez, former *seringueiro* turned ecological activist during the 1970s, who was murdered in 1988, remains a symbol and inspiration for the continued local engagement against human and ecological harm related to rubber tapping on an international scale.

Agwa does not explicitly discuss the colonial legacy of rubber exploitation in Brazil. Instead, it proposes a more allusive reading of plastic as a second-generation material that has today partially replaced rubber. The performance's interpretation joins historical references, present practices, and future speculations via art, through its nonchronological form and emphasis on collaboration. Dance scholar André Lepecki formulates this method compellingly in *Singularities. Dance in the Age of Performance*: "In not being individual, the repetition of a story . . . beyond personal survival, is singularity: that precipice and swerve that in persisting beyond the self, turns performance into the event it must always become: gathering of pasts and futures in the shared urgency of the collective now" (176). The performance is a medium, in the words of art theorist and art historian Catherine Soussloff, "to counteract the reenactment of an actual history in the future . . . to re-present its fictional potential *as* art in the present" (583, emphasis in original).[17] Via the insistent, stubborn surfacing of plastic objects that at times are used by the dancers, at other times act on and condition the human body, *Agwa* ponders contemporary perversions of globalized consumerism and anthropogenic fossil exploitation.

The performance invites its audience to engage with things "on the perceptual, cognitive, and emotional level" toward an "embodied experience of speculative future environments" (Weik von Mossner 15), based on current habits of excess. The aesthetic mode of dance engages in an ecological rethinking of the human–nonhuman interaction based on motion beyond conventions shaped by modern episteme. It unfolds in collaboration. Moreover, it resists conceptual and anthropocentric closure. This "refusal to stay put" (Bal 37) is precisely one of the fundamental characteristics of the metaphor. According to Bal, this figure of speech "does not simply change meaning but enriches it, and confronts [its readers and audiences] with dilemmas of understanding that activate them" (32). Hence, plastic may well reference different contexts for different members of the audience, among which the exploitation of raw materials from the Brazilian rainforest is one. It may be also a pejorative adjective referring to an object perceived as fake (Meikle 6), oftentimes in direct contraposition to an assumed "realness" of the human body. In addition, it may be interpreted as a means of imitation and substitution of previous materials or a metonym of human yearning for intentional shaping. Finally, it may be conceived as a translucent material that, when looked through it, visually alters that which is behind it. The dancers in *Agwa* gesture at all these possible uses, connotations and meanings of plastic. Their final motion of throwing the cups up in the air and throwing themselves on the dance floor, while the cups are falling down on them, powerfully closes *Agwa*'s cautionary tale of overlooking plastic's furtive yet at the same time all-too-present presence on stage and, by extension, in our lives: if ignored, it might fall back on us, literally, in stop motion, interaction, and play in and beyond a dance performance.

While the main metaphor referenced by *Agwa* is water and our dependency on plastic and everyday cultural practices of creating and disposing waste, *Pixel* examines the impact of immaterial electromagnetic impulses in the human body and environment. It is an interactive and intermedial choreography for and by humans and light dots that includes three videos of movements of pixels, along with a live animation enacted remotely from the electronic booth at the far back of the performance site that interacts with the dancers on stage. Unlike a dance centered on the power of the human body as the sole agent of creativity—or its opposite, a mechanical ballet moved by absent or invisible forces—*Pixel* explores webs of multiple agencies and interrelations of dancing bodies, things, and immaterial entities.

The swarms of light, created by the digital artists in collaboration with the dancers, shape an interactive environment and perform alongside the human bodies: at times, they cover the floor and transform it into a moving tissue of waves and grid structures that separate the dancers from one another. They slop onto the back wall of the theater, just as water would wash along a quay, snow down on the bodies, swirl up, form walls and semitranslucent veils that limit and alter the dancers' motions. Merzouki, his dancers, cochoreographers, and sound and light artists create situations where bodies may be illuminated in part, hence seeming fragmented (at times, only the dancers' hands or feet are visible) or moving once in plain sight, framed by an impressive assembly of flashing lights, then invisibly, so that their audience can see them only at intervals. This gives the impression that the bodies are flying from one side of the stage to another.[18] We, viewers in the darkness of the hall, are asked to reflect on our habits of looking that may include the overlooking of that which is in plain sight and apparent seeing, which is an imaginative filling in the blanks of that which is not shown to us.

Asked during an after-show Q&A about the presence of pixels in the work, video artist Adrien Mondot responded that art, for her, brings "alchemy to the everyday" (Mondot). The visual artists' website adds: "We want to deform perception . . . cross the daily boundaries of reality, and reveal things that are not 'possible'" (Bardainne and Mondot). In their performance, the dancers and the sound and virtual artists propose an artistic reexamination of light dots that estranges them from the daily virtual use of pixels and activates a new perspective on reality and its representations.

With its focus on electromagnetic waves, the piece proposes a reconsideration of the interplay of physical with seemingly immaterial forces, including elusive qualms about control, contamination, or radiation. At certain moments, it is a significantly immaterial matter that determines the dancers' motions to the point of immobilizing them. *Pixel* portrays this in a scene where a growing wave of dots forms on the right margin of the stage and rolls toward the dancers in a menacing way; the dancers run away from it to huddle together at the dark left margin. In another scene, light dots form moving circles that become vortexes on the dance floor, within which the dancers move in a turning motion. These illuminated isles seem to possess the power to accelerate the human bodies that now move in a dervish-like fashion, evoking perhaps electromagnetic pollution. In yet another passage that begins with the performance of a quartet of dancers in the middle of the dance floor, a grid structure builds on all four corners of the stage that slowly tightens around the dancers, whose range of motion becomes increasingly contained. At this moment, one dancer takes a brave leap into the sea of moving white spots, which then opens up for him to break free. In contrast to him, the other three dancers seem afraid: they get closer together, until they form a motionless bundle of bodies that is being figuratively carried away by the dots of light: the dancers' bodies fall on the floor and follow the wave-like motions of the dots that steadily "wash" them toward the side edge of the stage, where they disappear in the dark. In these scenes, the performance propels a rethinking of our entanglement with matter beyond limiting dualistic divides such as those of the visible versus the invisible, human versus nonhuman, subject versus object, and biosphere versus semiosphere.

The word *pixel*, an amalgamation of "picture" and "element," is a basic unit or "single point on a visual display device such as a monitor. Graphic devices display pictures, which can consist of thousands of pixels arranged in rows and colors" ("pixel," Ince. As the smallest controllable element in a digital image, a pixel is a "minute area . . . of uniform illumination" (*OED*) and thus a tiny space of representation that connects an invisible, abstract context with a visible reality. A pixel, this fundamental component of our everyday lives in digital environments, is too small to be seen. Trained to perceive the overall shape of an image, we ignore the cellular units of visual representation; until, as shown in the performance, they grow, become visible, group together, and act against the human body. In their entirety, on one (our usual) level of reality, the light dots function as a constellation. Up close, on another, correlating, level of reality, individual pixels are miniature spaces of representation of their own. Their joint force, in the form of an assemblage of light that becomes visible to the dancers and audience, is the result of a "swarm intelligence and collaboration" among pixels that behave similarly and move in sync with one another (Rolling 5).

The tiny areas of visualization assemble to form a mosaic of an electromagnetic body that moves on stage to interact with the bodies of the dancers. *Pixel* represents and, moreover, embodies encounters of "more-than-human" materiality in "a constant process of shared becoming [that] tell[s] us something about the 'world we inhabit'" (Iovino and Oppermann 1). This combined (at times antagonistic) performing of human and nonhuman agencies characterizes the performance as an artistic encounter in a political space that invites a critical engagement related to the range of interaction between human bodies and swarms of electromagnetic waves that span the spectrum from collaboration to attack.

Pixel develops further *Agwa*'s creative exploration of a shared spatiality. While in *Agwa*, echoing the capoeira space of the *jogo*, real space is used "in four dimensions, defining a sphere with circular movements, like a gyroscope in constant motion" (Almeida 132),[19] in *Pixel*, too, the light dots and human bodies carve a space out of motion. But now this spatiality expands and includes virtual sites of representation, created out of electromagnetic impulses, thermic and luminous energy. During the performance, the difference between real and virtual space disperses (Figure 12.3).

Figure 12.3 *Pixel* (2015).

Source: # 14111804–19BF, reprinted with permission by photographer Benoîte Fanton.

The prefix *eco* [Greek οἶκος, "house" or "adobe"] of the word *ecology* now includes virtual "houses" and those that are invisible or, conversely, too visible to be seen from our everyday perspective or that present spaces of connection that develop similarly to or beyond the traditional chronology.

A metaphor has its beginning not in language but in a situation in need of expressing and understanding.[20] The situations that Compagnie Käfig examines in both performances are those of human hubris related to our creation of plastic waste, neoliberal practices, and ensuing situations of bodily and spatial control, dependency, and exclusion both in traditionally described real and virtual spaces. *Agwa*'s and *Pixel*'s allusive and metaphorical explorations of these topics confront their audiences with "dilemmas of understanding that activate them [metaphors, and in extension, situations]" (Bal 32) and help perceive that which is too present or discover that which remains largely unobserved as it develops on different levels or on alternative "intertemporal" axes (Franko 8). These strategies in both performances slow down and deepen perception. They redirect our awareness and revisit old facets, as well as exploring new facets of established metaphors, which is a move toward a conceptual plurality and mobility. This technique openly contrasts our contemporary politics of speed and spectacular entertainment (Nixon 11); it may convey that which is usually invisible, be it too present or subcutaneous, too small or dilated to be taken into consideration. The artistic medium of this ambitious interplay of material, immaterial, and transparent matter is the human body in motion in space. It is precisely the dancers' and our shared bodily perception that alerts to the challenges and interferences related to the more-than-material spaces we inhabit and that carves out a potential of agency that goes beyond the mere individual.

Notes

1. I dedicate this essay to Jesús Alonso-Regalado, fellow modern dance aficionado, Latin American studies research librarian at UAlbany, and ever enthusiastic and steadfast travel companion to the Jacob's Pillow Summer Dance Festivals for the last decade. Let's (watch) dance!
2. Both *Agwa* and *Pixel* have been performed worldwide for nine and four years, respectively. The casts have slightly varied during these years. Each performance has two casts; in addition, during tours, dancers at times switch roles (Mondot, "Post-Performance Talk"). In my analysis, I am referring to the casts of the performances that I saw; *Agwa*: Fidelis da Conceição, Cristian Faxola Franco, José Amilton Rodrigues Junior, Diego Gonçalves do Nascimento Leitão, Aguinaldo de Oliveira Lopes, Leonardo Alves Moreira, Wanderlino Martins Neves, Aldair Junior Machado Nogueira, Luiz Caetano de Oliveira, Diego Alves dos Santos, Alexsandro Soares Campanha da Silva; *Pixel*: Kader Belmoktar, Sofiane Tiet, Marc Brillant, Elodie Chan, Aurélien Chareyron, Yvener Guillaume, Antonin Cattaruzza, Ludovic Lacroix, Ibrahima Mboup, Paul Thao, and Médé Yetongnon.
3. Since Merzouki became director of the Centre choréographique national (CCN) at Créteil and Val-de Marne, his mission has been to choreograph for his own troupe, in addition to inviting independent and emerging dancers and companies from many parts of the world to collaborate with him, his dancers, and cochoreographers (*Compagnie Käfig*, "Projet artistique," web).
4. While Merzouki's pieces explore a wide range of performative traditions that include North African dances, the legacy of expressive dance of the first half of the twentieth century, and contemporary acrobatics, in the two pieces selected for this study, the main creative reference is the tradition of capoeira and related street arts from Brazil.
5. For a recent analysis of the function of the performing arts in the country, see *Performing Brazil: Essays on Culture, Identity and the Performing Arts*, edited by Kathryn Bishop-Sanchez and Severino Joao Medeiros Albuquerque.
6. I follow James Haywood Rolling's definition of swarm identity developed in "Swarm Intelligence and Collaboration": "A human swarm is a social network of individuals behaving for a time like-mindedly or self-similarly. While behaving together as a local swarm can at times lead to an unthinking 'mob mentality' or ill-reasoned 'crowd hysteria,' for the most part human swarm intelligence contributes to the ongoing development of distinct cultural practices, belief systems, and patterns of mutually beneficial social actions—our general altruistic intent to perpetuate our species" and to transmit knowledge and values (5).
7. Ioviello and Oppermann use Donna Haraway's term of *natureculture*s that transcends a simple juxtaposition of nature and culture and proposes instead a combined "mesh" of both, where "culture and nature become a hybrid

compound," which both see as "the necessary terrain of every critical analysis" (5).

8. Following Jane Bennett, "[T]he bodily disciplines through which ethical sensibilities and social relations are formed and reformed are *themselves* political and constitute a whole (underexplored) field of 'micropolitics' without which any principle or policy risks being just a bunch of words" (xii; emphasis in original).
9. Both performances explore the concept of transcorporeality, which Stacy Alaimo describes as follows: "As the material self cannot be disentangled from networks that are simultaneously economic, political, cultural, scientific, and substantial, what was once the ostensibly bounded human subject finds herself in a swirling landscape of uncertainty where practices and actions that were once not even remotely ethical and political matters suddenly become so" (187). Alamo highlights this critical rethinking of the notion of the human body as a discourse that counters those of "global capitalism and the medical-industrial complex, that reassert a more convenient ideology of solidly bounded, individual consumers" (187).
10. In addition to Butler's conceptualization, I adopt Mieke Bal's definition of political art from the initial chapter of *What One Cannot Speak. Doris Salcedo's Political Art,* according to which, a "work of art emphatically endorses the inescapable fact that it is part of the world in which it occurs" (9). For her, in the case of "certain artwork [that] is in and of and about the world . . . the two elements of the phrase *political art* cannot possibly be separated" (1). For a definition of space in an ecological framework, Emily Apter's notion of a critical habitat as "a concept that explores the links between territorial habitat and intellectual habitus; between physical place and ideological forcefield, between economy and ecology" has been a helpful critical approach for my study (23).
11. In their 2017 study "Production, Use, and Fate of All Plastics Ever Made," Roland Geyer, Jenna R. Jambeck, and Kara Lawender Law "estimate that 8300 million metric tons (Mt) of virgin plastics have been produced to date. As of 2015, approximately 6300 Mt of plastic waste had been generated, around 9% of which had been recycled, 12% was incinerated, and 79% was accumulated in landfills or the natural environment. If current production and waste management trends continue, roughly 12,000 Mt of plastic waste will be in landfills or in the natural environment by 2050" (1).
12. To give an example, the United States, the country where I am writing this essay, had until March of 2018 sent about half of all its plastic scrap to China. After China's drastic reduction of imports of foreign scrap materials, the United States currently recycles around 10% of its discarded plastics, burns some 15% in waste-to-energy facilities, and dumps some 75% of its plastic waste in landfills (O'Neill 1–5).
13. A compelling narrative equivalent to the photographic portrayal of plastic scrap in the ocean can be found in Charles Moore and Cassandra Phillips' *Plastic Ocean. How a Sea Captain's Chance Discovery Launched a Determined Quest to Save the Oceans.* One example, among many others reads, as follows: "Here on deck of *Alguita* drips a broken blue plastic crate holding items that might have been culled in a children's neighborhood treasure hunt. A toothbrush, a toy car, rubber sandal, a comb, bottle caps, a Popsicle stick, a shopping bag, all plastic. But they are faded and worn from years, maybe decades, in salt water. Most have been nibbled and look like chewed dog toys. They'd seem benign if I hadn't found them where they least belong, in the middle of the Pacific Ocean, about six hundred miles north of Honolulu" (57). Captain Moore discovered the Great Pacific Garbage Patch.

14. This zooming in on a precise constellation is a strategy to counter the risk of universalization, which would put the topic of creating waste out of reach for critical engagement.
15. According to John Tully, the name *caoutchouc* was likely a corruption of the Amerindian word *cauchu* meaning "weeping wood" (20).
16. In *The Devil's Milk: A Social History of Rubber,* Tully describes one of the most gruesome killings, the Putumayo genocide of 1912 along the Putumayo River on the borders of Peru and Colombia as follows: "A British House of Commons report published in 1912 estimated that some 32,000 mainly Huitoto tribesmen, women, and children had been murdered or worked to death within a five-year period, leaving a scant 8,000 survivors to wander in the ruins of their world" (86).
17. My analysis is here indebted to Soussloff's reading of historical reenactment in art. In the essay "A Proposition for Reenactment. *Disco Angola* by Stan Douglas," Soussloff describes Canadian artist Douglas's photographic series of dance related to the Angolan Civil War as the creation of an "intentional fiction of the past for purposeful use in the present" (583).
18. This combination of movement and light directly references David Parson's famous strobe dance *Caught* from 1982, where a soloist caught in intervals by strobe light at the top of repeated leaps, seems to fly in the air.
19. Following Almeida, "[t]his sphere envelops the energy of the fighters and the best Capoeirista controls the inner space. His opponent must be handled carefully, as if he were inside a bubble of gelatin that needs to be moved around intact. An abrupt movement of attack that is mistimed will shatter the harmony of the jogo" (132).
20. This definition of the figure is significant in the case of my study, as both performances abstain almost exclusively from using language—the titles consist in single words, the description of the pieces are extremely sparse—and develop in and out of encounters of human beings, matter, and electromagnetic impulses that are then explored in motion and interaction. However, dancers do communicate with one another via quick shouts to coordinate synchronized movement.

Works Cited

Agwa. By Mourad Merzouki. Choreographed by Mourad Merzouki and Kader Belmoktar, Compagnie Käfig, Biennale de la Danse de Lyon, 2008.

Alaimo, Stacy. "Oceanic Origins, Plastic Activism, and New Materialism at Sea." *Material Ecocriticism*, edited by Serenella Iovino and Serpil Oppermann. Indiana UP, 2014, pp. 186–203.

Almeida, Bira. *Capoeira: A Brazilian Art Form.* Sun Wave, 1981.

Apter, Emily. "The Aesthetics of Critical Habitats." *October*, vol. 99, 2002, pp. 21–44.

Arons, Wendy, and Theresa J. May, eds. *Readings in Performance and Ecology.* Palgrave Macmillan, 2012.

Bal, Mieke. *Of What One Cannot Speak: Doris Salcedo's Political Art.* U of Chicago P, 2010.

Bardainne, Claire, and Mondot, Adrien. "Pixel." *AM-CB.Net* (Artists Web Site), www.am-cb.net/en/projets/pixel. Accessed 3 Dec. 2018.

Barham, Bradford L., and Oliver T. Coomes. *Prosperity's Promise: The Amazon Rubber Boom and Distorted Economic Development.* Westview Press, 1996.

Barthes, Roland. "Le plastique." *Mythologies.* Seuil, 1970, pp. 171–73.

Beckman, Ericka. *Capital Fictions: The Literature of Latin America's Export Age*. U of Minnesota P, 2012.

Bennett, Jane. *Vibrant Matter: A Political Ecology of Things*. Duke UP, 2010.

Bishop-Sanchez, Kathryn, and Severino Joan Medeiros Albuquerque, eds. *Performing Brazil: Essays on Culture, Identity and the Performing Arts*. U of Wisconsin P, 2015.

Butler, Judith. *Precarious Life: The Powers of Mourning and Violence*. Verso, 2006.

Cash, Debra. "Fuse Dance Review: *Compagnie Käfig*: Arbitrary Exoticism." 8 Feb. 2014, http://artsfuse.org/100532/fuse-dance-review-compagnie-kafig-arbitrary-exoticism/.

Caught. By David Parson. Choreographed by Elizabeth Cornish and David Parsons, David Parsons Dance Company. Yonkers, NY, 1982.

Compagnie Käfig. "Projet artistique. Une fenêtre ouverte sur le monde." https://ccncreteil.com/ccn-cie-kafig/missions-et-projets/projet-artistique. Accessed 10 Sept. 2018.

Cotter, M., et al. "Biodiversity and Ecosystem Services: A Case Study for the Assessment of Multiple Species and Functional Diversity Levels in a Cultural Landscape." *Ecological Indicators*, 2017, pp. 111–117. *EBSCOhost*, doi:10.101016/j.ecolind.2016.11.038.

da Silva, Ivone S., et al. "Heavy Metal Distribution in Recent Sediments of the Tiete-Pinheirons River System in Sao Paulo State, Brazil." *Applied Geochemistry*, vol. 17, no. 2, 2002, pp. 105–6. EBSCOhost, doi:10.1016/S0883-2929(01)00086-5.

DeLoughrey, Elizabeth, and George B. Handley. "Introduction: Toward an Aesthetics of the Earth." *Postcolonial Ecologies: Literatures of the Environment*, edited by Elizabeth DeLoughrey and Georges B. Handley, Oxford UP, 2011, pp. 3–51.

"Ecosystem." *OED Online*. Oxford UP. www.oed.com/view/Entry/144833. Accessed 13 Dec. 2018.

Fanton, Benoîte. *Agwa Compagnie Käfig* (# 16110503–10BF, # 16110503–25BF). 2016. CNN, Créteil, France. *Benoîte Fanton. Photographe de Théâtre et de Danse*. www.benoitefanton.org/161105-agwa. Accessed 7 June 2019.

————. *Pixel. Création mondiale* (# 14111804–19BF). 2014. CNN, Créteil, France. *Benoîte Fanton. Photographe de Théâtre et de Danse*. CNN, Créteil. www.benoitefanton.org/15120301-pixel. Accessed 7 June 2019.

Franko, Mark. "Introduction: The Power of Recall in a Post-Ephemeral Era." *The Oxford Handbook of Dance and Reenactment*, edited by Mark Franko. Oxford UP, 2017, pp. 1–18.

Geyer, Roland, Jenna R. Jambeck, and Kara Lavender Law. "Production, Use, and Fate of All Plastics Ever Made." *Science Advances*, vol. 3, no. 7, 2017, pp. 1–6. http://advances.sciencemag.org/content/3/7/e1700782.full. Accessed 19 Aug. 2018.

Gonçalves do Nascimento Leitão Diego, and Cristian Faxola Franco. Interview by Philip Szporer. *Agwa and Correira: Post-Performance Interview*. Jacob's Pillow, MA, 28 June 2013, www.youtube.com/watch?v=STsfpTQ-JMA. Accessed 4 Oct. 2018.

Ince, Darrel. "Pixel." *A Dictionary of the Internet*. Oxford UP, 2013, <http://oxford reference.com.librpoxy.albany.edu/view/10.1093/acref/9780191744150.001.00 01/acref-9780191744150-e-2541>. Accessed 17 Nov. 2018.

Iovino, Serenella, and Serpil Oppermann. "Stories Come to Matter." *Material Ecocriticism*, edited by Serenella Iovino and Serpil Oppermann. Indiana UP, 2014, pp. 1–17.

Johnson, Mark. *The Meaning of the Body: Aesthetics of Human Understanding*. U of Chicago P, 2007.

Lebreton, L., et al. "Evidence That the Great Pacific Garbage Patch Is Rapidly Accumulating Plastic." *Scientific Reports*, vol. 8, no. 466, Mar. 2018. *EBSCOhost*, doi:10.1038/s41598-018-22939-w. Accessed 9 Oct. 2018.

Lepecki, André. *Singularities: Dance in the Age of Performance*. Routledge, 2016.

Meikle, Jeffrey, L. *American Plastic: A Cultural History*. Rutgers UP, 1997.

Merrell, Floyd. *Capoeira and Candomblé. Conformity and Resistance Through Afro-Brazilian Experience*. Iberoamericana/Vervuet, 2005.

Merzouki, Mourad. Interview by Lucia Camargo Rojas and Briana Prevost. "Uncaged: A Q&A with Choreographer Mourad Merzouki." *The Post and Courier*, 23 May 2013, www.postandcourier.com/news/uncaged-a-q-a-with-choreographer-mourad-merzouki/article_fddc476b-6620–56d3-ac6b-ebb1201dc2f7.html. Accessed 9 Oct. 2018.

Molzahn, Laura. "Dance Review: *Compagnie Kafig* [sic] at B-Real Chicago." www.chicagotribune.com/entertainment/theater/dance/ct-ent-0224-kafig-dance-review-20140222-story.html 22 Feb. 2014. Accessed 22 Oct. 2018.

Mondot, Adrien. "Post-Performance Talk." Fuerstenfeldbruck, 27 June 2018.

Moore, Charles, and Cassandra Phillips. *Plastic Ocean: How a Sea Captain's Chance Discovery Launched a Determined Quest to Save the Oceans*. Avery, 2011.

Morgan, Carmel. "Compagnie Käfig: *Correria, Agwa*." 8 Mar. 2014, https://criticaldance.org/compagnie-kafig-correria-agwa/. Accessed 3 Oct. 2018.

Nixon, Rob. *Slow Violence and the Environmentalism of the Poor*. Harvard UP, 2011.

O'Neill, Kate. "The Plastic Waste Crisis Is an Opportunity for the U.S. to Get Serious About Recycling at Home." *The Conversation*, 17 Aug. 2018, http://theconversation.com/the-plastic-waste-crisis-is-an-opportunity-for-the-us-to-get-serious-about-recycling-at-home-93254. Accessed 22 Oct. 2018.

Pixel. By Mourad Merzouki. Choreographed by Mourad Merzouki, Adrien Mondot and Claire Bardainne. Compagnie Käfig, Maison des Arts de Créteil, 2014.

"Pixel." *OED Online*. Oxford UP, www.oed.com/view/Entry/144833. Accessed 13 Dec. 2018.

Rigby, Kate. *Dancing with Disaster: Environmental Histories, Narratives, and Ethics for Perilous Times*. U of Virginia P, 2015.

Rolling, James Haywood. "Swarm Intelligence and Collaboration." *ArtEducation*, vol. 69, no. 5, Sept, 2016, pp. 4–6. EBSCOhost, doi:10.1080/00043125.2016.1201400.

Rosa, Cristina F. "Performing Brazilianness Through Dance: The Case of *Grupo Corpo*." *Performing Brazil: Essays on Culture, Identity and the Performing Arts*, edited by Kathryn Bishop-Sanchez and Severino Joao Medeiros Albuquerque. U of Wisconsin P, 2015, pp. 65–97. EBSCOhost.

The Sacrifice of Giants. By Camille Hanson. Choreographed by Camille Hanson, Teatro Victoria Madrid, 2016.

Santiago Muíño, Emilio. "De nuevo estamos todos en peligro. El petróleo como eslabón más débil de la cadena neoliberal." *Petróleo*, Arcadia Macba, 2018, pp. 15–75.

Soussloff, Catherine. "A Proposition for Reenactment. *Disco Angola* by Stan Douglas." *The Oxford Handbook of Dance and Reenactment*, edited by Mark Franko. Oxford UP, 2017, pp. 571–86.

Three Branch Song. By Merián Soto. Choreographed by Merián Soto. Hostos Center for the Arts and Culture, 2006.

Thomas, Chelsey. "*Compagnie Käfig*: Both Raw and Refined." *Dance Informa*, 28 Feb. 2015, www.danceinforma.com/2015/03/02/compagnie-kafig-both-raw-and-refined/. Accessed 13 Dec. 2018.

Tully, John. *The Devil's Milk: A Social History of Rubber*. Monthly Review P, 2011.

Weik von Mossner, Alexa. *Affective Ecologies: Empathy, Emotion, and Environmental Narrative*. Ohio State UP, 2017.

13 *Llubia Negra*

Fetishism of Form, Temporalities of Waste, and Slow Violence in Cartonera Publishing of the Triple Frontier (Paraguay, Brazil, Argentina)

Elizabeth M. Pettinaroli

Around two o'clock on the afternoon of April 4, 2009, a storm battered eastern Paraguay with winds gusting up to more than 50 miles an hour followed by a mysterious black rain, sowing massive confusion and panic. In a matter of minutes, the temperature plummeted some 40 degrees from 99 to 57 degrees Fahrenheit, and day was plunged into night. This puzzling phenomenon, known to Paraguayans ever since as the *lluvia negra* (black rain), kindled apocalyptic dread among the devout, who read the strange storm as a sign of divine disapproval of the lascivious proclivities of then-president Fernando Lugo; incited fears of a possible nuclear holocaust due to the rain's resemblance to the precipitation that fell on Hiroshima after the atomic bomb was dropped there; and even excited the transplanetary imaginings of some who suspected an extraterrestrial invasion.

Once the worst of the storm had passed, the focus turned to "intenta[r] descubrir lo que nos cayó encima" ("the attempt to identify what has fallen on us"), as expressed by the newspaper *ABC Color* (April 6, 2009). Official media rapidly mobilized a campaign of mockery toward catastrophic speculations and conspiracy theories and, contradicting the Instituto Nacional de Tecnología, Normalización y Metrología (INTN, National Institute of Technology, Standards and Measures), reporters insisted on blaming the phenomenon first on the recent eruption of the Llaima Volcano in Chile and subsequently on soot from forest fires near Lake Ypoá. Both hypotheses were improbable given the direction of the prevailing winds. Experts did point out that, if this unusual rain were to contain volcanic debris, the ashes would float to the top of water samples. A viscous substance was instead consistently found at the bottom of testing containers. The INTN responded, notably, that its study of water samples had ruled out volcanic ash as the culprit, verifying the presence of organic elements and high levels of sodium, sulfur, and potassium. Suspiciously, however, further chemical analyses were halted; the official explanation provided was a sudden lack of funding (*ABC Color* April 6, 2009). A few days later, the INTN issued its official finding that the phenomenon had been the result of smoke from nearby pastures fires, followed by a strong storm. The manifestly inadequate scientific assessment of the events of April 4 by governmental entities charged with investigating such matters contributed to a generalized public distrust. Speculation about secret chemical warfare testing, weather modification experiments, and other popular hypotheses ran wild. There was also widespread discussion of long-standing economic practices, from dumping of toxic industrial waste, to intensive chemical agribusiness, to flood-inducing infrastructural projects like the Itaipu dam. The mystery surrounding the event's causes also fed the practitioners of the distinctively Paraguayan form of humor known as *cachiai*, who had a field day declaring the return of the *chaparrones aislados* (isolated downpours), the phrase the late dictator Alfredo Stroessner had infamously forced meteorologists to use to describe the torrential downpours and massive flooding caused by the region's severe deforestation.

The volume *Llubia Negra: 11 narradores paraguayos y non-paraguayos* (2009), edited by poet Douglas Diegues, is the fruit of a cooperative project of the Yiyi Yambo *cartonera* publishing house. (See Figure 13.1). The novel spelling of the volume's title suggests a gesture meant to highlight the unique quality of the strong, damaging storms in the Triple Frontier region of Paraguay, Brazil, and Argentina. The stories gathered in the book are described evocatively by Diegues as "brotados como hongos de essa lluvia negra . . . como kakemono [sic] paraguayo de lo que se escribe em la tri border guaraní punk"[1] (3; "sprung up like mushrooms from that black rain . . . like a Paraguayan kakimono scroll of what gets written along that Guaraní punk tri-frontier"). Yiyi Yambo *cartonera* is part of a cultural and political phenomenon born in Latin America and now well into its second decade. Eloísa Cartonera, the original publishing house, emerged in Buenos Aires, Argentina, in 2003 as a collaborative space bringing together writing and publishing in an alternative to the aesthetic, ethos, and practices of the transnational publishing industry (Bilbija and Carbajal 6–10). Spurred by the dramatic contraction of the publishing market in the wake of the financial crisis of 2001, the model was replicated in most major cities of South America, is currently expanding to several European urban centers, and has emerged most recently in a number of cities in the United States.[2] Yiyi Yambo is inspired by Eloisa's foundational project, which reconsiders the materiality of cardboard, transforming it from refuse to riches in the form of unique, hand-painted book covers: the very name of the phenomenon embodies the material, *cartón* (cardboard), and evokes the precarious economic niche occupied by *cartoneros* (gatherers of discarded cardboard) on the Latin American urban scene. But, as with *cartonera* publishers as a whole, Yiyi Yambo does much more than creatively recycle cardboard. More broadly, it joins the manifold functions of publishing—editing, printing, binding,

Figure 13.1 Llubia Negra: 11 narradores paraguayos y non-paraguayos (2009).
Source: Photo by Elizabeth M. Pettinaroli.

marketing, distribution, dissemination, and so on—with an innovative literary praxis, and does so in a fashion organically tied to its "*origen fronterizo*" (borderland genesis) along the Triple Frontier (Diegues "Interview"). In so doing, Yiyi Yambo and its sister projects recover the nexus between aesthetic undertaking and socioenvironmental consciousness.

Cartonera publishing embodies the notion of "ecotone" that Rob Nixon defines as spaces in which life forms that ordinarily thrive in discrete conditions encounter one another and intermingle, opening new configurations of possibilities (30). (See Figure 13.2). From an environmental humanities perspective, Nixon analogizes literary studies that remain disconnected from world concerns to the separateness that ecotones disrupt; without the intermingling brought by ecotone, literary studies are condemned to a formalism severed from history and afflicted with a dramatically limited field of vision—a formalism that Anne McClintock has termed a "fetishism of form" (quoted in Nixon 31). This schism elides the political, historical, and environmental responsibility that is central to the imaginative capacity of literary and artistic creation. Nixon identifies this

Figure 13.2 Various publications by Yiyi Yambo *cartonera* publishing house, using local textiles and materials such as *ñandutí*—a traditional Paraguayan embroidered lace knitted by local women.

Source: Photo by Elizabeth M. Pettinaroli.

rupture as the greatest challenge facing writer-activists in their efforts to address the complex interface between aesthetic forms and movements of socioenvironmental change—the difficult task of representing that process of gradual, protracted destruction, almost imperceptible in its dispersal across space and time, for which Nixon coined the notion of "slow violence" (32). This devastation is brought by the excesses generated by the globalized neoliberal economy.

Challenging theories that envision the *lluvia negra* as a spectacular natural disaster and nothing more, and combating the "fetishism of form" in all its insidiousness, each of the volume's 11 authors deploys ecocritical lenses to discern in the event, rather, a phenomenon temporally and chronologically diffuse, one that unfolds on a scale unimaginably vast and utterly disproportionate to human experience. Timothy Morton has coined the notion of "hyperobject" to theorize the ontology of "things that are massively distributed in time and space relative to humans" and to explore both their impact on material, political, ethical, artistic human experience and their coexistence with the nonhuman (1). Emerging from a trifrontiered region, the stories of *Llubia Negra* explore the dialectic relationship between environment and historic processes and expose the occlusions that complicate the representation of slow violence. The "fungiform" theories in the volume's short stories by Gabriela Alemán's "Lluvia negra, polvo gris" (Black Rain, Gray Dust), by Ever Román's "Osobuco" (Ossobuco), and by Diegues's "Perdidos en la black rain guaranga" (Lost in the Guaraní Black Rain)[3] challenge the temporalities of toxic waste (Nixon 46–7) and, as part of a *cartonera* collective, embody a contestation to the fetishism of form and to inherited notions of materiality. The stories address the region written in the margins of the "*aburrido oficialismo*" (3; "tedious official culture"), raising an ecocritical reproach against it.

Yiyi Yambo's collective initiative to publish *Llubia Negra* in the wake of the mysterious event reminds us of the place of the material world and its agency in creativity, freeing us from the limitations of restricting such agency exclusively to the human subject. It invites us to consider the vital capacity of nonhuman entities to act as quasi agents or forces that run alongside and within humans and to challenge our awareness about our relationships with the material world. Jane Bennett explores the vibrant materiality in things and their agency in relation to our ecological awareness. In particular, she proposes the presence of "a life" in matter with vitality of pure immanence, "a restless activeness, a destructive-creative force-presence" (an "actant" and a source of action, following Bruno Latour's coinage), and conceives of rain within these realms of being (53–4). During the author's 2017 visit to the Paraguayan capital, Asunción, casual inquiries to locals about the event unleashed a veritable cascade of oral narratives shaped by retrospection and shared with spontaneity and generosity, as well as with utter puzzlement. Ever Román recalls the overwhelming shock he experienced while witnessing the storm, the feeling of being in the presence of "*lo fantástico*" (the fantastic), sensations that infused him and indeed the entire group of authors with an urgency to grapple with the event: "Tuvimos algo así como un es para escribir algo" (Zalgade; "We had a reaction something like: It's worth writing something about this"). The materiality of Paraguay's black rain wields agency over locals and, commanding the will of writers, propels relationships and assemblages that give rise to new networks of actors exercising collective creativity.

Biosemiotics has been offered as a cross-disciplinary approach to shed light on our understanding of human creativity and materiality. Hubert Zapf proposes an analogy between creative processes in nature and in culture, as they both share an element of agency in addressing changes in the environment, in realigning patterns of life, and in reimagining the world and creating new meanings (53). The logical process of "abduction" (rather than the logical operations of induction or deduction), as elucidated by the semiotician Charles Sanders Peirce, is appropriate here to think about a complex degree of unconscious reasoning that entails imaginative reinterpretation and that brings formerly separate domains together into new amalgamations (quoted in Zapf 53). An ecocritical perspective invites us to explore the connections between the semiotic emergences of the Paraguayan black rain and the creativity of the collective. Imaginatively, these authors bring together both rain and cardboard as actants that propel a reinterpretation of the event from the familiar "*lluvia*" (rain) to *llubia* (its unconventional spelling, marking the event's shocking distinctiveness) and that transform generalized pluvial precipitation into a particular, localized phenomenon of the Triple Frontier region.[4] Moreover, together they create a system capable of recreating and maintaining itself, a process that evokes the concept of auto-poesis that Zapf adapts from the notion of auto-poesis as used by Chilean biologists Humberto Maturana and Francisco Varela and that Latour deploys by analogy in the aesthetic realm (54). In a dynamic alliance, Yiyi Yambo's carto-poesis spearheads a *cartonera* publishing project to create novel meanings while fostering a space for creativity in a drastically contracting publishing market.

The stories selected from *Llubia Negra* for this study partake of three qualities that Tim Morton recognizes as essential properties of hyperobjects:[5] nonlocality, Gaussian temporality, and viscosity. Examples of hyperobjects include a variety of vast entities from an oil field to a black hole; from the Florida Everglades to the solar system; and from the sum of all nuclear materials on Earth to "the very long lasting product of direct human manufacture, such as Styrofoam or plastic bags" (1). The way our authors conceive of that *lluvia negra* and transform it into a literary *llubia negra*—a massive event of indecipherable cause—clearly places it into this category. With regard to the first quality, nonlocality, Morton underscores that hyperobjects have no local, direct manifestation because they remain in what is known as a high-dimensional phase space (a space in which all possible states of a system are represented) that lends them a spectral quality (38–42). In other words, the object is massive, and its cause cannot be identified in any single, identifiable point in space. In addition, there is never proof of a direct causal link to their impact but rather an oblique inference through association, correlation, and probability afforded at best by statistics—a Humean causal system. The second quality, Gaussian temporality, refers to the altering of spatiotemporal conventions that generates space-time vortices through which a hyperobject envelops or surrounds us while being massively distributed in time (55–60). Viscosity—the third quality—is the property by which a hyperobject sticks to other entities and becomes enmeshed in them (27–30). All three qualities contribute to the relative invisibility and temporal dispersion that Nixon identifies as the greatest challenges to the perception of and response to slow violence.

The authors in the volume identify the phenomenon through these three properties, eliding direct discussion of its possible causes; the narrative engine of these stories has little to do with the exercise of searching for technocratic explanations that neatly resolve the mystery. In her story "Lluvia negra, polvo gris" (Black Rain, Gray Dust), Ecuadorian writer Gabriela Alemán (who lived in Paraguay as an adolescent) engages the logical process of "abduction" by inserting the toxic rain in Asunción into a new transtemporal and transnational framework; this framework projects and transforms the local events into a multipolar landscape. With irony, Alemán recalls the theories that erroneously propounded a volcanic origin for the *lluvia negra* and shifts the scene to seven years earlier and 3,000 miles away in Ecuador, where the ashes from the Reventador volcano's 2002 eruption spread far to the west, reaching the capital city and clothing it in a blanket of gray. Through a web of recombinations, Alemán merges the *lluvia negra* from Asunción with the *polvo gris* (gray dust) of Quito, fashioning that strange familiarity or familiar strangeness that is the hallmark of the uncanny.

In a shift that renews the picaresque and transports it from its origin on the Iberian Peninsula to the Global South,[6] Alemán's protagonist, a young, nameless, unemployed *quiteña* woman, embarks on a series of travails that reminds us of the sequence of calamities faced by Lázaro in the foundational work *Lazarillo de Tormes*. Her journey from Quito to Cuenca in pursuit of a possible job entails a chain of complications full of vague and ambiguous situations whose import becomes clear only after the volcanic dust settles, enabling her to find new meaning in the topography of her apartment. Slowly grasping the occluded relationships that imperceptibly shape local events, she experiences a sudden flash of understanding: "Ahora es fácil, fácil ver donde [sic] empezó" (30; "Now it's easy, easy to see where it started"). Alemán's *quiteña* struggles to make sense out of the dispersion of facts in space and time—much like the environmental writers' striving to lay bare the pervasive temporal dissociations occluding the causal relationships that underlie slow violence. The series of episodic tragedies begins with the sudden appearance of a long absent father scarcely seen in years who, to her puzzlement, arrives at her house to take care of her pet *axolotl* while she is away on her job-hunting trip (30).[7] There ensues a bizarre job interview in Cuenca with a technocratic "*curepa*" ("Argie"),[8] who scornfully deploys a cacophony of pseudo scientific terms meant to evoke the latest, cutting-edge human resources practices (32); an encounter with "*ecuatorianos icónicos*" (33; "iconic Ecuadorians"), sporting pan flutes hanging from their necks and canvas "*caza gringas*" (33; "gringa bait") sandals in a fashionably minimalist urban pub saturated with tourists immersed in first-world existential dilemmas; unanswered telephone calls from Cuenca to her father in Quito; and a dehumanizing night spent under a bridge, with dinner scavenged from the city dump of a Cuenca no longer able to fulfill the dreams of prosperity of its inhabitants (35). Finally, hitchhiking her way back to the capital, she reaches Quito only to find it buried in volcanic ash and her father and the *axolotl* dead, each in a different corner of her apartment (38).

The volcanic ash, analogous to Paraguay's toxic *lluvia negra*, reorients the main character's gaze, yielding a bleak epiphany: history, reinterpreted from a new perspective, as a tragic, episodic chain of events unleashed by slow violence: "Al final todo tiene que ver con la economía" (30; "In the end, it's all about the economy"). This new environmental picaresque unearths occluded intersections and exposes the complex network of ostensibly disjointed violence brought about by neoliberal globalization.[9] Like beads on the rosary of a penitent, the *quiteña* protagonist's long series of misfortunes composes a chain of calamities that overshadows the volcanic eruption in all its spectacular power, relegating it to a secondary plane. Instead, it gives pride of place to the insidious experiences that compound the effects of toxicity, everyday alienation, precarious employment, commodification of communal identities, the extractivism of the tourism industry—all part of a process that exacerbates the protagonist's vulnerability. Recalling the mechanical nature of the interview leads her to discern the trickery of the *curepa* and, concluding that there was never a job behind the interview, to contemplate the exclusionary processes and restrictive categorization of humanity of the "unlivable lives" that Judith Butler also identifies as "precarious" (33). A minimalist pub now emerges in the cityscape as a saturation of the local with Westernized aesthetics and touristic consumerism, brought on by the foreign-language teaching industry. The marketing of a purported indigenous authenticity in the region renews the eighteenth- and nineteenth-century quest for the noble savage. Epitomizing indigeneity with their *caza gringas* sandals, "iconic nationals" indulge the anxieties of foreign tourists who, preoccupied with their soul-searching, remain oblivious to the ecological impact of the tourism in which they themselves participate and that they placidly believe sustainable. These projects feed the fetishization of poverty by First World expats and students who project their social and ecological anxieties onto the Global South (that inclination to "overworlding" that DeLoughrey, Didur, and Carrigan see as imposed on the region). Reviewing the sequence of calls, she suddenly realizes that the reason for her father's unanswered phone is not the volcanic eruption but a delinquent account. Technological modernization is available only through exorbitant fees and the extractive practices of multinationals which leave locals destitute and vulnerable, fostering inequitable access to modernity's infrastructural networks (Nixon 42). These occluded truths, Nixon reminds us, are concealed by capitalism's "tendency to abstract in order to extract," the same tendency that hides the sources and consequences of environmental violence (41).

Reconstructing the topography of her apartment, its territoriality reconfigured by different layers of volcanic ash (a powdery viscosity), invites the narrator to theorize another possible ending in which father and *axolotl* lie "felices, inmersos en la centrífuga del tiempo" (39; "happy, immersed in the centrifuge of time") while observing the spectacle of lava and volcanic ashes, dying in communion "con los ojos encandilados al tiempo" (39; "their eyes dazzled by time"). Through a Gaussian, vortex-like temporality, the new perspective unveils a different landscape in which to essay further possibilities for imaginable time, one in which old linear frameworks are now deemed unreliable. The new spatiotemporal hermeneutics unearth the temporal dispersion occluded by geographies of concealment constitutive of slow violence. These concealing emplacements obfuscate manners of causation and bolster competing regimes of truth that, as Nixon states, affirm or discount the severity of damage from environmental toxicity (47).[10] Alemán's amalgamation of locations (Ecuador in 2004, Paraguay in 2009) through allusion to a centrifuge viewpoint contributes a kaleidoscopic perspective that brings attention to the repetition of patterns of experience both deep and far-reaching but that are spectacle deficient. The new perspective addresses the representational vacuum in the administration of difference that separates official victims and nonvictims. This official calculus turns a blind eye to the suffering of those not considered indigent enough because they fall outside the prevalent political and scientific logics of causation used to define the official category of sufferers: namely, those chronically underemployed individuals who, like Alemán's protagonist, swell the nameless ranks of citizens enduring a quasi invisible, toxic existence. Ashes in Quito and *lluvia negra* in Asunción merge to give materiality to the obstruction that impedes the gaze (both retrospective and prospective) that might unveil the forces pervading and shaping the local.

Moreover, Alemán amalgamates landscapes and connects the disasters in a futuristic prophecy that envisions a terrible ordeal of vulnerabilities resulting from the slow violence throughout the past. John Berger's admonition comes to mind: "Prophecy now involves a geographical rather than historical projection; it is space and not time that hides consequences from us. To prophesy today it is only necessary to know men [and women] as they are throughout the world in all of its inequality" (40). This historical connectivity offers new epistemologies through which it is possible to grasp what has not been codified as expressible and to galvanize urgency for change. The revelation is announced in Alemán's aphoristic phrase: in the end, it's all about the economy.

In his story "Osobuco", Paraguayan writer Ever Román explores the difficulty of narrating the experience of a post–*lluvia negra* Asunción covered in a jet-black, viscous patina in which there remains—as Román chooses to say it in Guaraní—only "*mba' eve*": nothing (89). This "nothingness" does not fit common postapocalyptic representations of destruction and chaos but rather is theorized through a series of sections that alternate the main character's comical, episodic quest to procure a piece of roast beef and, on the other hand, conjectures about the value of such literary genres as TV melodrama and cyberpunk. Ramiro Biedermann,[11] our second ecopicaresque hero, struggles to obtain the coveted roast beef to feed his family and, in a grotesque dispute with a butcher and a mob of cranky elderly customers, resigns himself to buying a lower-quality cut of osso buco. The populace reminds our *ecopícaro* of the value that the humble cut of meat has, for it can be repurposed time and again by boiling it to make soup. The degradation from roast beef to the less desirable ossobuco stands as a metaphor for the deterioration of Asunceno bodies oblivious to the effects of post–black rain toxicity. In turn, these scenes are interspersed with speculations about generic forms to interpret a society seemingly oblivious to the history of slow violence in Asunción. The metaliterary meditation crafted by Román in the form of what he coins "*teorías peregrinas*" ("wandering theories") (Zalgade) essays a critique of the fetishism of form, a probing that Nixon finds at the center of writer-activists working on the representation of slow violence. Reflecting on TV melodrama, the protagonist hypothesizes, "Cuando las telenovelas terminan, nadie recuerda que el desarrollo de la historia estuvo plagado de tragedia, con cegueras, amnesias, muertes, huérfanos y el Mal Absoluto desenvolviéndose a sus anchas en cada rincón" (89; "When telenovelas end, nobody remembers that the unfolding of the story was riddled with tragedy, blindness, amnesia, deaths, orphans—and Absolute Evil riding high and at its ease in every quarter"). Form, he contemplates, imposes a type of aesthetically inflicted amnesia that erases the pernicious effects of slow violence from the process of history making and displaces them as economic externalities.[12] Death as solution is recontextualized in the *mba' eve*, the void or nothingness, in which the characters end up losing themselves at the end of Román's story. Along similar lines, he finds that the apocalyptic novel as a whole lacks the ironic optimism of telenovelas and that placing death as an omnipresent theme "no aporta absolutamente nada" (90; "contributes absolutely nothing"). This renewed *mba' eve* continues to permeate Román's critique of form.

Román's musings through his "wandering theories" leads him to speculate that the cyberpunk aesthetic, instead, addresses the vital concerns he shares with Nixon in representing socioenvironmental transformation. In its aesthetic rebelliousness, cyberpunk abandons the tenets of the Enlightenment and joins the rational with the irrational, new with old, and mind with body; it does so in perplexing ways, while integrating structures of high technology with the lawlessness of street subcultures. The technology is of little interest to Román, however. Rather, he is concerned with how the notions of time, reality, materiality, community, and space are radically reimagined in this genre. Moreover, cyberpunk devises a new language and visual imaginary to depict and mediate contemporary culture. Ultimately, cultural transformations in this genre are centered on the human body as a stage that, through the fusion of the biological and technological, disappears and is eventually reduced to "lifeless meat" (Dani Cavallaro xii, xv). Imbued with this drive to reimagine the world, Román uses the jocular incursion into the butcher shop to explore new forms of degraded corporeality, much like the decline from roast beef to ossobuco. These downward shifts usher in new forms of intersubjectivity in the face of hyperobjects such as the black rain of 2009. This process of dissolution, now made visible through the cyberpunk aesthetic, points to an emerging ontological vacuum whose terrors Román allays by fashioning a particularized Guaraní expression of nullity: *mba' eve*.

This exploration of "nothingness" in the Paraguayan capital also serves Román as a forum to denounce those forces promoting inaction and confusion, those political and social actors whom the authors in this volume, with Nixon, identify as supervising the "production of doubt" (39). Asuncenos exist in a state of aporia and a reign of irresolution under a jet-black patina of *óleo negro* (95; "black oil") that blankets the city after the mysterious rain of April 4, 2009, covering everything and rendering the cityscape with a viscous uncanniness. Even the traditional Asunceno orange trees that adorn the cityscape, standing as a metaphor and metonymy for the body politic, are drenched in *lluvia negra*.[13] These traces of unreality echo Tim Morton's observation about the social and experiential quality of hyperobjects, particularly their pervasive "viscosity" from which escape is impossible (30). Attempts to extricate oneself, to wriggle free through cognitive processes or efforts of the will, are futile, for the substance or its toxicity eventually becomes fused with the self. Immersed in the urban desolation, passersby make their way, resigned and submissive, zombie-like victims "ciegos al guión oficial" (94; "blind to the official script") and surrounded by the scars of the recent looting. This aporia in the face of the imperceptible, toxic violence described by Román is internalized and somatized in dramas of mutation (from roast beef to ossobuco), particularly in the bodies of those whom Kevin Bales terms "disposable people," individuals who remain invisible, impervious to diagnosis, and therefore untreatable. From a narrative perspective, this invisible and mutagenic theater unfolds slowly, uncertainly, in open-ended fashion, avoiding neat, self-contained conclusions imposed by conventional narratives of winners and losers. They hide behind well manicured, orderly plots complete with orthodox eschatological endings.

Moreover, slow violence challenges both the bodies of those who "[s]ufren y no dicen ni mu" (94; "suffer and utter not even a peep") and their place, their very being in space and (particularly) in time. The viscous character of hyperobjects, Morton reminds us, fosters temporal plots in which time emanates or "leaks" from objects rather than forming a continuum in which the objects float (33). That jet-black patina intrudes and hinders the human capacity to discern the operations that inscribe or erase etiologies within the particular "regimes of truth" mentioned earlier. In particular, Román denounces the effect of this patina on the temporal imagination of an Asuncena society informed by kitsch melodramas permeated by the slow violence of rhetorical cleaning operations that impose sanitary beginnings and endings. In these types of narratives, the author finds a sort of temporal aporia in which "la historia no avanza, está estancada en la pasión, entendida ésta en la acepción cristiana" (99; "history fails to advance and instead remains trapped, stagnating in a 'passion' understood in the Christian sense"). Through the analysis of the possibilities for plot that genres yield, Román explores the temporal, Gaussian undulation that Morton finds in hyperobjects, which organize modes of thinking inherited from Christian Neoplatonism with its envisioning of time (and space) as absolute containers (65–6). Nonetheless, the author cautions, ironically and comically, that the genesis of the aporia of the Asuncenos ultimately lies not in Christian dogma—since the citizenry's ignorance of such doctrines is virtually complete—but rather in "las telenovelas que son el dietario de la vida, como si fueran las tablas de Moisés" (99; "the *telenovelas* that constitute life's ledger, almost in the manner of Moses's stone tablets").

Alternatively, Román theorizes that if Paraguayan society had a predilection for consuming sci-fi, the reaction to the apocalyptic process would be different because it would enact syncretism in an altogether different way. In that case, two messianisms—one theocratic, the other technological—would converge in "un ser humano nuevo, rapiñero y feroz, despierto a la época" (99; "a new human being, rapacious and fierce, cynically attuned to the era"). The new historical agent's suspicious character—like our *ecopícaro* Biedermann—echoes the call to beware that Eduardo Gudynas issues against the fallacy of ecomessianism and the dangers of an "inordinate faith and certainty, that is without reason or proof that environmental and ecological ideas can serve as agents of change" (170). The end of times in Román's renewed apocalyptic plots would still result in death. However, the process would generate "una historia más activa que reflexiva" (99; a "history marked more by action than reflection"). This epistemological approach to the *telos* of history echoes the notion of messianicity as adapted by Jacques Derrida from Walter Benjamin and understood as an emancipatory promise that emerges in relation to an unrealizable future.[14] The connection with this retrospective Derridean vision enables Román to explore qualitative difference, disjunction, even in historical time perceived as linear. However, his theorization differs from that of the French critic, who conceived of the future as a heterogeneous temporality, while in Román's ecocritique, despite the congeries of form, the end remains the same.

In Douglas Diegues's "Perdidos en la black rain guaranga," this *llubia negra* resembles "petróleo" (107; crude oil) and falls heavily on the city, causing flooding on a diluvial scale in the poorest sections of Greater Asunción. Viscosity, in this case, has a leveling property that transforms bodies and minds. The phenomenon produces an intoxicating (and toxic) effect on all realms of materiality, embracing locals and tourists, ants and robots, fashion models and scholars, Muslims and Jews, among many others, all wallowing in the sticky darkness. This pervasive patina imposes alternative logics and transforms the local landscape into a Hollywood-like carnival scene where John Huston fondles a female hippo, and King Kong lusts after Jessica Lange while suggestively puffing a vast cloud of cannabis smoke over a bastardized, superficial "Asunciónlândia" (109). The hallucinogenic effects of the wafting fumes cut through the concealing powers of this hyperobject to unveil an occluded chapter in the history of the region: the path to a "Nueba Germania" where Elisabeth Nietzsche (sister of the philosopher Friedrich) and her radically anti-Semitic husband, Bernhard Förster, founded a "pure Aryan" colony in 1887. *Portunhol selvagem,*[15] a distinctive term for the amalgamation of Argentine Spanish, Guaraní, and Brazilian Portuguese widely spoken in the Triple Frontier region, serves as a fitting vehicle to reveal the complexity that is disguised by the patina of official history. A telegraphic narrative summarizes the scant information found in the history books that, to date, have all followed the official line. A plan to recruit 140 families for this chimerical project ended up yielding only a tenth that many who moved to the locality of San Pedro. Once arrived, their survival was impeded by the failure of their European agricultural methods on Paraguayan soil. The practice of incest and vulnerability to tropical illnesses, gigantic insect pests, predators, and the intimidating power of mysterious local myths all contributed to the fiasco of a nineteenth-century project to bring alleged racial superiority to the New World.

Through an exercise of reinterpretation, on the other hand, an unofficial account emerges through the dispersed character of a story in which what is occluded seeps through a delirious narrative. The narrator debunks myths of racial purity, telling off the "cabrones primermundistas" (114; "first-world sons of bitches") in a rant that puts the local "nativas new-germánikas" (110, "native new-Germanic chicks"), those "cute" nieces of Förster and Nietzsche, forward as the most beautiful, mixed, indigenous, and savage *kuñataís* (111, Guaraní for "young women"). The fragmented, hybrid narrative recovers the material history that was sanitized, airbrushed from the historical panorama by foundational narratives of the Paraguayan nation and later on by Stroessner's regime. A farcical account reports on the gullibility of locals who, embracing the official master narratives, believe to this day that "[s]olo existía un paraiso, diziam, el paraiso capitalista" (114; "there only existed one paradise, they said, the capitalist paradise"). Veering to the present, the narrator denounces pretensions of high art by the director of the Bayreuth Festspielhaus who, through use of digital technology with her *vaka pó* (112; "cow hands"), declines his request to host a reading of the *Llubia Negra* anthology at the local opera house. The forum, she explains, was appropriate only for the operas of Wagner—the very composer whose *Parsifal* inspired Förster to undertake the founding of his Aryan utopia in Paraguay. A variety of distractions (shallow pornographic eroticism, the impact of Hollywood and its ideological machine) derail Paraguayans from historical reality. This reality entails problematic, underexamined chapters of the country's history spanning from the foundation of the first Nazi party outside Germany to the ecological havoc inflicted on the local environment by the Itaipu hydroelectric dam. In Diegues's Gaussian framework, Nueva Germania marks the end of history, where places lose meaning and finally perish, and where "em medio de todas las mentiras imbeciles, solo la llubia negra era verdadera" (114; "amidst all the idiotic lies, only the black rain was true"). Black rain's materiality, in all its vibrancy,[16] acquires historical agency, helping to unveil the occluded degradation brought by modernity on the Triple Frontier. (See Figure 13.3.)

Diegues also shares some of Román's preoccupations with the folly of form and literary genres and, in addition, adopts some of the conventions of Tupinipunk/Brazilian Cyberpunk—fragmentary prose, challenges to inherited signifiers, sarcasm and satire, modernity and postmodernity (Vázquez 213–14)—to denounce the veiled, *guaranga* history of the larger Triple Frontier. Moreover, he refashions some of Tupinipunk's conventions, pointing them in new directions and thereby creating what I call Triple-Frontier hyperpunk: a Triple-Frontier aesthetic sensibility that addresses the representational challenges of hyperobjects and their

Figure 13.3 Photo collage by Douglas Diegues, *Llubia Negra: 11 narradores paraguayos y non-paraguayos* (2009).

Source: Photo by Elizabeth M. Pettinaroli.

slow, insidious, toxic effects. Whereas Tupinipunk's notion of boundaries is ambiguous, Diegues's Triple Frontier hyperpunk aesthetic vanishes borders, making it an impossibility to delineate the boundaries of the hyperobject. Through the fusion of the three languages into *portunhol selvagem*, as well as by the merging of characters and localities from a diversity of realms and historical times (King Kong and Jessica Lange, Elisabeth Nietzsche and the *kuñataís*, Nueva Germania and the Itaipu hydroelectric dam), all converges in a dispersed, transfrontiered existence, immersed in the viscosity of the Guaraní black rain. The punk aesthetic that had shifted to the cyberworld in cyberpunk genre is now redeployed to a grotesque, urban setting to critique the occlusions in official narratives that promise modernity's limitless progress. Sex is no longer a utopic force of liberation (to some extent via the primitive) that Tupinipunk had borrowed from anthropophagic modernism (Vázquez 214). In Diegues's *Perdidos*, it is rather a Triple Frontier excess that calls attention to the dispersed, outlandish character of the official history and that serves as the key to debunking myths. Resisting the invention of a metalanguage to craft a historiography of the region, the new eschatological scheme Diegues puts forward in his interpretation of history is now based in the materiality of the *llubia negra* itself—the sole and ultimate truth. The defiant mockery of fetishism of form in "Lluvia negra, polvo gris," "Osobuco," and "Perdidos en la black rain guaranga" exposes the fallacy of crafting a reified aesthetics into an ineluctable *telos* driving history toward the future; instead, the stories unveil what Morton calls the "wicked problem" of hyperobjects: that, in the end, there are no good rational solutions (135)—what there is, is simply *mba' eve*.

Attunement to ecological awareness[17] through critiques of slow violence serves the authors of *Llubia Negra* in particular and the practitioners of *cartonera* publishing in general to transcend the occlusive effects of narratives fostered by inherited canons and genres, and of materialities produced by dispersed, global structures of production. (See Figure 13.4). Their ecologically informed aesthetic experience is grounded in a renewal of practice as well. Through a more capacious conceptualization of the literary, these independent publishing groups foster discrete spaces and places (the ecotones previously mentioned) that open new possibilities for collective, inclusive practices in which the frontiers between content, materialities, authorship, and readership melt. These fora yield novel creations that reject the fetishism of the form. As Marcy Schwartz observes, these environments are starting points for conversations and interactions, catalysts for participation and exchange (190) that foster a new ecology of reading (151). *Cartonera* publishers contest the spectral quality by which slow violence is rendered invisible and that economic theory elides from cost–benefit analysis in the form of mere externalities. The collectives empower a variety of social actors, including displaced workers who turn to gathering cardboard as a means of survival in the wake

Figure 13.4 Yiyi Yambo collective.
Source: Photo by Douglas Diegues.

of the recurring financial crisis in the region. Increasingly, they attract writers and artists who see in the *cartoneras* productive spaces to essay new meaning through creative practices and as valuable sites for activism against the contexts of precariousness that result from the functioning of the neoliberal economic system.[18] The kinds of geographically dispersed, disembodied, "virtual" social interactions so characteristic of digital culture are reinvented in the *cartonera* scene. Technology has helped these small-scale collectives deeply rooted in local places to forge transnational connections (Epplin 393–4). Their probing into the slow violence effected by policies and ideological frameworks geared solely toward the generation of private profit through production and circulation unveils the logics of coloniality still present in these economic processes (Bilbija and Carbajal 16). In their spirit and practice of cooperation and solidarity, *cartoneras* answer Walter Mignolo's call to seek paths to decolonization of the literary realm.

The reunion of the aesthetic, the political, and the environmental, Heffes observes, expands recycling from a merely conservational act to a social one that addresses the situation of scarcity of the displaced by engaging the potentialities of art (229–30). *Cartonera* manifestos reveal a multiplicity of principles and missions by which these collaboratives align common ecological interests with a continual reconceptualization of book and audience (Álvarez Oquendo and Madureira quoted in Schwartz). In his *Ecological Thought*, Tim Morton reminds us that ecology transcends the mere assessment of the effects of extraction on nature (global warming, for instance) and the development of palliative practices still based on the logics of extraction (solar power and other alternative energies). An ecological perspective entails the critical discernment of an expansive web of interconnection that is centerless and fringeless, in which the political is constitutive of the whole. In their critique of slow violence, the authors in *Llubia Negra* transcend the ecological and engage in projects with kindred practices informed by an ecocritical mission. Moreover, they theorize and enact *cartonera* collectivism as forward-thinking practices with the power to actualize new aesthetic forms and human relations that forge paths to new futures. Thus, *cartonera* publishing fosters innovative cultural ecosystems that bring together areas previously disjointed, continuously reimagining new intersections (that generative matrix brought about by ecotones) through the creation of collective communities with an ecocritical foundation and purpose itself at its core.

The stories in *Llubia Negra* propose new theorizations of rain that renounce the fantasy of scientific control, instead embracing the *llubia negra* in all its vibrant materiality to reimagine, with shared agency, a constellation of futures that displace official teleologies in all their occlusiveness. The volume's rainy eco- and carto-poesis offers liquid theorizations centered on reconsideration of the dialectic between environmental slow violence and the historical process, prompting us to rethink form and matter though new ecosystems. The stories show us that slow violence is not confined to the past and that its environmental consequences still dwell in the bodies and spaces of the Paraguayan, Argentine, and Brazilian Triple Frontier. The incisive *cachiai* humor of the people of the region has allowed them to recognize that Stroessner's "*chaparrones aislados*" ("isolated downpours") return in many guises—sometimes as rain, sometimes as dust, sometimes as smoke, and even sometimes as *mba' eve*. Against the backdrop of that wry folk cynicism, and sharing in it, the *cartonera* movement nevertheless seeks in its ecocritical vibrancy to imagine ways forward from *mba' eve* to possibility.

Notes

1. The quotation is characteristic of Diegues's writing and of much linguistic and literary production along the *Triple Frontera*: while plainly a Spanish sentence, it shows an admixture of Portuguese words and/or orthography (e.g.,

em instead of *en*, *essa* instead of *esa*), English words ("border" for *frontera*, "punk"), and even the Japanese *kakemono* "a hanging scroll."

2. In addition to cities in Latin America, *cartonera* publishers can be found in Finland, France, Germany, Spain, and Sweden, as well as in Mozambique. Memphis Cartonera is the most recent addition to the collectives that have been established in the United States. For the cartonera publishers database, please see https://uwdc.library.wisc.edu/collections/arts/eloisacart/.
3. Diegues resignifies the word *guaranga* into a spatial designation for the Guaraní region of Paraguay, which he labels *guaranítika*, and into a positive notion that describes the qualities of the Guaraní natural world. He recognizes that the word is commonly associated with Guaraní peoples to convey their presumed lack of formal education and western manners. The new term abandons the negative Spanish connotations which include "vulgar," "salacious," and "uncouth." The specific concept of a *lluvia negra guaranga* is used to describe the powerful storms typical of the region, which cause deaths and material damage.
4. Bruno Latour's actor-network theory conceives of reality as an intersection between nonhuman entities and human agents collaborating to form assemblages and collectives (63–9).
5. Morton's notion of "hyperobject" is a concept that navigates the fields of philosophy, science, literature, the arts (visual and conceptual), and popular culture in order to theorize the ecological.
6. The notion of the Global South is used in a postnational sense to refer to spaces and peoples negatively affected by contemporary capitalist globalization. The term also conveys the idea of a deterritorialized geography to account for oppressed peoples within the borders of wealthier countries. The notion of "global" is meant to unhinge the South from a one-to-one relationship to geography. Ann Garland Mahler provides a longer treatment of the concept and an extensive bibliography.
7. An *axolotl* is a type of salamander found in the mountain lakes of Mexico and the western United States that generally lives and breeds in the larval form without metamorphosing.
8. *Curepa* is a pejorative term used in the region to designate a person from Argentina.
9. Rob Nixon coins the term *environmental picaresque* to label Indra Sinha's reimagining of the genre in his *Animal People*.
10. I adopt a notion of emplacement derived from Edward Soja's "social production of space and the restless formation and reformation of geographical landscapes" (11) and expand it to explore the delusive aspects of such active deployment of space and time in particular historical and geographic contextualizations.
11. When asked about the origin of "Osobuco," Román reminds readers that stories have a literary aspect; here, it allows him to essay a philosophical and sociological reflection on one hand, while at the same time expressing a nostalgic component that prompted him to incorporate a critique of Asuncena society through his use of local names (Zalgade). The Biedermanns are a wealthy, "patrician" family in Asunción.
12. His reference to "*la historia*" can be understood not only to refer to the telenovela's "story" but also to "history" writ large.
13. Orange trees in Paraguay in general and in Asunción in particular became part of the local landscape due to urban forestation efforts that took place in the early twentieth century. This feature of the city has become an iconic element of the region, widely celebrated in poetry and popular music. See "Canto al Paraguay" by Heriberto Altinier and "Noches Blancas" by Mauricio Cardozo

Ocampo, among other compositions. Many candidates running for office announce orange and other native tree-planting campaigns under the banner of recovering Asunción's lost beauty and identity, yet no such project has gained political or ecological traction. The difficulty has been dual: on one hand, the ephemeral nature of electoral promises, and on the other, changes in the microclimate of the historical section of downtown that have made cultivation of the aromatic orange trees more challenging.

14. Owen Ware reminds us that Benjamin's understanding of messianism is "past-gazing," while Derrida's is disjointed, heterogeneous, and affirming of both past and future (100). Michael Levine offers a more extensive treatment of "weak messianic power" in the context of figurations of time.
15. The first word in the phrase is a portmanteau of *portugués* (Portuguese) and *español* (Spanish), with the substitution of the Portuguese grapheme *nh* for Spanish *ñ*; the second word, *selvagem*, is Portuguese for "savage" or "wild" (Spanish would be *salvaje*).
16. Here, we recall Jane Bennett's view of matter as vibrant and restlessly creative (54).
17. This stasis goes beyond a notion of attunement that privileges presence or observation, which Kant finds essential to aesthetic experience.
18. Their celebration of bibliodiversity, as Schwartz notes, ranges from international writers like Mario Bellatin and Ricardo Piglia to the literary work of the displaced and the incarcerated. In a personal interview, for instance, Diegues shared that Yiyi Yambo selects "autores que nos gusten, literatura de vanguardia en contextos fronterizos, y autores 'arribeños'" ("authors we like, cutting-edge literature in borderland contexts, and arribeño [upland] writers").

Works Cited

ABC Color. "No se sabe que [sic] tenía la 'lluvia negra' que cayó sobre Paraguay." 6 Apr. 2009, www.abc.com.py/edicion-impresa/locales/no-se-sabe-que-tenia-la-lluvia-negra-que-cayo-sobre-paraguay-1161795.html. Accessed 24 May 2019.

———. "Lluvia negra contenía basura y ceniza." 7 Apr. 2009, www.abc.com.py/edicion-impresa/locales/lluvia-negra-contenia-basura-y-ceniza-1162117.html. Accessed 24 May 2019.

Bales, Kevin. *Disposable People: New Slavery in the Global Economy*. U of California P, 2004.

Bennett, Jane. *Vibrant Matter: A Political Ecology of Things*. Duke UP, 2010.

Berger, John. *The Look of Things: Essays*. Viking Press, 1974.

Bilbija, Ksenija, and Paloma Celis Carbajal, eds. *Akademia Cartonera: A Primer of Latin American Cartonera Publishers/Un ABC de Las editoriales cartoneras en América Latina*. Parallel Press/U of Wisconsin P, 2009.

Butler, Judith. *Precarious Life: The Powers of Mourning and Violence*. Verso, 2006.

Cavallaro, Dani. *Cyberpunk and Cyberculture: Science Fiction and the Work of William Gibson*. Athlone Press, 2000.

DeLoughrey, Elizabeth M., et al., eds. *Global Ecologies and the Environmental Humanities: Postcolonial Approaches*. Routledge, Taylor & Francis Group, 2015.

Derrida, Jacques. *Specters of Marx: The State of the Debt, the Work of Mourning, and the New International*. Routledge, 1994.

Diegues, Douglas, ed. *Llubia Negra: 11 narradores paraguayos & non-paraguayos*. Yiyi Yambo Cartonera, 2009.

———. *Personal Interview*. Conducted by Elizabeth M. Pettinaroli. 29 July 2017.

Epplin, Craig. "New Media, Cardboard, and Community in Contemporary Buenos Aires." *Hispanic Review*, Autumn, 2007, pp. 385–98.

Gudynas, Eduardo. "Fallacy of Ecomessianism: Observations from Latin America." *Global Ecology: A New Arena of Political Conflict*, edited by Wolfang Sachs, Zed Books, 1993, pp. 170–8.

Heffes, Gisela. *Políticas de la destrucción, poéticas de la preservación: apuntes para una lectura (eco)crítica del medio ambiente en América Latina*. Beatriz Viterbo Editora, 2013.

Latour, Bruno. *Reassembling the Social: An Introduction to Actor-Network-Theory*. Oxford UP, 2005.

Levine, Michael G. *A Weak Messianic Power: Figures of a Time to Come in Benjamin, Derrida, and Celan*. Fordham UP, 2013.

Mahler, Anne Garland. "The Global South in the Belly of the Beast: Viewing African American Civil Rights Through a Tricontinental Lens." *Latin American Research Review*, vol. 50, no. 1, 2015, pp. 95–116.

Maturana, Humberto, and Francisco Varela. *Autopoiesis and Cognition: The Realization of the Living*. D. Reidel Publishing, 1980.

Mignolo, Walter. *The Idea of Latin America*. Blackwell, 2005.

Morton, Timothy. *Hyperobjects: Philosophy and Ecology After the End of the World*. U of Minnesota P, 2013.

———. *The Ecological Thought*. Harvard UP, 2010.

Nixon, Rob. *Slow Violence and the Environmentalism of the Poor*. Harvard UP, 2011.

Schwartz, Marcy E. *Public Pages: Reading Along the Latin American Streetscape*. U of Texas P, 2018.

Soja, Edward W. *Postmodern Geographies: The Reassertion of Space in Critical Social Theory*. Verso, 1989.

Vázquez, Karina. "Brazilian Cyberpunk and the Latin American Neobaroque." *Luso-Brazilian Review*, vol. 49, 2012, pp. 208–24.

Ware, Owen. "Dialectic of the Past/Disjuncture of the Future: Derrida and Benjamin on the Concept of Messianism." *Journal for Culture and Religious Theory*, vol. 5, no. 2, 2004, pp. 99–114.

Zalgade, Darío. "Ever Román: La literatura paraguaya se preguntó poco acerca de sí misma. Los escritores de ahora se están haciendo cargo de esta pregunta." *Oculta Lit* (blog), 4 July 2017.

Zapf, Hubert. "Creative Matter and Creative Mind: Cultural Ecology and Literary Creativity." *Material Ecocriticism*, edited by Serella Iovino and Serpil Opperman. Indiana UP, 2014, pp. 51–66.

Contributors

Diana Aldrete is Visiting Assistant Professor of Hispanic Studies in the Language and Culture Studies Department at Trinity College, Hartford, Connecticut. She earned her PhD from the University at Albany, SUNY. Her interest in human rights in Latin American literature led her to a dissertation focusing on the representation of the female body in texts concerning the feminicides in Ciudad Juarez, Mexico. Her areas of research include contemporary Mexican literature and culture, twentieth- and twenty-first-century Latinx/Queer representations, and transnational feminist studies in Latin America. She is currently working on her first manuscript expanding on her dissertation, which examines the questioning of justice in literary texts on feminicidal violence and antifeminicidal activism in Mexico.

Laura Barbas-Rhoden is Professor of Spanish at Wofford College. She is the author of two books, *Ecological Imaginations in Latin American Fiction* (2011) and *Writing Women in Central America* (2003). She has published numerous articles on literature and the environment in journals such as *Interdisciplinary Studies in Literature and the Environment*, *Revista de crítica literaria latinoamericana*, and *Studies in Twentieth and Twenty-First Century Literature*. She serves on the Executive Council for the Association for the Study of Literature and the Environment (ASLE).

Ida Day is Assistant Professor of Spanish at Marshall University. She specializes in Latin American literature and culture. Her current research includes ecocriticism and indigenous studies. She published book chapters in *Global Issues in Contemporary Hispanic Women's Writing* (2012) and *Ecological Crisis and Cultural Representation in Latin America* (2016), as well as several articles related to the field. Ida Day received her PhD in Hispanic Studies from the University of Georgia, her MA in Spanish from Texas A&M University, and her MA in English Philology from Nicolaus Copernicus University, Poland.

Carlos Gardeazábal Bravo, PhD, teaches at Colby College. His current book project, "Human Rights and the Politics of Empathy in Twenty-First

Century Latin American Literature," is centered on the work of Evelio Rosero, Juan Gabriel Vásquez, Alonso Sánchez Baute, Claudia Hernández, Rodrigo Rey Rosa, and Horacio Castellanos Moya. His research interests lie at the intersections of contemporary Latin American literature, human rights narratives, and the cultural politics of emotion, ecocriticism, and critical theory. Carlos has presented on these issues at national and international conferences, including LASA, ACLA, and NeMLA. He has published articles in academic journals such as *Chasqui*, *Teatro-Revista de estudios culturales*, and *Folios*.

Gisela Heffes is a writer and Professor of Latin American literature at Rice University (Houston), where she also teaches creative writing in Spanish. She has published the anthology *Judíos/Argentinos/Escritores* (1999), two monographs: *Las ciudades imaginarias en la literatura latinoamericana* (2008) and *Políticas de la destrucción/Poéticas de la preservación. Apuntes para una lectura (eco)crítica del medio ambiente en América latina* (2013). She also has edited *Poéticas de los (dis) locamientos* (2012) and *Utopías urbanas. Geopolítica del deseo en América latina* (2013) and was the guest editor for the special issue of *Revista de Crítica Literaria Latinoamericana* on "Ecocrítica" (2014). She is currently finishing, with Professor Jennifer French of Williams College, *The Latin American Eco-Cultural Reader* and is also working with Professor Carolyn Fornoff of Lycoming College on an edited volume, *Pushing Past the Human in Latin American Cinema*, which theorizes Latin America's rich cinematic production on environmental issues.

Adrian Taylor Kane is Professor of Spanish and Chair of the Department of World Languages at Boise State University. He is the author of *Central American Avant-Garde Narrative: Literary Innovation and Cultural Change* (2014) and editor of *The Natural World in Latin American Literatures: Ecocritical Essays on Twentieth-Century Writings* (2010).

Ilka Kressner is Associate Professor of Hispanic Studies at the University at Albany, SUNY. Her scholarship and teaching examine Spanish American literature, film, and visual arts and explore conceptions of space in art, intermediality, and ecocriticism. Her book *Sites of Disquiet: The Non-Space in Spanish American Short Narratives and Their Cinematic Transformations* (2013) analyzes representations of alternative spaces in Spanish American short narratives and their adaptations to the screen. She has coedited *Walter Benjamin Unbound* (*Annals of Scholarship*, Vols. 21:1 and 2, 2015) and has published scholarly articles in, among others, *Bulletin of Hispanic Studies*, *MELUS*, *Confluencia*, *Hispanic Journal*, *Chasqui*, *Hispanófila*, and *Revista Chilena de Literatura*.

Ana María Mutis is Assistant Professor at Trinity University in San Antonio, Texas. She has published articles on Latin American narrative, poetry, and film in academic journals such as *Hispanic Research Journal*, *Revista de Crítica Literaria Latinoamericana*, *Revista Iberoamericana*, *Revista Canadiense de Estudios Hispánicos*, and *Revista Hispánica Moderna*, among others. She coedited, with Elizabeth M. Pettinaroli, the volume *Troubled Waters: Rivers in Latin American Imagination* (2013). She also cowrote, with Elizabeth M. Pettinaroli, a chapter in *History of Colombian Literature* (Ed. Raymond Williams, 2016). She is currently working on the intersections between violence and the environment and their representation in film and literature.

Thaiane Oliveira is Professor of the Postgraduate Program in Communication at the Fluminense Federal University. She is Coordinator of the Forum of Editors and Communication of Science and Chair of the latmetrics network, which promotes discussions on alternative metrics and open science in Latin America. She is also Leader of the Laboratory of Research in Science, Innovation, Technology and Education (Cite-Lab) and Laboratory of Experiences of Engagement and Transformation of the Audience (LEETA). In addition, she is a member of the Literacies of Global South project, hosted by the University of Tübingen, Germany. She has been researching topics related to disputes about the circulation of science in the geopolitics of knowledge and epistemic crisis in the digital age.

Elizabeth M. Pettinaroli is Associate Professor of Spanish at Rhodes College and founder and editor of Memphis Cartonera, a cooperative, sustainable publishing house dedicated to fostering literacy through literature and art in Memphis, Tennessee. Her scholarship examines the elaboration of notions of space and place in competing visions of globality, as well as ecocriticism and the environmental humanities. Pettinaroli's research includes natural history, the intersection of literature and cartography, literary history, and cooperative publishing. She coedited the volume *Troubled Waters: Rivers in Latin American Imagination* (2013) with Ana María Mutis and has contributed chapters to the *History of Colombian Literature* (2016). She has published in journals such as *Colonial Latin American Review*, *Journal of Spanish Cultural Studies*, *Dieciocho: Hispanic Enlightenment*, and *Journal of Medieval Iberian Studies*.

Jacob G. Price received his PhD candidate in Hispanic Literature/Cultures from Rutgers, The State University of New Jersey. His current project, "Environmental Under the Gun: Environmental Reactions to the Cold War in Central America," explores emerging environmentalist thought that counters the literal and ideological warfare brought by the Cold War to Central American conflicts in the latter half of the twentieth century. He has published in scholarly journals such as *Ometeca, Millars* and has also cofounded the renewal of the *Yzur* literary journal at Rutgers.

Charlotte Rogers received her PhD in Spanish from Yale University, where she specialized in literatures of the tropics. Her first book, *Jungle Fever: Exploring Madness, Medicine and Writing in 20th Century Tropical Narratives*, was published by Vanderbilt University Press in 2012. Her book manuscript about the legacy of the legend of El Dorado in contemporary fiction about the South American forest is under contract with University of Virginia Press. She has published in *Revista de Estudios Hispánicos, Revista Canadiense de Estudios Hispánicos*, *MLN*, *Hispania*, *Bulletin of Spanish Studies*, *Cuadernos hispanoamericanos*, and *Bulletin of Hispanic Studies*. She currently holds a Mellon Humanities Fellowship at the University of Virginia (where she is Assistant Professor of Spanish) for her new book project about creative responses to hurricanes in the Greater Caribbean.

Lauren Woolbright is a scholar of games, environmentalism, and social justice who has published a video game as well as video essays and text articles on the topic of video games and environmental issues. She holds the position of Assistant Professor of New Media Studies at Alma College, where she teaches game design, interactive media, and environmental communication. She is the 2018–2019 Chair of the Council for Play and Game Studies for the Conference on College Composition and Communication and cofounder of the journal *One-Shot: A Journal of Critical Play and Games*.

Index

Note: Figures are indicated by page numbers in *italics*.

abduction 261–262
Acevedo-Muñoz, Ernesto 100, 110n2
Acosta, Alberto 28n11, 29n14, 71n7, 125n2
Acosta, José de 9–10, 28n11, 29n14
accumulation: of capital 13, 16, 24, 147, 151–152, 154, 156, 158–160; of toxins 128, 133
actor-network theory 274n4
Adorno, Rolena 14
Afro-diasporic spirituality 180–184, 186
agency 6, 14, 17, 16–17, 19–20, 23, 36, 59, 101–102, 106, 108, 129, 140, 170, 205, 237, 240–241, 245, 251, 261, 270, 273
agrochemicals 12, 22, 39–40, 42, 44–45, 48–50, 51n4, 52n17, 66–70, 71n17, 155–156
Aguilar Camín, Héctor 18, 147–154, 160
Agwa (Water) (Merzouki) 229
Aguirre, Emiliano 28n7
Alaimo, Stacy 68, 252n9
Alemán, Gabriela 262–265
Almeida, Bira 245
Alonso-Regalado, Jesús 251n1
Altinier, Heriberto 274n13
Alves, Maria Thereza 29n20
amaranth 57
Anderson, Mark 10
Andrade, Pablo 29n16
animals, trauma and 165–167
Animal's People (Sinha) 76
Anker, Elizabeth 163
Anthropy, Susan 201
Año del desierto, El (Mairal) 55, 58–59
Anthropy, Anna 201
apocalypse 64, 74–76, 119, 162, 167, 199, 203, 209, 211, 266, 268
Apter, Emily 252n10
Arancibia, Florencia 44
Arendt, Hannah 30n22
Armbruster, Karla 147
Arons, Wendy 238
art, material 16–18
AUC *see* United Self-Defense Forces of Colombia (AUC)
Augé, Marc 78–79

Bacon, Francis 10
Baer Hester and Ryan Long 100, 103, 110n2
Bal, Mieke 20, 247, 252n10
Baldenegro, Isidro 207–208
Baltimore, Maryland 6–7
Barthes, Roland 244
Baudelaire, Charles 78
Bauman, Zygmunt 78
Beck, Ulrich 62
Beckman, Erika 128–129, 142n5
Beilin, Katarzyna 51n4, 57
Bellatin, Mario 275n18
Benbrook, Charles M. 51n3
Bennett, Jane 239, 261, 252n8, 275n16
Bentancor, Orlando 8
Berry, Thomas 122
biocapitalism 60, 62, 71n7
biosemiotics 261–262
Biron, Rebecca 110n2
Bleeding Border (video game) 205
Boff, Leonardo 106
Bolaño, Roberto 74–87, 87n3
Bolívar, Simón 58
Bolsonaro, Jair 220

Borders (video game) 204
Bracero Program 88n14
Branche, Jerome 193n2
Brandt, Barbara 153
Braziel, Jana Evans 193n3
Brazil 220, 227–231, *228–229*
Bristow, Tom 16, 30n24
Brundtland Report 226
Buell, Lawrence 4–5, *56–57*
Buen Vivir 12–14, 16, 59, 120, 125n2, 126n6
Bullet Boy (video game) 204
Burke, Edmund 123
Burtynsky, Edward 68
Butler syndrome 225–226
Butler, Judith 166, 241, 252n10, 264

Caballero, Vander 203
"Caballo ahogado, El" (The Drowned Horse) (Cuadra) 130–131
Cabellos Damián, Ernesto 114, 120
Cadena, Marisol de la 120
Cajete, Gregory 121
Can Poetry Save the Earth? (Felstiner) 115
Cantos de Cifar y del mar dulce (Cuadra) 128–135
capital fictions 128
Capital Fictions (Beckman) 129
capitalism 6, 7, 10, 12–13, 22, 48, 49, 59, 74, 75, 76, 77, 78, 80, 82, 84, 125, 129, 132, 135, 137–138, 140–141, 142n5, 148, 154, 157–158, 160, 221, 252n9, 264
capoeira 237–238, 240, 245, 249
Cardenal, Ernesto 128–129, 134–139, 141, 142n5
Cardozo Ocampo, Mauricio 274n13
Carroll, Noël 52n12
Carruthers, David 86, 88n9
Carson, Rachel 56–57, 64, 69
cartonera publishing 20–21, 258–261, 271–274, 274n2; *see also* ecocritical publishing projects
carto-poesis 20, 262, 273
Cash, Debra 240–241
Castle of Otranto (Walpole) 52n6
Castro-Gómez, Santiago 98, 223
Cecire, Natalia 50
"Cementerio de los pájaros, El" (Bird Cemetery) (Cuadra) 131–133
children: and slow violence 6; as monsters 20, 22, 41–43, 46–49; in environmental discourse 50; in horror fiction 52n8, poisoned bodies 46–47, 60, 156–157
Ciudad Juárez 80, 83
classification 221–224
Clement, Jennifer 154–160
Clark d'Lugo, Carol 86
climate change 7, 25, 40, 57, 116, 203, 215, 220, 226–227, *228*, *228–229*, *229*
closed spaces 41–42
Cohen, Jeffrey Jerome 45, 46, 51, 52n12
Coherence of Gothic Conventions, The (Sedgwick) 41,42
Cold War 15, 128–141, 176n4
collateral damage 24, 74, 147, 154
Colombia 162–176, 176n4
colonialism 9–11
coloniality 97–99, 218–219
Compagnie Käfig 236–251, *242–243*, *250*
consumption 120
corporate social responsibility (CSR) 27n1
Correa-Cabrera, Guadalupe 157
Cossío, Dinorah 56, 70n3
Costa Lima, João Vicente 227
Costikyan, Greg 200
Costo humano de los agrotóxicos, El (Piovano) 68
Cromo (television series) 5–6
Crosby, Sara L. 216n8
CSR *see* corporate social responsibility (CSR)
Cuadra, Pablo Antonio 128–134, 139–141
Cuarón, Alfonso 95–96
cyberpunk 266, 270–271; *see also* Triple Frontier hyperpunk

Dabove, Juan Pablo 59
Daly, Tara 29n21
dance 236–251, *242–243*, *250*
Daughter of the Lake (film) 114–118, *118*, 119–121, 125
declassification 221–224
decolonial: ecocriticism 96–98, 218–219, 223, 225, 231; thinking 223; turn 13, 97
decolonization 224–227, 272
degradation 25, 49, 75–76, 82, 117, 193n3, 270

degrowth 114, 120, 125n2, 126n6
DeLoughrey, Elizabeth 192
Del Valle, Ivonne 9, 28n10
De procuranda indorum 9
Derechos de la naturaleza (Gudynas) 71n14
descriptive geography 27n2
Detroit, Michigan 84, 89n17
Diamond, Jared 124–125
Diegues, Douglas 258, 268–270, 274n3
Distancia de rescate (Schweblin) 39–60, 62–64
D'Lugo, Carol Clark 86
docility 89n15
Domínguez, Nora 44, 45, 52n10
dromocolonization 6
Dussel, Enrique 107
dust 83

Earth Beings: Ecologies of Practice Across Andean Worlds (Cadena) 120
Echeverría, Ignacio 87, 87n3
ecocritical publishing projects 273; *see also* cartonera publishing
ecocriticism 261–262; decolonial 218–233, *228–229*; in Latin America 13–14; spatial justice and 97–98
Ecological Thought (Morton) 273
ecomutation 65
ecomaternalism 43
ecospirituality 187–189
ecotone 259–260
ecopícaro 265, 268; *see also* ecopicaresque
ecopicaresque 275
ecosystem: science 218; culture 273
eco-zombie 49–51
Eimbcke, Fernando 95–96
Ejércitos, Los (Rosero) 162–176, 176n3
Electronic Monuments (Ulmer) 216n4
ELN *see* National Liberation Army (ELN)
Eloísa cartonera 258, 268
Embry, Karen 52n16
endogenous 14, 25, 209–211, 214, 221–222, 224–225
Environmental History of Latin America (Miller) 126n8
environmental justice 87n2, 183
environmentalism of the poor 183; writer-activists 260; *see also* cartonera publishing
epistemic disobedience 218–219
Erin Brockovich (film) 71n17
Escobar, Arturo 29n19, 97, 104, 107
Estrecho dudoso, El (The Doubtful Strait) (Cardenal) 138
Estrellas federales, Las (Incardona) 59–63, 67
extraction 8–9; *see also* mining; neoextractivism
extractivism 11–12, 28n11, 29n14, 118–119, 125n2, 175, 264; *see also* neoextractivism

Facundo (Sarmiento) 55, 58
Fallout (video game series) 203–204
FARC *see* Revolutionary Armed Forces of Colombia (FARC)
Felstiner, John 115
Femicide Machine, The (González) 89n16
Ferré, Rosario 181
fetishism of form 17, 259–261, 271
Fiebre del loco, La (film) 126n9
Flint, Michigan 71n17
Flor de piel, A (Salcedo) 29n20
Flowers of Evil, The (Baudelaire) 78
Fordism 89n17
formalism 7
Forns-Broggi, Roberto 110n2
Foucault, Michel 81–82, 89n15
Franco, Jean 78, 90n20
Freud, Sigmund 42
Franko, Mark 237
Frigola, Nuria 120
Fritsch, Kelly 62
Funes, Reinaldo 182, 193n7
Funes Monzote, Reinaldo 182

Gaad, Greta 180
Galeano, Eduardo 11, 116–118
Galtung, Johann 177n7
García, Gustavo 105
García Gutiérrez, Antonio 218–219, 225, 233n3
García Moreno, Laura 166
Gaussian temporality 262, 270
geography, descriptive 27n2
Gibson, Charles 28n9
Giddens, Anthony 67
Gilmore, David 52n12
Ginés de Sepúlveda, Juan 9
Giudicelli, Christophe 212
Glifosatos 2

Global North 220
Global South 218, 221–222, 263–264, 274
globalization 12, 76, 96
glyphosate 2, 3, 13, 39, 44, 51n4, 51n5, 57, 60, 68, 70n4
gold 119–120
González Dinamarca, Rodrigo Ignacio 52n8, 52n13
González Rodríguez, Sergio 75, 89n16
Gothic literature 40–45, 50, 52n6
green revolution 227
Grineski, Sara E. 81, 88n14
Gudynas, Eduardo 11, 12, 29n13, 59, 71n14, 268
Guzmán, Patricio 114

Half-Real: Video Games Between Real Rules and Fictional Worlds (Juul) 202
Hallstead, Susan 59
Handley, George 192
Hanson, Camille 238
Haraway, Donna 251n7
Hartmann, Tilo 202
Hathaway, Mark 106
haunting 167–175
Healy, Kathleen 42
Heholt, Ruth 167
Heise, Ursula 4–5, 68–69, 141n1
Hell, Julia and Andreas Schönle 75
Henighan, Stephen 138
herbicide 2, 4, 39, 44, 51n3, 57, 60, 155–159; *see also* agrochemicals
Hind, Emily 111n3
Historia moral y natural de las Indias (Acosta) 9–10
"Historical Roots of Our Ecological Crisis, The" (White) 150
horror fiction: children in 52n8
Hoyos, Héctor 75
Huesos en el desierto (Bolaño) 75
Humboldt, Alexander von 57–58
Humean causal system 262
Hurley, Kelly 40
Hurricane San Ciriaco 184–186, 194n12
hyperobjects 32, 69, 203, 260, 262, 266, 268–269, 271 274n5

IARC International Agency for Research on Cancer 2, 39, 70n4
immateriality 69, 229, 237
Incardona, Juan Diego 59–63, 67, 70
indigenous games 199
Ingold, Tim 8
interspecies war 57
Iovino, Serenella 17, 239–240
Irizarry, Guillermo B. 183, 190, 194n20

Jordan, Chris 68
Joven, Julián 65–67, 70
justice: environmental 87n2; spatial 77; spatial, ecocriticism and 97–98; spatial, film and 99–109; spatial, in Latin America 14–16; spatial, modernity and 97–98
Juul, Jesper 202

Kane, Adrian Taylor 135
Kawésqar language 124–125
Kilomba, Grada 222
Klein, Naomi 119, 126n7
Kressner, Ilka 52n10
Kristeva, Julia 76
Kuhn, Thomas 225, 227

Language Matters (film) 125
Lake Tacarigua 57–58
Lake Valencia 57–58
languages, native, disappearance of 124–125
Lasarte, Francisco 133
Las Casas, Bartolomé de 9, 28n3
Latin America: ecocriticism in 13–14; political ecology in 14–16; slow violence in 8–13, 97–99; spatial justice in 14–16
Latinx 180, 202–204
Latour, Bruno 261, 262, 274n4
Lauro, Sarah Juliet 49, 52n14, 52n16–52n17
Leff, Enrique 98
Leone, Lucía de 56, 57
Lepecki, André 246
Levine, Michael 275n14
Life Exposed: Biological Citizens After Chernobyl (Petryna) 68
Llubia Negra: 11 narradores paraguayos y non-paraguayos (Diegues, ed.) 258–259, *259*, *270*
lluvia negra 257–258
"Lluvia negra, polvo gris" (Black Rain, Gray Dust) (Alemán) 262–265

Lobanyi, Jo 168
Long, Ryan and Hester Baer 100, 103, 110n2
López Parada, Esperanza 74
Luberza Oppenheimer, Isabel 180, 193n1–193n2
Ludmer, Josefina 55–56, 70n2
Lugo, Fernando 257
Lugones, Leopoldo 62

Mahler, Ann Garland 221, 274n5
Maguire, Geoffrey and Rachel Randall 105
Mairal, Pedro 55, 58–59
maquiladoras 80–81, 88n14
Marcone, Jorge 132
Márquez, Alan 212
Martínez, Juliana 163, 167, 172, 176n2
Martinez Alier, Joan 15, 218
Masiello, Francine 17
material art 16–18
materiality 271, 273
maternity 42; and ecocriticism 43; *see also ecomaternalism*, and the Gothic genre 43–44, 52n8; politicization of motherhood 44–45
Matter of Empire, The (Bentancor) 8
Maturana, Humberto 262
May, Theresa J. 238
Mbembe, Achille 223
McClintock, Anne 7, 259
McDowell, Megan 52n7
McGregor, Sherilyn 43, 52n9
McNeish, John Andrew 12
Meikle, Jeffrey L. 243
memory 100, 114, 121–123, 174, 178n9, 203
Mendez, Chico 246
Méndez Becerra, Penélope 89n18
Méndez Panedas, Rosario 188, 195n18
Menne, Jeff 111n3
Merrell, Floyd 238
Merzouki, Mourad 236, 238–239, 251n3
Messianism 268, 275n14
Metaphors 225
Mexico 74–87, 100–101, 126n7, 205–206
Mignolo, Walter 4–5, 13–14, 16, 30n23, 218, 222, 272
Miller, Shawn William 126n8, 132
mining 6, 117–120; *see also* extraction
Mitchell, Mozella 194n15
modernity 17, 22–23, 62, 74–75, 78–79, 85–86, 88, 97, 100, 107, 129, 133, 135, 138, 168, 218, 239, 264
Mondot, Adrien 248
Monsanto Company 39, 57
monster 45–51, 52n12, 52n15, 71n13, 205, 209, 214
Montaldo, Graciela 59, 65
Moore, Charles 252n13
Moore, Jason 154
Moraña, Mabel 45, 46, 48, 52n12
Moreira, José Roberto 226
Moreiras, Cristina 168
Morton, Timothy 203, 262, 267–268, 273, 274n5
Mothers of Ituzaingó 44
Mulaka (video game) 199–200, 205–215, *213*, *215*
Muriente Pérez, Julio A. 182
Mutis, Ana María 170

NAFTA *see* North American Free Trade Agreement (NAFTA)
National Liberation Army (ELN) 177n4
native languages, disappearance of 124–125
naturecultures 251n7
neoextractivism 11, 65
neoliberalism 12, 17, 49, 74, 76, 89n16, 96, 102, 129, 154, 164, 219, 227, 231, 233
Never Alone (video game) 216n4
Neves, Fabrício 227
Nicaragua 128–141
Nichols, Bill 115
Nixon, Rob 2, 4–5, 39–40, 49, 69, 74, 76, 80, 83, 88n8, 96, 102, 104, 115–117, 119, 132, 147, 156, 159–160, 164, 171–172, 177n7, 183, 215, 218, 236, 244–245, 259–260, 274n9
nonlocality 262
non-places 79
Non-Places: Introduction to an Anthropology of Supermodernity (Augé) 78–79
North American Free Trade Agreement (NAFTA) 80, 84, 100–101

Nuestra Señora de la Noche (Santos-Febres) 180–192
"Nueva ecología" (New Ecology) (Cardenal) 134–140

occlusion, strategies of 6–7, 270
Oloff, Kerstin 48, 52n15
Open Veins of Latin America (Galeano) 116–117
Operation Bootstrap 181, 190
Oppermann, Serpil 17, 239–240
Oslender, Ulrich 168
"Osobuco" (Osso buco) (Román) 265–268, 274n11
Ospina, William 3, 26
Our Common Future 226
overworlding 264

pampa 55–56, 71n6
Papo & Yo (video game) 203
Parson, David 253n18
Partido Revolucionario Institucional (PRI) 100–101
Pearl Button, The (film) 114–117, 121–125
Peirce, Charles Sanders 261
Pequeño enfermo, Un (Joven) 65–67
"Perdidos en la black rain guaranga" (Lost in the Guaraní Black Rain) (Diegues) 268–269
Peronism 61, 71n11
"Pescador" (Fisherman) (Cuadra) 133–134
pesticides *see* agrochemicals
Petersen, Amanda L. 168, 175
Petryna, Adriana 68
Pettinaroli, Elizabeth M. 52n10
Phan Hein 142n4
Phillips, Cassandra 252n13
Piglia, Ricardo 275n18
Pinter, Harold 177n5
Pinto, Rodrigo 171
Piovano, Pablo Ernesto 68
Pixel (Merzouki) 229
Plastic Ocean: How a Sea Captain's Chance Discovery Launched a Determined Quest to Save the Oceans (Moore & Phillips) 252n13
Plumwood, Val 180
Poblete, Juan 106, 109
Poe, Edgar Allen 216n8
Poemas nicaragüenses (Cuadra) 134
poetic documentary 115
poetry 115–116, 128–130
political ecology: in Latin America 14–16
Poniatowska, Elena 83
portunhol selvagem 269, 271, 275n15
Portuguese river 180–182, 189
Posmentier, Sonia 183–184, 194n10
Postal (video game) 216n3
postcolonial 192, 214, 223
Postnatural world 67–68
Pourgouris, Marinos 169
poverty 79–80, 183
Powers of Horror (Kristeva) 76
Prayers for the Stolen (Clement) 154–160
praxis 18, 102, 239, 259
precariousness 223–224, 272
PRI *see* Partido Revoluncionario Institucional (PRI)
Puerto Rico 180–192

Quijano, Aníbal 97, 218

Ramos Otero, Manuel 181
Randall, Rachel and Geoffrey Maguire 103
Ray, Sarah Jaquette 69
Readings in Performance and Ecology (Arons and May, eds.) 238
recycling; *see also* cartonera publishing
Red Alert! Saving the Planet with Indigenous Knowledge (Wildcat) 214
regimes of control 96, 110n1
resilience 225–226
resistance 225–226
Resplandor de la madera, El (The Radiance of Wood) (Aguilar) 147–154, 160
Ress, Mary Judith 193n4
"Rethinking the World from the Perspective of *Buen Vivir*" (Acosta) 125n2
Revolutionary Armed Forces of Colombia (FARC) 2, 164, 177n4
Ribas-Casasayas, Alberto 168, 175
Richard, Nelly 168
riskspaces 68–69
Rivera de la Cuadra, Patricia 86
Rivera, José Eustasio 142n5
Rock of Ages (video game) 204
Rodríguez, Natalia 59, 63–64, 70

Rolling, James Haywood 249, 251n6
Román, Ever 265–268, 274n11
Romero, Silvia 207
Rosa, Cristina 245
Rosario Natal, Carmelo 183, 189
Rosas, Juan Manuel de 61–62
Rosero, Evelio 162–176
Rothberg, Michael 173
Roundup 57
Rudd, Alison 45
ruin 74
Ruiz, Pedro 2–4, 26–27; *see also* *Glifosatos* 2
rural turn 55–56

Sacrifice of Giants, The (performance) 238
sacrifice zones 119
Salcedo, Doris 29n20
Saldaña-Portillo, Maria 101, 110n2
Salzinger, Leslie 89n15
San Ciriaco Hurricane 184–186, 194n12
Sandinistas 128, 134–135, 137–138
Santa Cruz, Alonso de 9
Santería 180, 194n14; *see also* Afridiasporic spirituality
Santos-Febres, Mayra 18, 180–192, 193n6
Sarmiento, Domingo F. 55, 58, 61–62
Schwartz, Marcy 271, 275n18
Schwartz, Stuart 184, 194n12
Schönle, Andreas and Julia Hell 75
Schweblin, Samanta 39–60, 62–64, 70
Scientific networks 219
Sedgwick, Eve Kosofsky 41,42
Seeds of Change (Alves) 29n20
Seeking Spatial Justice (Soja) 77
Segato, Rita Laura 89n19
Sense of Place and Sense of Planet (Heise) 141n1
Serrano, Edgar 206, 208–209
Servín, Enrique 212
Shaw, Deborah 110ns, 111n3
Sheldon, Rebekah 50, 156–157
Sibara, Jay 69
Silent Spring (Carson) 56–57, 64
Sinha, Indra 76, 274n9
Signates, Luiz 232
slow violence 2–3; of agrochemical pollution 48–49, 69–70; defined 4–5, 69, 164, 218; environmental history and 181–184; impact of 5–7; as invisible 169; alternative responses 226, 271–273; in Latin America 8–13, 97–99; in the Triple Frontier 257–259; memory and 178n9; in Nicaragua 132–133; resistance to 189–190; science and 219–220; theorization of 97–99; trauma and 167–175; water and 115
Slow Violence and the Environmentalism of the Poor (Nixon) 2–3, 39, 116
Soderbergh, Steven 71n17
Soja, Edward 14, 76–77, 97, 274n9
Solanas, Pino 71n16
Sommer, Doris 18
Soto, Domingo de 28n4
Soto, Merián 238
Sousa Santos, Boaventura de 222–223
Soussloff, Catherine 246–247, 253n17
soy 51n4, 57
spatiality 97–99, 102, 106, 249
spatial justice 77; coloniality and 97–98; ecocriticism and 97–98; film and 99–109; in Latin America 14–16
spectral topographies 15, 24, 163, 167
spectrality 163, 168, 172–175
spiritual ecology 23, 114, 121–122, 125
Spiritual Ecology: A Quiet Revolution (Sponsel) 114
Sponsel, Leslie E. 114
Steinberg, Samuel 109
Steinmetz, George 84, 89n17
structural violence 169, 174, 177n7
Suárez Findlay, Ellen 194n17
sublime 123
Summerfield, Derek 173
Super Mario Odyssey (video game) 205–206
supermodernity 78–79
Suryanarayanan, Sainath 51n4, 57
Sze, Julie 76, 88n7

Tarahumara people 199–200, 206–208, 212–214
Techeira, Lucas 68
Temporada de patos (Duck Season) (film) 95–96, 99–100, 104–109
temporality: children and 50
Thiong'o, Ngugi wa 224, 232
This Changes Everything: Capitalism vs. The Climate (Klein) 119

Thomas, Chelsea 241
Three Branch Song (dance) 238
Total Eclipse of the Heart (Ruiz) 2–4, 26–27
toxic discourse 56, 57, 66
toxicity 64–67, 133, 158, 209, 214, 264–267, 271
trauma 165–175
Triple Frontier hyperpunk 270
Tropic of Orange (Yamashita) 88n7
Tully, John 246, 253n15–253n16
Tupinipunk/Brazilian cyberpunk 270
2666 (Bolaño) 74–87, 87n3

Ulmer, Greg 216n4
United Nations Conference on Sustainable Development 228
United Self-Defense Forces of Colombia (AUC) 164, 177n4
urban islands 55
Uribe Vélez, Álvaro 177n4
Úselo y tírelo (Use It and Throw It Away) (Galeano) 118

Varela, Francisco 262
video games 201–202
Vi mutar, La (Rodríguez) 59, 63–65
violence *see* slow violence
viscosity 262, 267–268
Vizcaino, Guillermo 206
von Mossner, Weik 170–171
Von Werlhof 97
Vorágine, La (Rivera) 142n5
"Voyage, The" (Baudelaire) 78
Vuelos de victoria (Flights of Victory) (Cardenal) 128, 134–135, 137, 140–141

Waldron, John 96
Wallace, Kathleen R. 147
Walpole, Horace 52n6
Ware, Owen 275n14
waste 7, 12, 17, 68, 77, 79, 81–82, 84–85, 117, 129, 133, 141, 243–244, 246–247, 250, 257–258
water, as protagonist 126n8
water crisis 114–115
Weik van Mossner, Alexa 237
Weldt-Basson, Helene 193n2
White, Lynn, Jr. 150
White, Steven F. 134
WHO *see* World Health Organization (WHO)
Wildcat, Daniel 214
Williams, Raymond 96
Wimmer, Natasha 88n4
wood 30n22
Wood, Andrés 126n9
Work of Art in the World, The: Civic Agency and Public Humanities (Sommer) 18
"World games" 199
World Health Organization (WHO) 70n4
worldism 220, 221
World Until Yesterday, The: What Can We Learn from Traditional Societies (Diamond) 124
Wulf, Andrea 57
Wynter, Sylvia 194n10

Yamashita, Karen Tei 88n7
Yang, Sharon Rose 42
Y tu mamá también (And Your Mama, Too) (film) 95–96, 99–104, 110n2, 111n3
Yiyi Yambo cartonera 258, 272

Zapatista uprising 100–101, 126n7
Zapf, Hubert 261
Zavala, Osvaldo 89n19
Žižek, Slavoj 169
zombie 47–51, 52n15, 267
Zurita, Raúl 122–123

For Product Safety Concerns and Information please contact our EU representative GPSR@taylorandfrancis.com
Taylor & Francis Verlag GmbH, Kaufingerstraße 24, 80331 München, Germany

www.ingramcontent.com/pod-product-compliance
Ingram Content Group UK Ltd.
Pitfield, Milton Keynes, MK11 3LW, UK
UKHW021300190926
475701UK00009B/166
* 9 7 8 1 0 3 2 0 8 4 8 2 4 *